FUNDAMENTALS
OF AIR POLLUTION

ARTHUR C. STERN

Department of Environmental Sciences and Engineering
The School of Public Health
The University of North Carolina
Chapel Hill, North Carolina

HENRY C. WOHLERS

Consultant
Villanova, Pennsylvania
formerly

Department of Environmental Engineering and Science
Drexel University
College of Science
Philadelphia, Pennsylvania

RICHARD W. BOUBEL

Department of Mechanical Engineering
Oregon State University
Corvallis, Oregon

WILLIAM P. LOWRY

Department of Geography
University of Illinois
Urbana, Illinois
formerly

Department of Regional Planning
University of Pennsylvania
Philadelphia, Pennsylvania

FUNDAMENTALS

OF

AIR

POLLUTION

ACADEMIC PRESS   New York and London

A Subsidiary of Harcourt Brace Jovanovich, Publishers

ACADEMIC PRESS, INC.
111 Fifth Avenue, New York, New York 10003

United Kingdom Edition published by
ACADEMIC PRESS, INC. (LONDON) LTD.
24/28 Oval Road, London NW1

LIBRARY OF CONGRESS CATALOG CARD NUMBER: 72-82650

PRINTED IN THE UNITED STATES OF AMERICA

Dedicated to Our Wives

CONTENTS

Part I THE ELEMENTS OF AIR POLLUTION

Chapter 3.   Air Pollution

Chapter 4.   Scales of the Air Pollution Problem

Chapter 5.   The Air Pollution System

Chapter 6.   History of the Problem

Chapter 7.   Air Quality

Chapter 8.   The Chemistry of Air Pollution

Part II   THE EFFECTS OF AIR POLLUTION

Chapter 9.   Effects of Air Pollution on Inert Materials

Chapter 10.   Air Pollution Effects on Vegetation and Animals

Part III THE METEOROLOGY OF AIR POLLUTION

Chapter 16. Sun, Earth, Atmosphere, Weather, and Climate

Chapter 17. Radiation in the Atmosphere

Chapter 18. Atmospheric Thermodynamics

Chapter 19.   Atmospheric Precipitation

Chapter 20.   Motion in the Atmosphere

Chapter 21.   Effects of Pollutants on the Atmosphere

Chapter 22.   Studies of Air Pollution Climatology

Part IV THE CONTROL OF AIR POLLUTION

Chapter 23. Sources of Air Pollution

Chapter 24. Emission Inventory

Chapter 25. Source Sampling

Chapter 26. Engineering Control

FUNDAMENTALS
OF AIR POLLUTION

Chapter 1

INTRODUCTION

I. The Authors' Point of View

The authors are collectively a chemist (HCW), a meteorologist (WPL), and two mechanical engineers (RWB; ACS). This 1:1:2 ratio has some relevance in that it tends to approximate the ratio of those professionally involved in the field of air pollution. In the environmental protection and management field, the experience of the recent past has been that physicists and electrical engineers have been most attracted to the radiation, nuclear, and noise areas; biologists and civil engineers to the aquatic and solid waste areas; chemists, meteorologists, chemical and mechanical engineers to the area of air pollution and its control. These remarks are not intended to read all others out of the party (or out of this course). The control of air pollution requires the combined efforts of all the professions mentioned in addition to the input of physicians, lawyers, and social scientists. However, the professional mix of the authors, and their expectation of a not-too-dissimilar mix of students using this textbook, forewarns the tenor of its contents and presentation.

Although this book consists of four parts and four authors, it is not to be considered four short books put together back-to-back to make one large one. By and large the several parts are the work of more than one author. Obviously the meteorologist member of the author team is principally responsible for the part of the book concerned with the meteorology of air pollution, the chemist author for the chapters on chemistry, and the engineer authors for those on engineering. However, as you will see, no chapters are signed, and all authors accept responsibility for the strengths and weaknesses of the several chapters and for the book as a whole.

By virtue of its division into four sections, this text may be used in several ways. Part I, by itself, provides the material for a short course to introduce a diverse group of students to the subject—with the other three parts serving as a built-in reference book. Parts I and II, which define the problem, can provide the basis for a semester's work, while Parts III and IV, which resolve the problem, provide the material for a second semester's work. Part III may well be used separately as the basis for a course on the meteorology of air pollution; and the book as a whole may be used for an intensive one-semester course.

The viewpoint of this book is first that most of the students who elect to receive some training in air pollution will have previously taken courses in chemistry in high school and college, and that those few who have not would be well advised to defer the study of air pollution until they catch up on their chemistry. Because of this there is only one chapter on chemical principles in this book.

The second point of view is that the engineering design of control systems for stationary and mobile sources requires a command of the principles of chemical and mechanical engineering beyond that which can be included in a one-volume textbook on air pollution. Before sallying forth into the field of engineering control of air pollution, a student should, as a minimum, master courses in internal combustion engines, power plant engineering, the unit processes of chemical engineering, engineering thermodynamics, and kinetics. However, this does not have to be accomplished before taking a course based upon this book, but can well be done simultaneously with or after so doing.

The third point of view is that *no one*, regardless of his professional background, should be in the field of air pollution control unless he or she sufficiently understands the behavior of the atmosphere, which is the feature that differentiates *air* pollution from the other aspects of environmental protection and management. This requires a knowledge of some basic meteorology in addition to some rather specialized air pollution meteorology. The viewpoint presented in the textbook is that very few of the students using it will have studied basic meteorology previously, and that, if this course serves its purpose, it is not mandatory that they do so simultaneously or later. It is hoped that exposure to air pollution meteorology at this stage will excite a handful of students to later delve deeper into the subject, but, again, it is not mandatory that all students do so. Therefore, a relatively large proportion of this book has been devoted to meteorology because of its projected importance to the student.

The authors have tried to maintain a universal point of view so that the material presented would be equally applicable in all the countries of the world. Although a deliberate attempt has been made to keep American provincialism out of the book, it has inevitably crept in through the exclusive use of English language references and suggested reading lists, and the preponderant use of American data for examples, tables, and figures. The saving grace in this respect is that the principles of chemistry, meteorology, and engineering are universal.

TABLE 1-1

Metric and English Equivalents

Linear measure	Square measure
1 in. = 2.54 cm	1 in.^{2} = 6.4516 cm^2
1 ft = 3.048 dcm	1 ft^2 = 9.2903 dcm^2
1 yd = 0.9144 m	1 yd^2 = 0.83613 m^2
1 mile = 1.6093 km	1 mile^2 = 2.5900 km^2
1 mm = 0.03937 in.	1 mm^2 = 0.00155 in.^{2}
1 cm = 0.3937 in.	1 cm^2 = 0.15500 in.^{2}
1 dcm = 3.937 in.	1 dcm^2 = 0.10764 ft^2
1 m = 3.2808 ft	1 m^2 = 1.1960 yd^2
1 km = 0.62137 mile	1 km^2 = 0.38608 mile^2

Cubic measure	Weight
1 in^3 = 16.387 cm^3	1 grain = 0.064799 gm
1 ft^3 = 0.028317 m^3	1 ounce (avoirdupois) = 28.350 gm
1 yd^3 = 0.76455 m^3	1 pound (avoirdupois) = 0.45359 kg
1 mile^3 = 4.16818 km^3	1 ton (short) = 0.90718 metric ton
1 cm^3 = 0.061023 in^3	1 gm = 15.432 grains
1 dcm^3 = 61.023 in^3	1 gm = 0.035274 pounds (avoirdupois)
1 m^3 = 35.315 ft^3	1 kg = 2.2046 pounds (avoirdupois)
1 km^3 = 0.23990 mile^3	1 metric ton = 1.1023 tons (short)

Temperature	Liquid measure
$\dfrac{°C}{°F - 32} = \dfrac{5}{9}$	1 gallon = 3.7854 liters
$°K = °C + 273.16$	1 ml = 0.033814 fluid ounces
$°R = °F + 459.7$	1 liter = 0.26417 gallon

II. Units and Nomenclature

The units in this book are metric, temperatures are in Centigrade. A conversion table from the inch-pound-degree Fahrenheit system to the metric (°C) is provided (Table 1-1). Units and nomenclature, not in Table 1-1, are presented in Appendix F.

III. Problems

One of the things that distinguishes a textbook from a reference book is the inclusion in the former of questions and problems at the end of each chapter. The student should not expect to find all the information needed for a satisfactory answer or solution in the text. Part of the reason for including lists of suggested reading is to refer the student as to where the information required might be found. The following sections of this chapter suggest other information sources.

IV. Bibliographical Material

There is a difference between the function of the Suggested Reading list and the numbered Reference list. The inclusion of the former is intended to instruct the student: There is much more information on this subject than could be incorporated in this brief textbook chapter. In fact, there is pertinent material that was not even mentioned in the chapter. If you want to avail yourself of it, here is where you will find it.

The inclusion of numbered references is intended to tell the student one of several things: (a) The piece of data, table, or figure referenced is not our work, but rather the work of the person or persons whose names are referenced. They, not we, did the hard work of obtaining the data, assembling the table, or drawing the figure, and to them should go the credit. (b) The material referenced follows a line of reasoning originated by or best expressed by the referenced person(s), and we, the authors, wish to acknowledge this fact by appropriate referencing.

V. Other Literature Resources

Searches of the air pollution literature are facilitated by the existence of excellent bibliographic resources in the air pollution field. In general, one should start a literature search with the most recent listing of references and work backward in time both through earlier lists, and through references in the bibliographies of the books or papers selected from the most recent list. One stops searching for additional references when the time allotted for reference searching runs out, or when the

TABLE 1-2

English Language Air Pollution Bibliographic Sources
Published in the United States

Bibliography	Publisher	Monthly since
Air Pollution Abstracts	Office of Air Programs (1)	Feb. 1971
Air Pollution Titles	Center for Air Environment Studies (6)	Jan. 1965[a]
NAPCA Abstract Bulletin	National Air Pollution Control Administration (1)	Jan. 1970[b]
APCA (*Air Pollut. Contr. Ass.*) *Abstr.*	Air Pollution Control Association (2)	June 1955[b]
Air Pollution Bibliography Vols. I and II	Library of Congress (3, 4)	
Air Pollution: A Bibliography	Bureau of Mines (5)	

[a] Bimonthly; November–December issue is cumulative for year.
[b] Ceased publication in January 1971.

references uncovered will take all the time allotted for study of references. Of course, if there are no such time constraints, these rules do not apply.

One uses different resources for different languages. What follows refers to a search in the English language. One would begin with a search through the abstracts published by the Air Pollution Technical Information Service (APTIC) of the Office of Technical Information and Publications of the Office of Air Programs (1) of the U.S. Environmental Protection Agency (Table 1-2). The name of the abstracts and of the publishing agency have had some changes since the initiation of their publication by APTIC, as various governmental reorganizations occurred incident to the transfer of this activity from the Public Health Service, U.S. Department of Health, Education, and Welfare to the Environmental Protection Agency.

The next source to be searched started in January, 1955 and ceased publication in December, 1970. It, too, changed names slightly during its lifetime, but for most of the time it was known as APCA Abstracts and was published by the Air Pollution Control Association (2) (of the United States and Canada).

Prior to the publication of APCA Abstracts and to a slight extent overlapping its early years, there were a series of Library of Congress Abstracts, published in both card and book (3, 4) form. Preceding the Library of Congress Abstracts there was a 1954 Bibliography published by the U. S. Bureau of Mines (5).

In addition, since 1966, the Center for Air Environment Studies of Pennsylvania State University (6) has published a key word in context (KWIC) Index of Air Pollution Titles, listing the titles of air pollution books, papers and reports.

The foregoing are all government-supported bibliographic ventures. A recent development has been the advent of several private bibliographic publications in the environmental protection and management field (7, 8).

In addition to the abstracts on pollution discussed above, there are abstracting services in the chemical, meteorological, engineering, and medical fields (9–12).

For a more complete discussion of "Air Pollution Literature Resources," the student is advised to refer to the chapter of that name in Volume III of "Air Pollution" (13). Much of the early air pollution literature, otherwise difficult to obtain, is available on microfilm in the Bay Area Microfilm Library (14).

VI. The Authors' Philosophy

As persons who have dedicated all or significant parts of their professional careers to the field of air pollution, the authors believe in its importance and relevance. We believe that as the world's population increases, it will become increasingly important to have an adequate number of well-trained persons professionally engaged in air pollution control. If we didn't believe this, it would have been pointless for us to have written this textbook.

We recognize that, in terms of short-term urgency, many nations and communities may rightly assign a lower priority to air pollution control than to problems of population, poverty, nutrition, housing, education, water supply, communicable disease control, civil rights, mental health, aging, or crime. Air pollution control is more likely to have a higher priority for a person or community of persons already reaping the benefits of society in the form of adequate income, food, housing, education, and health care than for persons who have not and may never reap these benefits.

However, in terms of long-term needs, nations and communities can ignore air pollution control only at their peril. A population can subsist, albeit poorly, with inadequate housing, schools, police, and care of the ill, insane, and aged; it can also subsist with primitive water supply. The ultimate determinants for survival are its food and air supplies. Conversely, even were society to succeed in providing in a completely adequate manner all of its other needs, it would be of no avail if the result were an atmosphere so befouled as not to sustain life. The long-term objective of air pollution control is to allow the world's population to meet all its needs for energy, goods, and services without sullying its air supply.

References

1. Air Pollution Technical Information Service, Office of Technical Information and Publications, U.S. Environmental Protection Agency, P.O. Box 12055, Research Triangle Park, North Carolina 27709.
2. Air Pollution Control Association, 4400 Fifth Avenue, Pittsburgh, Pennsylvania 15213.
3. Gibson, J. R., Culver, W. E., and Kurz, M. E. "The Air Pollution Bibliography," Vol. I. Library of Congress, Washington, D.C., 1957.
4. Jacobius, A. J., Gibson, J. R., Wright, V. S., Culver, W. E., and Kassianoff, L. "The Air Pollution Bibliography," Vol. II. Library of Congress, Washington, D.C., 1959.
5. Davenport, S. J., and Morgis, G. G. Air pollution; a bibliography. *U.S. Bur. Mines Bull.* 537, (1954); [U.S. Government Printing Office, Washington, D.C., 1954] 448 pp.
6. Center for Air Environment Studies, The Pennsylvania State University, University Park, Pennsylvania 16802.
7. Institute for Scientific Information, 325 Chestnut Street, Philadelphia, Pennsylvania 19106.
8. Pollution Abstracts, Pollution Abstracts, Inc., La Jolla, California 92037.
9. Chemical Abstracts, American Chemical Society, Washington, D.C.
10. *Meteorol. Geoastrophys. Abstr.*, American Meteorological Society, Washington, D.C.
11. Engineering Index, Engineering Index Inc., New York, New York.
12. Index Medicus, National Library of Medicine, Bethesda, Maryland.
13. Stern, A. C. Ed., "Air Pollution," 2nd ed. 3 vols. Academic Press, New York, 1968.
14. Bay Area Air Pollution Control District, 939 Ellis Street, San Francisco, California 94109.

Suggested Reading

"Air Conservation." Report of the Air Conservation Commission of the American Association for the Advancement of Science. AAAS Publ. No. 80. Washington, D.C., 1965, 335 pp.
"Restoring the Quality of Our Environment." Report of the Environmental Pollution Panel. President's Science Advisory Committee. U.S. Government Printing Office, Washington, D.C., 1965, 317 pp.

Questions

1. Discuss the part to be played by social scientists in air pollution control.
2. List the more important professional journals in the air pollution field.

3. List the more important abstract journals covering the fields of chemistry, meteorology, engineering, and medicine.

4. Discuss the extent that a nation or a community that has problems of population, poverty, nutrition, housing, etc., is justified in spending money on air pollution control.

5. What are the subject areas that must be mastered for an advanced degree in meteorology? Discuss their relevance to the training of an air pollution meteorologist.

6. Discuss the proposition that public interest in air pollution did not arise until a large segment of the public had flown in airplanes and had seen polluted air from aloft.

7. Discuss the gaps in your academic training and professional experience that need to be filled in to best prepare you for work in the field of air pollution.

8. Convert to the metric system:
   (a) Density: 1 pound/ft^3
   (b) Viscosity: 1 pound/inch-second
   (c) Weight: 1 grain
   (d) Acceleration: 1 ft/sec^2

9. Convert to English units:
   (a) Diameter: 1 μ
   (b) Absolute temperature: 1°K
   (c) Dust loading: 1 μg/m^3
   (d) Gravitational constant: g

Part I
EFFECTS OF AIR POLLUTION

Chapter 2

ECOLOGY

I. Air–Water–Soil Interrelations

The relationships among air, water, soil, and living matter are called "ecological relationships," and their study is called "ecology." The end result of essentially all atmospheric scavenging mechanisms is to cause the pollutant to leave the air by sedimentation, washout, or impaction. All three mechanisms transfer pollutants from the air to the surface of the earth, i.e., its soil, water, vegetation, or buildings. If a pollutant that has been deposited on the ground is in solution, it can be carried through the root system of plants into the plants themselves, or it can be leached out of the deposit, perhaps after action by soil microbes. From that point it can be carried to underground aquifers, or surface runoff to lakes, streams, and eventually the sea. If this water, or these plants or plant parts are used for human or animal food or drink, the pollutant will enter the food chain.

Even when the pollutant is insoluble in rainwater or groundwater, it may come to rest on a plant, fruit, or vegetable that is used as a foodstuff by either man or animals. After ingestion it may prove soluble in stomach acids, thereby entering the human or animal bloodstream. Once the pollutant becomes a constituent of a lake, stream, or sea—even as an undissolved particle suspended therein—it can enter the marine food chain to accumulate in some form of sea food to be used as a food or foodstuff of man or terrestrial animals. It is not necessary that a particle

11

be metabolized by an organism in order to enter the food chain. An organism can be eaten by a larger one, and eventually enter the stomach of a fish. The fish can be eaten for food or be ground into fish meal for poultry feed. In the latter case, the toxicant will arrive on your dining room table in a chicken or turkey.

Perhaps the prime example of the transport of a pollutant through ecological pathways is the fate of the tetraethyl lead used as an antiknock agent in automotive gasoline. When leaded gasoline is burned in an internal combustion engine, some lead-containing particles are discharged from the tail pipe of the car as very fine particles that become nuclei upon which condensation of water droplets can occur, the remainder plates out on the walls of the tail pipe and muffler. Backfires, bursts of acceleration, or sudden jolts will break these platelets loose and discharge them from the tail pipe onto the road. Road traffic will grind them into fine dust, which becomes airborne by the turbulence created by vehicular motion and wind action. Since this dust is composed of relatively large particles, it will settle out of the air quite rapidly, mostly along both sides of the highway or street. In the former case it will settle onto vegetation or soil; in the latter case, onto paved sidewalks and rooftops. After each rainfall, the runoff washes into the nearest stream. The result is the build-up of relatively high lead levels in the stream and in the oceans into which they eventually discharge. The build-up of lead concentrations in the sea, since the introduction of lead as an antiknock agent, has been documented by lead analysis of the annual increments to polar ice obtained from cores taken from the ice (1) (Fig. 2-1).

Thus from one source, lead can get into the air as both fine and coarse particulate matter. While it is still airborne, i.e., before it leaves the air by washout or sedimentation, these particles can be breathed by man and animals. In man, the fine

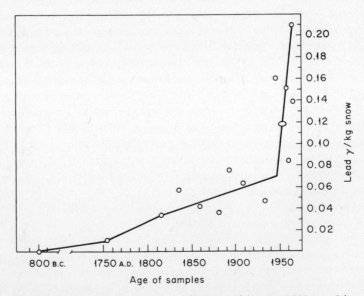

Fig. 2-1. Increase in lead content of snow and ice since 800 B.C. (1).

particles may reach and be retained by the lung from which their lead content can find its way into the bloodstream and the skeletal structure. The larger particles will be retained by mucus in the nose and the upper respiratory tract. Except for the very small amount of mucus rejected by nose-blowing, expectorating, coughing, and sneezing, the remainder is automatically ingested and is swept into the gastro-intestinal tract. Some people who are occupationally exposed to the dust or fumes of lead, as, for example, workers in a factory manufacturing lead storage batteries, will be exposed to larger quantities of airborne lead than the general populace.

II. Total Body Burden

Lead is a toxicant in the sense that excessive accumulation in the body can ultimately result in lead poisoning, which has a broad spectrum of manifestations ranging from colic, to palsy, to brain damage; and eventually to death. The total body burden concept is that the body cannot distinguish among the sources of a toxicant. The body accumulates the toxicant from all sources and is affected by the total accumulation, dose, or dose rate. The basic body burden of any substance is the level found in the average healthy person in an unpolluted environment. Substances such as lead are toxicants when present in the body in excessive quantity, yet they are found in unpolluted environments as normal constituents of plant and animal tissues. As a matter of fact, this is true for most elements: When too much is contained in the body, the effect is intoxication or poisoning; but when there is too little, the result is what are called deficiency diseases—the manifestation of an inadequacy of that element in the body.

The principal lead intake for a human, who is not occupationally exposed to lead, is from ingestion of food and water. Very little of his lead intake comes from inhalation. In the centuries before the air became infused with lead from vehicular exhaust, human exposure to excessive lead arose from obtaining drinking water through lead pipes; from storing of foodstuffs in lead-containing vessels, from lead-rich paints and cosmetics, and from spreading lead-containing insecticides on crops. By the time air became tainted with lead from vehicular exhaust, most of these other sources of lead were on the wane, and in fact had almost disappeared in most countries. Since that time the principal source of increased lead exposure above "background" has been associated with increases of lead in our water and food supplies due to the soil and water leaching processes previously described. Thus, although the present principal source of lead is an air pollutant, the total body burden of lead for most people arises mainly from the ingestion of lead, rather than from its inhalation.

This example was not cited to cause alarm but rather to explain a concept and a process. Actually tetraethyl lead is in the process of being phased out. It is totally banned in some countries of the world and is voluntarily disappearing from gasoline in the United States. Although it was banned in some countries because of a determination that its continued use would result in an excessive lead body burden in the populace, this was not the case in the United States. Studies in the United States prior to the time when the use of lead in fuels started to decrease had not revealed total body burdens of lead in the populace high enough to cause alarm (2). The decision to switch to lead-free or low-lead gasolines in the United States was prompted by the fact that continued use of lead would make it difficult for automobile manufacturers to use the types of catalytic converters they deemed necessary to meet Federal standards for carbon monoxide, hydrocarbons, and oxides of nitrogen in automobile exhaust.

III. Recycling

In solid waste disposal, the principal options are to burn, bury, compost, or recycle. When wastes are burned, a potential air pollution problem is created by the introduction to the air of the products of complete and incomplete combustion. Chemicals from land fills, dumps, and lagoons can leach into aquifers and streams. Composting gives rise to odors and wind-blown dust. Recycling garbage by feeding to hogs gives rise to problems of odor as well as of food-borne diseases. More sophisticated forms of solid waste recycling, which remain to be developed, may well require controls to prevent both air and water pollution.

Sewage and industrial liquid-waste treatment plants have the potential for odor emission and for other forms of air pollution if the resulting sludge is burned or dumped at sea. Many forms of air pollution control yield a solid or liquid waste, the disposal of which may create a water pollution problem. Conversely, the solution not infrequently proposed for some water pollution problems is to burn the effluent rather than discharge it to a stream.

Thus it will be seen that there are ecological trade-offs in solving pollution problems. The total environmental pollution problem is not solved by changing an air pollution problem to one of water pollution and *vice versa*. In the long run, total environmental pollution needs to be minimized by so balancing the interrelationships among releases to the air, water, and land as to optimize the overall control achieved.

IV. Population

Data from the U. S. National Air Sampling Network (3) (Table 2-1) show a marked dependence of suspended particulate matter in the air on the population of the area involved. Wohlers *et al.* (4) compared air pollutant concentrations with pollutant emissions and population (Fig. 2-2). The data demonstrate the effect of increasing city population upon emissions and concentrations for sulfur dioxide and nitrogen oxides. Similar relationships exist for other pollutants and also between population and visibility, corrosion of metal, deterioration of fabrics, and the soiling of exterior painted surfaces (4).

Another way of showing that population generates pollution is to look at pollution levels along a line transecting a populated area, which starts in the rural area on one side, passes first through the suburbs, then through the center city, then through the suburbs on the other side of the center city, and finally into the rural

TABLE 2-1

Distribution of Cities by Population Class and Particulate Concentration, 1957 through 1967 (3)

Population class	Average particulate concentration, µg/m³										
	Less than 40	40–59	60–79	80–99	100–119	120–139	140–159	160–179	180–199	More than 200	Total
Over 3 million	—	—	—	—	—	—	1	—	1	—	2
1–3 million	—	—	—	—	—	—	2	1	—	—	3
0.7–1 million	—	—	1	—	2	—	4	—	—	—	7
400–700,000	—	—	—	4	5	6	1	1	1	—	18
100–400,000	—	3	7	30	24	17	12	3	2	1	99
50–100,000	—	2	20	28	16	12	6	5	1	3	93
25–50,000	—	5	24	12	12	10	2	1	2	3	71
10–25,000	—	7	18	19	9	5	2	3	1	—	64
Under 10,000	1	5	7	15	11	2	1	2	—	—	44
Total	1	22	77	108	79	52	31	16	8	7	401

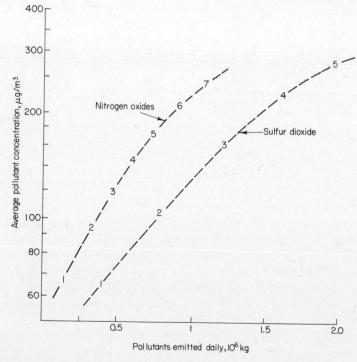

FIG. 2-2. Pollutant emissions versus concentration and population for American and Canadian cities. Numbers represent population in millions (4).

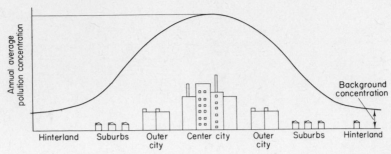

FIG. 2-3. Annual average pollution concentration over an urban area.

area on the other side of town. It will be seen (Fig. 2-3) that on an annual average basis (taking into account winds from all directions) pollution concentration is highest in the center city, lower in both the suburbs, and lowest in the surrounding rural area. The concentration in the surrounding rural area is called the background concentration, with respect to that in the city.

V. Nondegradation

Guardians of natural resources are called conservationists. Those concerned with ecosystems are frequently called environmentalists. Both groups are dedicated to a policy of preservation of natural resources and ecosystems. Conservationists recognize that preservation of natural resources is not equivalent to nonutilization of these resources. Without utilization of its natural resources society would grind to a halt. In the particular case of water flowing in streams, the use of the water

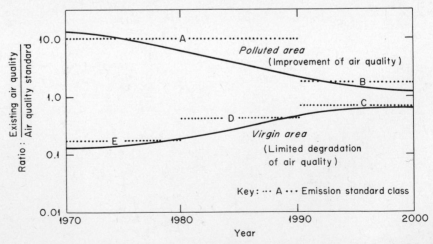

FIG. 2-4. Future air pollution control requirements of presently polluted and presently virgin areas.

body for navigation, hydraulic power production, fishing, water sports, and drinking water does not make it unfit for subsequent use downstream. Conservationists and environmentalists have evolved a "nondegradation" policy to the effect that no future use of any reach of a stream shall degrade it below its present status of fitness. This means that any reach of a stream, unused further upstream by man, must be maintained in its pristine state and that any liquids dumped into it by man must be essentially as pure as and at the temperature of the stream itself.

A similar nondegradation policy has not yet been enunciated for air. If it were, its strict interpretation would keep our population pinned down to the places we have already inhabited and thereby polluted. It would prevent us from exploring and inhabiting new lands, because these acts would bring into these new lands the polluting habits we now use to support life, i.e., burning fuels to cook, heat, provide energy, transportation, and communication. In practice, the future of air pollution control will involve simultaneously (a) the reduction of pollution from built-up areas to some acceptable lower level, and (b) limited increase of pollution in virgin areas the world will be forced to inhabit to accommodate its expanding population (Fig. 2-4). The test of success will be whether we can reduce the former and limit the latter at the same time.

References

1. Murozumi, M., Chow, T. J., and Patterson, C. Chemical concentrations of pollutant lead aerosols, terrestrial dusts and sea salts in Greenland and Antarctic snow strata. *Geochimi. Cosmochim. Acta.* **33,** 1247–1294 (1969).
2. Engel, R. E., Hammer, D. I., Horton, R. J. M., Lane, N. M., and Plumlee, L. A. "Environmental Lead and Public Health." Air Pollution Control Office Publication No. AP-90. Environmental Protection Agency, Research Triangle Park, North Carolina, 1971.
3. "Air Quality Data for 1967 [rev. ed, 1971]," Office of Air Programs Publication No. APTD 0741, Environmental Protection Agency, Research Triangle Park, North Carolina, 1971.
4. Wohlers, H. C., Katz, E., Malfara, L., and Plater-Zyberk, J. Air pollution as a function of population. Proceedings of the Institute of Environmental Sciences, Anaheim, California, April 1969.

Suggested Reading

Odum, E. "Ecology." Holt, New York, 1969.
Kormondy, E. "Concepts of Ecology." Prentice-Hall, Englewood Cliffs, New Jersey, 1969.

Questions

1. The text discusses the ecological fate of lead. Discuss in similar manner the ecological fate of mercury and of DDT.
2. What is the role of the chlorides and bromides incorporated in the ethyl fluid used to provide the tetraethyl or tetramethyl lead content of leaded gasolines? How does their presence in gasoline affect automotive exhaust pollution emissions?
3. What levels of lead in blood, urine, and bone are considered normal and what levels are considered indicative of disease?

4. Food containers are likely candidates for recycling. Discuss the relative ease and the means for recycling metal, glass, plastic, and paper containers.
5. Figure 2-3 is on an annual average basis. Draw similar curves on a monthly average basis for the two months of the year with the greatest dissimilarity of weather, and for two days, one of good and one of poor urban ventilation.
6. List arguments for and against a nondegradation policy for the air.
7. How would you design a laboratory experiment to show air–water–soil ecological interrelationships?
8. How does the pollution in your community, or the nearest one to your community that has adequate air quality data, compare with what you would have predicted from Table 2-1 and Figure 2-2?
9. Discuss the relative potential for environmental pollution of disposal of refuse by incineration and by sanitary landfill.

Chapter 3

AIR POLLUTION

I. Pollution*

Pollution, *n*: act of polluting, state of being polluted; defilement; impurity.
Pollute, *vt*: [L. *pollutus*, past part. of *polluere*, to pollute] to make or render unclean;
to defile, desecrate, profane-syn. see CONTAMINATE.
Air, *n*: [OF, *air*, fr. L. *aer*, fr. Gr. *aer*, air, mist] 1. the invisible, odorless, and tasteless
mixture of gases which surrounds the earth . . . .
Atmosphere, *n*: [*atmo* + Gr. *sphaira* sphere] 1. the whole mass of air surrounding the
earth . . . .

II. The Atmosphere

On a macroscale (Fig. 3-1) as temperature varies with altitude, so does density
(1). In general, the air grows progressively less dense as we move upward from the
troposphere through the stratosphere and the chemosphere to the ionosphere. In
the upper reaches of the ionosphere the gaseous molecules are few and far between
as compared with the troposphere.

The ionosphere and chemosphere are of interest to space scientists because they

---

* Definitions are from Webster's Seventh New Collegiate Dictionary'' (7th edition). Merriam,
Springfield, Massachusetts, 1967.

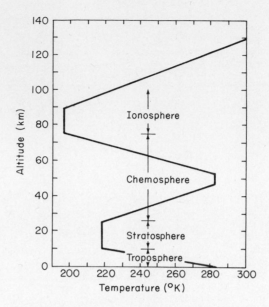

Fɪɢ. 3-1. The regions of the atmosphere.

must be traversed by space vehicles en route to or from the moon or the planets, and they are regions in which satellites travel in earth orbit. These regions are of interest to communications scientists because of their influence on radio communications, and they are of interest to air pollution scientists primarily because of their absorption and scattering of solar energy, thereby influencing the amount and spectral distribution of solar energy and cosmic rays reaching the stratosphere and troposphere.

The stratosphere is of interest to aeronautical scientists because it is traversed by airplanes; to communication scientists because of radio and television communications; and to air pollution scientists because global transport of pollution, particularly the debris of aboveground atomic bomb tests and of volcanic eruptions, occurs in the stratosphere; and because absorption and scattering of solar energy also occurs in the stratosphere.

The troposphere is the region in which we live and is the region to which this book is primarily devoted.

III. Unpolluted Air

The gaseous composition of unpolluted tropospheric air is given in Table 3-1. Unpolluted air is a concept, i.e., what the composition of the air would be if man and his works were not on earth. We will never know the precise composition of unpolluted air because by the time we had the means and the desire to determine its composition, man had been polluting the air for thousands of years. Now even at the most remote locations, at sea, at the poles, in the deserts, and mountains, the air may be best described as dilute polluted air. It closely approximates unpolluted

TABLE 3-1

The Gaseous Composition of Unpolluted Air (Dry Basis)

	ppm (vol)	$\mu g/m^3$
Nitrogen	780,900	8.95×10^8
Oxygen	209,400	2.74×10^8
Water	—	—
Argon	9,300	1.52×10^7
Carbon dioxide	315	5.67×10^5
Neon	18	1.49×10^4
Helium	5.2	8.50×10^0
Methane	1.0–1.2	$6.56–7.87 \times 10^2$
Krypton	1.0	3.43×10^3
Nitrous oxide	0.5	9.00×10^2
Hydrogen	0.5	4.13×10^1
Xenon	0.08	4.29×10^2
Organic vapors	ca. 0.02	—

air, but differs from it to the extent that it contains vestiges of diffused and aged man-made pollution.

The real atmosphere is more than a dry mixture of permanent gases. It has other constituents—vapor of both water and organic liquids; and particulate matter held in suspension. Above their temperature of condensation, the vapor molecules act just as permanent gas molecules in the air. The predominant vapor in the air is water vapor. Below its condensation temperature, if the air is saturated, it changes from vapor to liquid. We are all familiar with this phenomenon since it appears as fog or mist in the air, and as condensed liquid water on windows and other cold surfaces exposed to the air. The quantity of water vapor in the air varies greatly from almost

TABLE 3-2

The Gaseous Composition of Unpolluted Air (Wet Basis)

	ppm (vol)	$\mu g/m^3$
Nitrogen	756,500	8.67×10^8
Oxygen	202,900	2.65×10^8
Water	31,200	2.30×10^7
Argon	9,000	1.47×10^7
Carbon dioxide	305	5.49×10^5
Neon	17.4	1.44×10^4
Helium	5.0	8.25×10^2
Methane	0.97–1.16	$6.35–7.63 \times 10^2$
Krypton	0.97	3.32×10^3
Nitrous oxide	0.49	8.73×10^2
Hydrogen	0.49	4.00×10^1
Xenon	0.08	4.17×10^2
Organic vapors	ca. 0.02	—

complete dryness to supersaturation, i.e., between zero and 4% by weight. If Table 3-1 were to be recast to a wet air basis at a time when the water vapor concentration is 31,200 parts by volume per million parts by volume of wet air (Table 3-2), the concentration of condensible organic vapors is seen to be so low compared to that of water vapor that for all practical purposes the difference between wet air and dry air is its water vapor content.

Gaseous composition in Tables 3-1 and 3-2 has been expressed as parts per million by volume—ppm (vol). (When a concentration is expressed simply as ppm, one is in doubt as to whether a volume or weight basis is intended.) To avoid confusion caused by different units, air pollutant concentrations in this book are generally expressed as micrograms per cubic meter of air ($\mu g/m^3$) at 25°C and 760 mm Hg, i.e., in metric units. To convert from units of ppm (vol) to $\mu g/m^3$, it is assumed that the ideal gas law is accurate under ambient conditions. A generalized formula for the conversion at 25°C and 760 mm Hg is

$$1 \text{ ppm (vol) pollutant} = \frac{1 \text{ liter pollutant}}{10^6 \text{ liter air}}$$

$$= \frac{(1l/22.4) \times MW \times 10^6 \, \mu g/g}{10^6 l \times 298°K/273°K \times 10^{-3} \, m^3/l}$$

$$= 40.9 \times MW \, \mu g/m^3 \tag{3-1}$$

where MW equals molecular weight. For convenience, conversion units for common pollutants are shown in Table 3-3.

TABLE 3-3

Conversion Factors between Volume and Mass Units of Concentration (25°C, 760 mm Hg)

	To convert from	
Pollutant	ppm (vol) to $\mu g/m^3$, multiply by:	$\mu g/m^3$ to ppm (vol), multiply by ($\times 10^{-3}$):
Ammonia (NH_3)	695	1.44
Carbon dioxide	1800	0.56
Carbon monoxide	1150	0.87
Chlorine	2900	0.34
Ethylene	1150	0.87
Hydrogen chloride	1490	0.67
Hydrogen fluoride	820	1.22
Hydrogen sulfide	1390	0.72
Methane (carbon)	655	1.53
Nitrogen dioxide	1880	0.53
Nitric oxide	1230	0.81
Ozone	1960	0.51
Peroxyacetylnitrate	4950	0.20
Sulfur dioxide	2620	0.38

A minor problem arises for nitrogen oxides. It was common practice to add concentrations of nitrogen dioxide and nitric oxide in ppm (vol) and express the sum as "oxides of nitrogen." In metric units, conversion from ppm (vol) to $\mu g/m^3$ must be made separately for nitrogen dioxide and nitric oxide prior to addition.

IV. Particulate Matter

Neither Table 3-1 nor 3-2 lists among the constituents of the air the suspended particulate matter that it always contains. The gases and vapors exist as individual molecules in random motion. Each gas or vapor exerts its proportionate partial pressure. The particles are aggregates of many molecules, sometimes of similar molecules, often of dissimilar molecules. They age in the air by several processes. Some particles serve as nuclei upon which vapors condense. Some particles chemically react with atmospheric gases or vapors to form different compounds. When two particles collide in the air, they tend to adhere to each other because of attractive surface forces, thereby forming progressively larger and larger particles by agglomeration. The larger a particle becomes, the greater its weight and the greater its likelihood to fall to the ground, rather than continuing to remain airborne. The process of particles falling out of the air to the ground is called sedimentation. Washout of particles by snowflakes, rain, hail, sleet, mist, or fog is a common form of agglomeration and sedimentation. Still other particles leave the air by impaction onto and retention by the solid surfaces of vegetation and buildings. The particulate mix in the atmosphere is dynamic with the continual injection into the air from sources of small particles, their creation in the air by vapor condensation or chemical reaction among the gases and vapors in the air; and their removal from the air by agglomeration, sedimentation, or impaction.

Before the advent of man and his works, there must have been particles in the air from natural sources. These certainly included all the particulate forms of condensed water vapor; the condensed and reacted forms of natural organic vapors; salt particles resulting from the evaporation of water from sea spray; wind-borne pollen, fungii, molds, algae, yeasts, rusts, bacteria, and debris from live and decaying plant and animal life; particles eroded by the wind from the beach, desert, soil, and rock; particles from volcanic and other geothermal eruption, and from forest fires started by lightning; and particles entering the troposphere from outer space. As mentioned earlier the true natural background concentration will never be known because when it existed man was not there to measure it, and by the time man started measuring particulate matter levels in the air, he had already been polluting the atmosphere with particles resulting from his presence on earth for several thousand years. The best that can be done now is to assume that the particulate levels at remote places—the middle of the sea, the poles, and the mountains—approach true background concentration. The very act of man's going to a remote location to make a measurement implies some change in the atmosphere of that remote location attributable to the means man used to get himself there and to maintain himself while obtaining the measurements (Heisenberg Uncertainty Principle!). Particulate matter is measured on a dry basis, thereby eliminating from the measurement not only water droplets and snowflakes but also all vapors both aqueous and organic that evaporate or are dessicated from the particulate matter during the drying process. Since different investigators and investigative procedures employ different drying procedures and definitions of dryness, it is important to know the procedures and definition employed, when comparing data.

There are other ways of measuring particulate matter than by weight per unit volume of air. They include a count of the total number of particles in a unit volume of air; a count of the number of particles of each size range; the weight of particles in each size range and similar measures based on surface area and volume of the particles rather than on their number or weight. Some particles in the air are so small that they cannot be seen by an optical microscope, individually weighing so little that their presence is masked in gravimetric analysis by the presence of a few large particles. The mass of a spherical particle is

$$w = \tfrac{4}{3}\pi\rho r^3 \tag{3-2}$$

where w is the particle mass (gm); r is the particle radius (cm); and ρ is the particle density (gm/cm^3).

The size of small particles is measured in microns (μ). One micron is one-millionth of a meter or 10,000 Å (angstrom units)—the units used to measure the wavelength of light (visible light is between 3000 and 8000 Å) (Fig. 3-2) (2). Compare the weight of a 10-μ particle near the upper limit of those found suspended in the air and a 0.1-μ particle which is near the lower limit. If both particles have the same density (ρ), the smaller particle will weigh one-millionth the weight of the larger one. The usual gravimetric procedures can scarcely distinguish a 0.1-μ particle in the presence of a 10-μ particle. To measure the entire size range of particles in the atmosphere several measurement techniques must therefore be combined, each most appropriate for its size range (Table 3-4). Thus the smallest particles—those only slightly larger than a gas molecule—are measured by the electric charge they carry and by electron microscopy. The next larger size range is measured by electron microscopy or by the ability of these particles to act as nuclei upon which water vapor can be condensed in a cloud chamber. (The water droplets are measured rather than the particles themselves.) The still larger size range is measured by electron or optical microscopy; and the largest size range is measured gravimetri-cally, either as suspended particles separated from the air by a sampling device, or as sedimented particles falling out of the air into a receptacle near the ground.

The methods noted above tell something about the physical characteristics of atmospheric particulate matter but nothing about its chemical composition. One can seek this kind of information for either individual particles or for all particles *en masse*. It is beyond our present capability to chemically analyze an individual 0.1-μ particle, and a sufficient number of identical particles cannot be selectively collected from the air for analysis. Analysis of particles *en masse* means analysis of a mixture of particles of many different compounds. How much of each element or radical, anion, or cation is present in the mixture can be determined, but not how they are combined as chemical compounds or individual particles. Specific organic compounds may be separated and identified but, to date, only a small number of the

TABLE 3-4

Particle Size Ranges and Their Methods of Measurement

μ	Ions	Nuclei	Visibility	Suspended or settleable nonairborne	Dispersion aerosol	Condensation aerosol	Pollen or spores	Diffusion or settling
10^{-4}–10^{-3}	Small	—	Electron microscope	Suspended	—	Gas molecules	—	Diffusion
10^{-3}–10^{-2}	Intermediate and large	Aitken nuclei	—	Suspended	—	Vapor molecules	—	Diffusion
10^{-2}–10^{-1}	Large	Aitken and condensation	Electron microscope	Suspended	—	Fume–Mist	—	Diffusion
Air pollution $\left\{\begin{array}{l}10^{-1}$–$10^{-0} \\ 10^{0}$–$10^{1} \\ 10^{1}$–$10^{2}\end{array}\right.$	— — —	Condensation nuclei — —	Microscope: electron and optical Microscope: optical Eye sieves	Suspended Suspended and settleable Settleable	Dust–Mist Dust–Mist Dust–Mist	Fume–Mist Fume–Mist Mist–Fog	— — Pollen and spores	Diffusion and sedimentation Sedimentation Sedimentation
10^{2}–10^{3}	—	—	Eye sieves	Nonairborne	Dust–Spray	Drizzle–Rain	—	Sedimentation
10^{3}–10^{4}	—	—	Eye sieves	Nonairborne	Sand–Rocks	Rain	—	Sedimentation

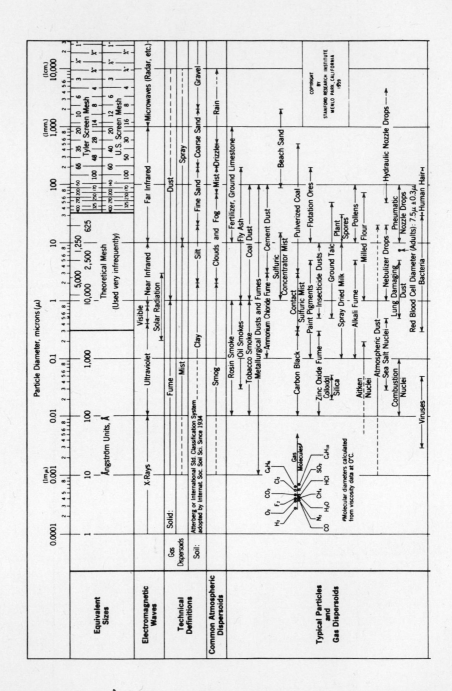

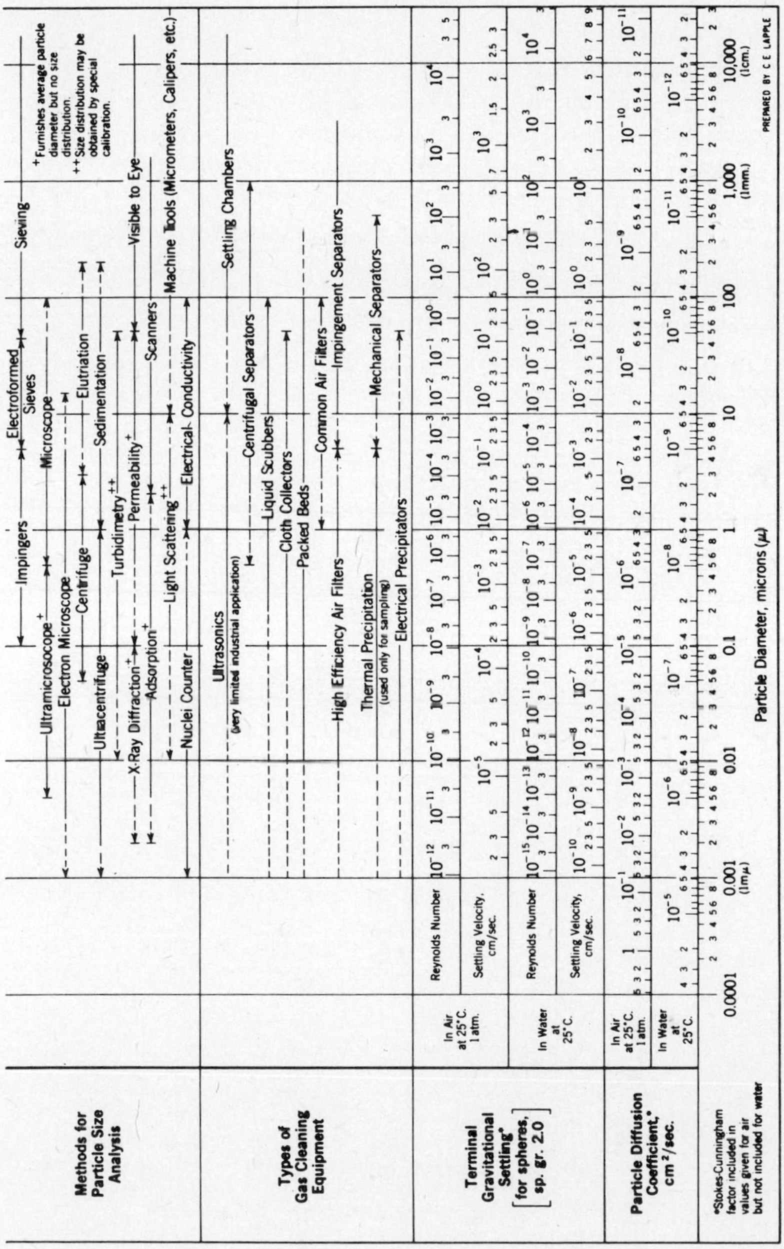

FIG. 3-2. Characteristics of particles and particle dispersoids. Reproduced by permission of Stanford Research Institute, Menlo Park, California, 1959.

infinite number of possible organic compounds that may be in the air have been identified.

V. Sources and Sinks

The places from which pollutants emanate are called sources. There are natural as well as man-made sources of the permanent gases considered pollutants. These include plant and animal respiration, and the decay of what was once living matter. Volcanoes and naturally caused forest fires are other natural sources.

The places to which pollutants disappear from the air are called "sinks." Sinks include the soil, vegetation, structures, and water bodies, particularly the oceans. The mechanisms whereby pollutants are removed from the atmosphere are called scavenging mechanisms, and the measure used for the aging of a pollutant is its half-life—the time it takes for half of the quantity of pollutant emanating from a source to disappear into its various sinks (3) (Table 3-5). Fortunately most pollutants have a short enough half-life, i.e., days rather than decades, to prevent their accumulation in the air to the extent that they substantially alter the composition of unpolluted air shown in Table 3-1. One gas that does appear to be accumulating in the air to the extent that its percentage of the total is increasing year by year is carbon dioxide (Fig. 3-3) (4). Other gases may also be increasing in percentage, but such an occurrence has not yet been documented.

Oxidation, either atmospheric or biological, is a prime removal mechanism for inorganic as well as organic gases. Inorganic gases, such as nitric oxide (NO), nitrogen dioxide (NO_2), hydrogen sulfide (H_2S), sulfur dioxide (SO_2), and sulfur trioxide (SO_3), may eventually form corresponding acids:

$$NO + \tfrac{1}{2}O_2 \rightarrow NO_2 \tag{3-3}$$

$$4\,NO_2 + 2\,H_2O + O_2 \rightarrow 4\,HNO_3 \tag{3-4}$$

$$H_2S + \tfrac{3}{2}O_2 \rightarrow SO_2 + H_2O \tag{3-5}$$

$$SO_2 + \tfrac{1}{2}O_2 \rightarrow SO_3 \tag{3-6}$$

$$SO_3 + H_2O \rightarrow H_2SO_4 \tag{3-7}$$

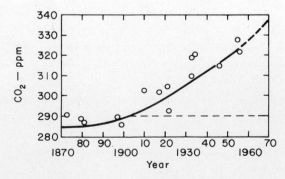

FIG. 3-3. Average CO_2 concentration in North Atlantic Region, 1870–1956(4). Dashed line: 19th century base value = 290 ppm.

Oxidation of SO_2 is slow in a mixture of pure gases but the rate is increased by light, NO_2, oxidants, and metallic oxides which act as catalysts for the reaction. The formed acids react with particulate matter or ammonia to form salts.

The speed of the reaction between nitric oxide and oxygen, normally slow (120 $\mu g/m^3$, $t_{\frac{1}{2}} = 1000$ hr), is increased manyfold ($t_{\frac{1}{2}} = 18$ sec) in the presence of ozone (O_3). The ozone-speeded reaction is important in photochemical air pollution. Background O_3 levels, approximating 50 $\mu g/m^3$, result from atmospheric mixing of high concentrations naturally occurring at 16 to 32 km above sea level. Interestingly, it may be postulated that nitrogen oxides added to the stratosphere by high-flying planes may react with and destroy this ozone layer, allowing shorter wavelength light to reach the earth's surface.

VI. Receptors

The receptor is that which is adversely affected by polluted air. The receptor may be a person or animal that breathes the air and whose health may be adversely affected thereby, or whose eyes may be irritated or whose skin made dirty. It may be a tree or plant that dies, or the growth, yield, or appearance of which is adversely affected. It may be some material such as paper, leather, cloth, metal, stone, or paint that is affected. Finally, some properties of the atmosphere itself such as, for example, its ability to transmit radiant energy may be affected; or it may be soil or water bodies that are adversely affected receptors.

VII. Transport and Diffusion

Transport is the mechanism that moves the pollution from a source to a receptor. The simplest source–receptor combination is that of an isolated point source and an

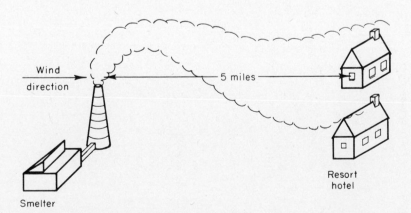

Fig. 3-4. Transport and diffusion from source to receptor.

TABLE 3-5

Summary of Sources, Concentrations, and Major Reactions of Atmospheric Trace Gases (June 1969) (3)

Contaminant	Major pollution sources	Natural sources	Estimated annual emissions		Atmospheric background concentrations	Calculated atmospheric residence time	Removal reactions and sinks	Remarks
			Pollution (tons)	Natural (tons)				
SO_2	Combustion of coal and oil	Volcanoes	146×10^6	None[a]	0.2 ppb[a]	4 days	Oxidation to sulfate by ozone, or after absorption, by solid and liquid aerosols	Photochemical oxidation with NO_2 and HC may be the process needed to give rapid transformation of $SO_2 \rightarrow SO_4$
H_2S	Chemical processes, sewage treatment	Volcanoes, biological action in swamp areas	3×10^6	100×10^6 [a]	0.2 ppb	2 days[a]	Oxidation to SO_2	Only one set of background concentrations available
CO	Auto exhaust and other combustion	Forest fires	275×10^6	75×10^6 [b]	0.1 ppm	<3 years[b]	None known, but large sink necessary	Ocean contributions to natural source probably low
NO/NO_2	Combustion	Bacterial action in soil (?)	53×10^6	NO 430×10^6 NO_2 658×10^6	NO: 0.2–2 ppb NO_2: 0.5–4 ppb	5 days	Oxidation to nitrate after sorption by solid and liquid aerosols, hydrocarbon photochemical reactions	Very little work done on natural processes

NH_3	Waste treatment	Biological decay	4×10^6	1160×10^6	6 ppb to 20 ppb	7 days	Reaction with SO_2 to form $(NH_4)_2SO_4$, oxidation to nitrate	No quantitative rate data on oxidation of NH_3 to NO_3, which seems to be dominant process in atmosphere
N_2O	None	Biological action in soil	None	590×10^6	0.25 ppm	4 years	Photo dissociation in stratosphere, biological action in soil	No information on proposed absorption of N_2O by vegetation
Hydro-carbons	Combustion exhaust, chemical processes	Biological processes	88×10^6	480×10^6	CH_4: 1.5 ppm non CH_4: <1 ppb	16 years (CH_4)	Photochemical reaction with NO/NO_2, O_3 large sink necessary for CH_4	"Reactive" hydrocarbon emissions from pollution = 27×10^6 tons per year
CO_2	Combustion	Biological decay, release from oceans	1.4×10^{10}	10^{12}	320 ppm	2–4 years	Biological absorption and photosynthesis, absorption in oceans	Atmospheric concentrations increasing by 0.7 ppm per year

[a] See also Kellogg, W. W., Cadle, R. D., Allen, E. R., Lazrus, A. L., and Martell, E. A., "The sulfur cycle," *Science* **175**, 587–596 (Feb. 1972). This suggests 1.5×10^4-tons/year volcanic SO_2.

[b] See also Maugh, T. H. II, "Carbon monoxide: Natural sources dwarf man's output," *Science* **177**, 338–339 (July 1972). This suggests that 75×10^6 may be too low by a factor of 40, and a 3-year residence time too high by a factor of 10.

isolated receptor. A point source may best be visualized as a chimney or stack emitting a pollutant into the air; the isolated point source might be the stack of a smelter standing by itself in the middle of a flat desert next to the body of ore it is smelting. The isolated receptor might be a resort hotel 5 miles distant on the edge of the desert. The effluent from the stack will flow directly from it to the receptor when the wind is along the line connecting them (Fig. 3-4). The wind is the means by which the pollution is transported from the source to the receptor. However, during its transit over the 5 miles between the source and the receptor, the plume does not remain a cylindrical tube of pollution of the same diameter as the interior of the stack from which it was emitted. Instead, as it travels over the 5-mile distance, turbulent eddies in the air and in the plume move parcels from the edges of the plume into the surrounding air, and move parcels of surrounding air into the plume. If the wind speed is greater than the speed of ejection from the stack, the wind will stretch out the plume until the plume speed equals wind speed. These two processes—mixing by turbulence and stretch-out of the plume, plus a third one—meandering (which means that the plume may not follow a true straight line between the source and the receptor, but may meander somewhat about that line as wind direction fluctuates from its mean value over the time of transit between the two points)—tend to make the concentration of the plume as it arrives at the receptor less than its concentration on release from the stack. The sum of all these processes is called diffusion. The process of diffusion becomes increasingly complex as the number of sources and receptors increases; as sources and receptors begin to group together into towns and cities; as some of the sources and receptors move as they do in moving vehicles; and, finally, as the weather and topography become more complex than a wind blowing in one direction over a flat desert for a prolonged period of time.

References

1. Turner, D. B. Meteorological fundamentals *in* "Meteorological Aspects of Air Pollution—Course Manual." Air Pollution Training Institute, Office of Manpower Development, Office of Air Programs. U.S. Environmental Protection Agency, Research Triangle Park, North Carolina.
2. Lapple, C. E. *Stanford Research Institute Journal* **5**, 95 (1961).
3. Robinson, E., and Robbins, R. C. "Sources, Abundance and Fate of Gaseous Atmospheric Pollutants." Final Report Prepared for American Petroleum Institute, Stanford Research Institute, Menlo Park, California, 1968. Supplementary Report (June 1969).
4. Callendar, G. C. *Tellus* **10**, 243 (1958).

Suggested Reading

"Guide to the Appraisal and Control of Air Pollution," 2nd ed. Prepared by the Program Area Committee on Air Pollution, American Public Health Association, New York, 1969, 80 pp.

Air Pollution Manual, Part I, Evaluation, 2nd ed. American Industrial Hygiene Association, Detroit, Michigan, 1972, 259 pp.

Questions

1. Prepare a graph showing the conversion factor from ppm (vol) to $\mu g/m^3$ for molecular weight compounds ranging from 10 to 200 at 25°C and 760 mm Hg as well as at 0°C and 760 mm Hg.

2. (a) Convert 0.2 ppm (vol) NO and 0.15 ppm (vol) NO_2 to $\mu g/m^3$ nitrogen oxides (NO_x) at 25 °C and 760 mm Hg.

   (b) Convert 0.35 ppm (vol) NO_x to $\mu g/m^3$ at 25°C and 760 mm Hg.

3. Prepare a table showing the weight in grams and the surface area in m^3 of a 0.1-, 1.0-, 10.0-, and 100.0-μ-diameter spherical particle of unit density.

4. What is the settling velocity in cm/sec in air at 25°C and one atmosphere for a 100 mesh size spherical particle, i.e., one which just passes through the opening in the sieve (specific gravity = 2.0).

5. How does the diameter of airborne pollen grains compare with the diameter of a human hair?

6. What are the principal chemical reactions that take place in the chemosphere to give it its name? How do they influence stratospheric and tropospheric chemical reactions.

7. What is the source and nature of the condensible organic vapors in unpolluted air?

8. Has the composition of the unpolluted air of the troposphere most probably always been the same as Tables 3-1 and 3-2? Will Tables 3-1 and 3-2 most probably define unpolluted air in the year 2075? Discuss your answer.

9. Describe the apparatus and procedures used to measure atmospheric ions and nuclei.

Chapter 4

SCALES OF THE AIR POLLUTION PROBLEM

It is important early on to recognize that there is not one air pollution problem but rather several distinct problems, all subject to the concepts previously discussed and each having its own characteristics. These problems may be distinguished by, among other things, their scale. There are four dimensions that establish scale (Table 4-1). The first is the horizontal dimension—how much of the earth's surface is involved. The second is the vertical dimension—how great a depth of the atmosphere is involved. The third is time—over what time scale does the problem develop and over what time scale may its control be resolved. The fourth is the scale of organization required for its resolution.

TABLE 4-1

Categories of the Air Pollution Problem

Category	Vertical scale	Temporal scale	Scale of organization required for resolution
Local	Height of stacks	Hours	Municipal
Urban	Lowest mile	Days	Regional
National	Troposphere	Months	State or provincial
Continental	Stratosphere	Years	National
Global	Atmosphere	Decades	International

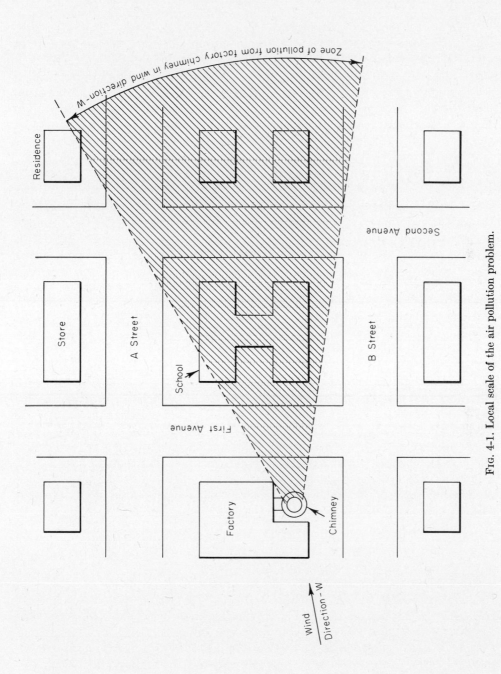

Fig. 4-1. Local scale of the air pollution problem.

I. Local

The smallest problem with regard to scale is the local problem (Fig. 4-1). Here what is meant is that a source and receptor are in close proximity—generally in plain sight of each other. An example would be the main street of a city, lined on both sides with multistory buildings and carrying extensive automotive traffic on a stop-and-go basis, with traffic lights on every corner. The sources are the automobiles. The receptors are their occupants, the pedestrians on the sidewalks, and the occupants of the adjacent buildings. The pollutants are automobile exhaust gases and particulate matter. The transport and diffusion mechanisms are those developed primarily by vehicular motion and convective air motion within the canyon formed by the building walls and the street, and secondarily by the weather pattern prevalent in the community. The horizontal scale of the problem is the street itself. The vertical scale is the height of the adjacent buildings. The time scale is measurable in minutes since traffic density can easily change by a factor of two or more within an hour. The time scale for control with no change in traffic flow is relatively long, but by restricting traffic the problem is resolvable in a matter of minutes.

The local scale of the air pollution problem is that for which the source or sources affecting a receptor can be identified without the necessity of specifically adding a tracer. Controlling the identified source will abate its adverse effects on the receptor. Generally, where a single source or group of sources can be traced to one specific receptor, it also can be traced to other receptors to which its effluent will be transported under different wind conditions. The most common form of local problem is identification of the source of a unique odor. Control consists of reducing the odorous effluent to the point at which it can no longer be smelled. Similarly, sources of unique particles or droplets falling in a neighborhood can be identified and abated. It can be readily demonstrated that street-level carbon monoxide concentration is proportional to traffic density along the street. Immediate abatement is to decrease traffic density; long-term abatement is to decrease emissions from vehicles.

Organizationally, local problems respond best to local resolution, i.e., at the lowest level of government at which there is both sufficient regulatory authority and technical expertise. This will usually occur at the municipal level.

II. Urban

Most urban areas consist of a center city surrounded by its suburbs, which in turn are surrounded by a nonurban hinterland. On a long enough averaging time, such as seasonal average, the pollution concentration over the region looks like Fig. 4-2, with highest concentrations in the center of the region and background concentrations in the nonurban hinterland. In an urban area a major air pollution problem is what happens when there is a cessation of ventilation of the city. Normally a city is ventilated by two mechanisms—horizontal wind flow to move pollution out of the community laterally; and vertical convection to move pollution from near the ground to upper levels of the atmosphere, and to bring cleaner air from

FIG. 4-2. Urban scale of the air pollution problem (1) showing theoretical adjusted winter suspended particulate concentrations. (—) State boundary; (—) county boundary in Maryland; (— - —) county boundary in Pennsylvania, Virginia, and Delaware.

aloft down to ground level. These two mechanisms normally occur in any city to a greater or lesser extent every day of the year. However, on certain days the meteorological situation is such that either or both mechanisms of ventilation fail to occur. Cessation of horizontal wind is called a calm. Suppression of vertical convection is usually the result of an atmospheric inversion. When both mechanisms fail for a protracted period of time, e.g., more than 36 hours, what is termed a "stagnation" has occurred. During a stagnation, pollution concentrations build up in the air over the region as if it were covered by an inverted box having as its sides the boundaries of the region, its bottom the earth's surface, and its top the inversion ceiling, into which pollution continues to pour at a substantially constant rate. The resulting increase of pollution concentration is called an "episode." People have died from acute exposure in air pollution episodes in major cities of the world.

Thus on an urban scale the horizontal dimension is the diameter of the urbanized area. The vertical scale is confined to the lowest kilometer of the atmosphere. The sources are all those within the area and the receptors are primarily the people inhabiting the area. The time scale is measurable in hours both for an episode to occur and for the necessary reduction in source emissions to be effectuated.

An urban complex usually comprises a number of separate city, town, village, and county governmental units. Therefore, the appropriate organizational scale for resolution of urban scale problems is regional. This will usually be in a regional organization for air pollution control, or at the state or provincial level where no such regional organization exists.

III. National

On the national scale our concern is with what is happening to the air in places normally considered unpolluted (Fig. 4-3). Air pollution is usually thought of as an urban problem that can be escaped by getting away from the city into the clean air of the hinterland. Unfortunately, the air of the hinterland is becoming increasingly contaminated by urban air. The nonurban air masses moving into cites to flush them is increasingly the diluted air from the same or another city. The only control that can possibly reduce nonurban background pollution concentration is the reduction of the urban pollution that causes it.

There is also a class of pollution sources found only in nonurban areas. These include the burning of fields and agricultural wastes, crop spraying, the tilling of soil, the disposal of animal excrement and of the wastes from processing animal and vegetable products. As the world population increases, food and fiber production will increase. The only way these kinds of pollutants can be prevented from in-

SUMMER

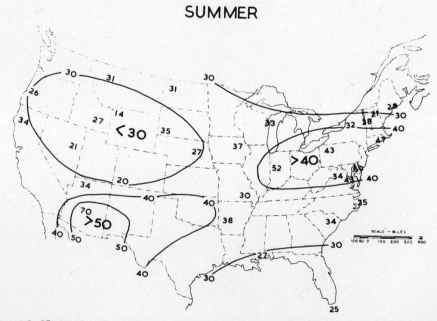

FIG. 4-3. National scale of the air pollution problem (2). Suspended particulates—nonurban National Air Sampling Network Stations, geometric means (1958–1964) summer. Values are in $\mu g/m^3$.

creasing in proportion is to change the practices that cause them. It was noted that diluted urban pollution forms the background level of nonurban pollution. Where there are appreciable nonurban sources, the converse situation occurs, i.e., the air advected into urban areas is diluted nonurban pollution.

Background concentration builds up slowly on a time scale measurable in months. Since nonurban areas are involved, organization for control at the state or provincial level is required.

IV. Continental

The continental-scale air pollution problem of greatest concern is the transport of pollution across international boundaries. The best documented example of this is the alleged transport of sulfur oxides from Great Britain and Germany to the Netherlands, Belgium, and Scandanavia where it is apparently washed out of the air as "acid rain," resulting in decreased pH of water bodies and soil (Fig. 4-4) (3). An earlier example on a smaller scale was the Trail smelter in Canada, whose plume of sulfur oxides crossed the border to cause extensive crop damage in the United States (4).

Transport on the scale postulated by the Scandanavian scientists studying the acid rain problem involves a large number of sources in one country combining to contaminate an extensive air mass which then moves to another country. This type of problem appears to be associated particularly with the use of tall stacks to release untreated gases from the burning of sulfur-containing fuel or the smelting of

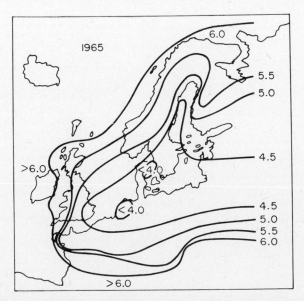

FIG. 4-4. Continental scale of the air pollution problem (3) in northwestern Europe. Isopleths of the acidity (pH) of precipitation; the hypothesis is that the acidity is caused by the SO_2 content of the atmosphere.

sulfur-containing ore. Since problems in this category involve more than one country, they are resolvable only by the action of nations through international treaty, compact, or mutual agreement, and on a time scale measurable in years.

V. Global

The global scale is that which transports pollution around the globe, as exemplified by global stratospheric transport of the radioactive debris from atmospheric testing of nuclear weapons (Fig. 4-5) and of nonradioactive debris from volcanic eruptions (Fig. 4-5) or those increases in pollutants that hypothetically may change the composition of the atmosphere, thereby changing the climate of the world. The earth could be cooled, conceivably to the point of initiating another ice age, if atmospheric contaminants sufficiently reduce the amount of solar energy that penetrates the earth's atmosphere. Conversely, the earth could be heated, conceivably to the point of melting presently existing polar ice, raising the sea level and inundating coastal cities, if contaminants sufficiently reduce the amount of the earth's heat that escapes from the earth into space as infrared radiation. The contaminants whose build-up in the atmosphere could hypothetically cause reduction of penetration of solar energy are suspended particulates that have the ability to absorb and scatter solar energy, thereby preventing it from reaching the earth. The contaminants whose build-up in the atmosphere could hypothetically cause reduction in infrared radiation from the earth are gases such as carbon dioxide which have the ability to absorb infrared energy and to be heated in so doing. (See also Chapter 17, Section F, and Chapter 21, Section E).

Measurement shows that the carbon dioxide and suspended particulate matter

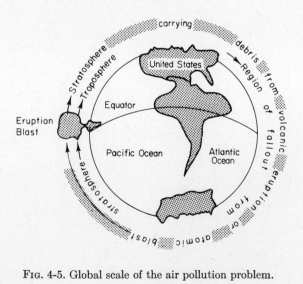

FIG. 4-5. Global scale of the air pollution problem.

content of the atmosphere have been increasing for the last several decades (Fig. 3-3 on p. 28), and that the temperature of the earth has decreased slightly over the same time period. This may be interpreted to mean that the opposite effects hypothesized for CO_2 and particulate matter increase may be counterbalancing each other, with the hypothetical effect of particulate matter to a slight extent prevailing.

The implications these matters have with respect to air pollution control are both unclear and disturbing. They imply that a major effort to decrease the particulate matter content of the atmosphere without a simultaneous decrease in the CO_2 content of the air might tip a precarious balance in one direction, and that the same could apply to any major effort at unilateral CO_2 reduction without concurrent particulate matter reduction. Since build-up of either particulate matter and gaseous contaminants alone is undesirable, it does not seem to be desirable to allow them both to build up because we are unsure of what will happen if we disturb the balance between them.

Because temperature is presently changing at such a slow rate, the world is not presently in a crisis situation where decisions have to be made and implemented rapidly. There is time to test the hypotheses from which the above-noted implications ensue, and by improvement of theory and experiment to refine our understanding of the effect of changes in atmospheric composition on global ecology. There is also time to find out how to reduce atmospheric carbon dioxide and particulate matter on a worldwide scale if it becomes imperative to do so. We certainly do not know how to accomplish this now, either scientifically or politically. Scientifically, we are embedded in a technology that burns fossil fuel to carbon dioxide for energy and see no clear path to a non-CO_2-producing technology. Likewise, we are a long way from a technology that will restore the particulate matter level of the atmosphere to that of a century or more ago. Finally, action to reduce the CO_2 and particulate level of the world's atmosphere must be a collaborative effort of all nations of the world; and as yet there is no effective international political machinery to accomplish this. The only international organizations to look to for help with our global problem are the United Nations and its affiliated organizations, the World Meteorological Organization (5), and the World Health Organization (6).

References

1. Report for Consultation on the Metropolitan Baltimore Intrastate Air Quality Control Region, U.S. Dept. of Health, Education and Welfare, Public Health Service, National Air Pollution Control Administration, April 1969 (47 pp).
2. McCormick, R. A. Air pollution climatology *in* "Air Pollution" (A. C. Stern, ed.), 2nd ed., Vol. I, p. 317. Academic Press, New York, 1968.
3. Oden, S. *Forskning och Framsteg* 1 (1969).
4. Dean, R. S., and Swain, R. E. Report Submitted to the Trail Smelter Arbitral Tribunal, U.S. Dept. of the Interior, Bureau of Mines Bulletin No. 453, 304 pp. U.S. Government Printing Office, Washington, D.C., 1944.
5. *Write to:* 41 Avenue G. Motta, 1211 Geneva 20, Switzerland.
6. *Write to:* 20 Avenue Appia, 1211 Geneva 27, Switzerland.

Suggested Reading

Singer, S. F. Ed. "Global Effects of Environmental Pollution," 218 pp. Springer-Verlag, New York/Reidel, Dordrecht, 1970.

Perloff, H. S., Ed. "The Quality of the Urban Environment," 332 pp. Resources for the Future Inc., Washington, D. C., 1969.

Questions

1. Automobiles are identified in the text in connection with the local scale of the air pollution problem. Discuss the contribution of automotive exhaust emission to the larger scales of the problem.
2. The text says that sources of unique particles or droplets falling in a neighborhood can be identified. How can such identification be accomplished?
3. Stagnations are discussed in the text in the urban context. Discuss the possibility and implications of nonurban stagnations.
4. How does corridor development between major cities affect the air pollution problem?
5. Discuss the air pollution problem associated with animal feed lots.
6. How can international organizations contribute to the control of air pollution on the continental and global scales?
7. Discuss krypton-85 as a global air pollutant.
8. The constraints on the horizontal and vertical dimensions of the scales discussed are geographic and physical, respectively. Discuss the constraints on the time scale involved.
9. Explain the variations in concentration of background pollution shown in Fig. 4-3.

Chapter 5

THE AIR POLLUTION SYSTEM

I. Primary and Secondary Pollutants

The gas and vapors emitted to the atmosphere in appreciable quantity by man-made sources tend to be relatively simple in chemical structure—carbon dioxide, carbon monoxide, sulfur dioxide, and nitric oxide from combustion processes; hydrogen sulfide, ammonia, hydrogen chloride, and hydrogen fluoride from industrial processes. The solvents and gasoline fractions that evaporate are paraffins, olefins, and aromatics of relatively simple structures. Substances such as these, emitted directly from sources, are called primary pollutants. They are certainly not innocuous, as will be seen when their adverse effects are discussed in later chapters. However, the primary pollutants do not, of themselves, produce all of the adverse effects of air pollution. Chemical reactions may occur among the primary pollutants and the constituents of the unpolluted atmosphere (Fig. 5-1). The atmosphere is a vast reaction vessel into which are poured reactable constituents, and in which are produced a tremendous array of new chemical compounds, generated by gases and vapors reacting with each other and with the particles in the air. The pollutants manufactured in the air are called secondary pollutants; they are responsible for most of the smog, haze, and eye irritation, and for many of the forms of plant and material damage attributed to air pollution. In air pollution parlance, the primary pollutants that react are termed the precursors of the secondary pollutants. With

43

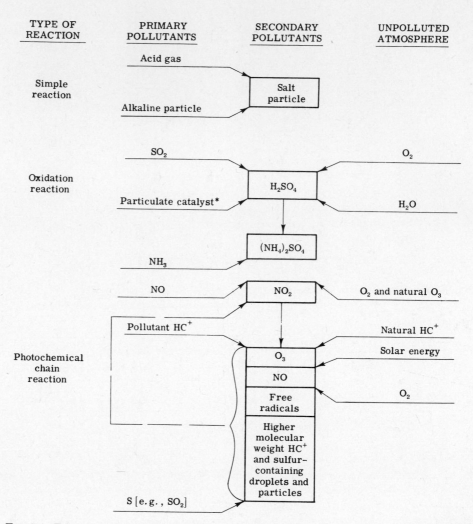

Fig. 5-1. Primary and secondary pollutants. *Reaction can occur without catalysis (HC+, hydrocarbons).

the knowledge that each secondary pollutant arises from specific chemical reactions involving specific primary reactants, it is within our power to control secondary pollutants by controlling how much of each primary pollutant is allowed to be emitted.

II. Strategy and Tactics

Since primary pollutants have the dual role of causing adverse effects in their original unreacted form, and of chemically reacting to form secondary pollutants, it can be seen that the main concern of air pollution control consists of reducing the emission of primary pollutants to the atmosphere. Air pollution control has two

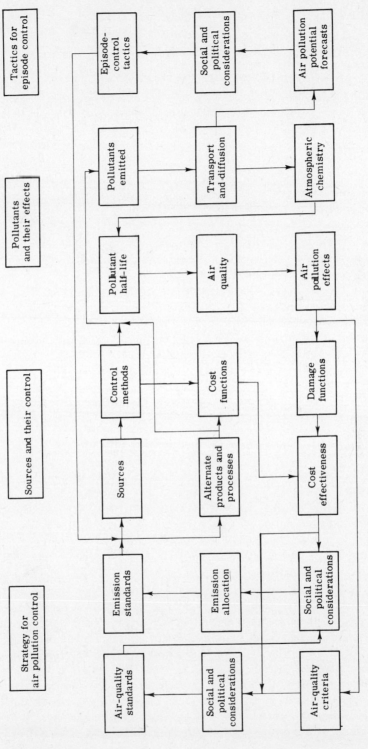

FIG. 5-2. A model of the air pollution system.

major aspects—strategic and tactical. The one is the long-term reduction of pollution levels in all scales of the problem from local to global. This aspect is called strategic, in that long-term strategies must be developed. Goals can be set for air-quality improvement five, ten, or fifteen years ahead and plans made to achieve these improvements. There can be a regional strategy to effect planned reductions at the urban and local scales; a state or provincial strategy to achieve reductions at the state, provincial, urban, and local scales; and a national strategy to achieve them at national and all lesser scales. The continental and global scales require an international strategy for which an effective instrumentality has yet to be developed.

The other major aspect of air pollution reduction is the control of short-term episodes on the urban scale. This aspect is called tactical, because, prior to an episode, a scenario of tactical maneuvers must be developed for application on very short notice to prevent an impending episode from becoming a disaster. Since an episode usually varies from a minimum of about 36 hours to a maximum of 3 or 4 days, temporary controls on emissions, much more severe than are called for by the long-term strategic control scenario, must be rapidly effectuated and maintained for the duration of the episode. After the weather conditions that gave rise to the episode have passed, these temporary episode controls can be relaxed and controls can revert to those required for long-term strategic control.

The mechanisms by which a jurisdiction develops its air pollution control strategies and episode control tactics are outlined in Fig. 5-2. Most of the boxes in the figure have already been discussed—sources, pollutants emitted, transport and diffusion, atmospheric chemistry, pollutant half-life, air quality, and air pollution effects. To complete an analysis of the elements of the air pollution system it is necessary to explain the several boxes not yet discussed.

III. Episode Control

The distinguishing feature of an air pollution episode is its persistence for several days allowing continued build-up of pollution levels. Consider the situation of the air pollution control officer who is expected to make the decision of when to invoke the stringent control restrictions required by the episode-control tactics scenario (Fig. 5-3 and Table 5-1). If he invokes these restrictions and the episode does not mature, i.e., the weather improves and blows away the accumulated pollution without allowing it to accumulate for another 24 hours or more, he will have required for naught a very large expenditure by the community and a serious disruption of the community's normal activities. He will have also used up part of his credibility in the community. If this happens more than once he will be accused of crying wolf, and when a real episode occurs his warnings will be unheeded. If however, the reverse situation occurs, i.e., he does not invoke these restrictions and an episode does mature, there can be illness and deaths in the community that could have been averted had he invoked the scheduled restrictions.

In making his decision whether or not to initiate episode emergency plans, the control officer cannot rely alone upon measurements from air-quality monitoring stations because even if pollutant concentration levels show a rise toward acute

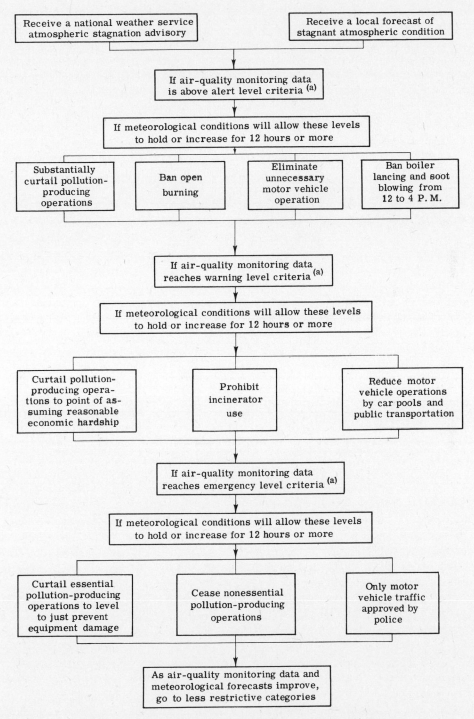

FIG. 5-3. Air pollution episode control scenario. [a]See Table 5-1.

TABLE 5-1

United States Alert, Warning, and Emergency Level Criteria[a]

Alert level criteria

SO_2
    800 μg/m³ (0.3 ppm), 24-hour average

Particulate
    3.0 COH's or 375 μg/m³, 24-hour average

SO_2 and particulate combined
    Product of SO_2 ppm, 24-hour average, and COH's equal to 0.2; or product of SO_2 μg/m³, 24-hour average and particulate μg/m³, 24-hour average, equal to 65 $\times$ 10³

CO
    17 mg/m³ (15 ppm), 8-hour average

Oxidant (O_3)
    200 μg/m³ (0.1 ppm), 1-hour average

NO_2
    1130 μg/m³ (0.6 ppm), 1-hour average; 282 μg/m³ (0.15 ppm), 24-hour average

Warning level criteria

SO_2
    1600 μg/m³ (0.6 ppm), 24-hour average

Particulate
    5.0 COH's or 625 μg/m³, 24-hour average

SO_2 and particulate combined
    Product of SO_2 ppm, 24-hour average, and COH's equal to 0.8; or product of SO_2 μg/m³, 24-hour average, and particulate μg/m³, 24-hour average, equal to 261 $\times$ 10³

CO
    34 mg/m³ (30 ppm), 8-hour average

Oxidant (O_3)
    800 μg/m³ (0.4 ppm), 1-hour average

NO_2
    2260 μg/m³ (1.2 ppm), 1-hour average; 565 μg/m³ (0.3 ppm), 24-hour average

Emergency level criteria

SO_2
    2100 μg/m³ (0.8 ppm), 24-hour average

Particulate
    7.0 COH's or 875 μg/m³, 24-hour average

SO_2 and particulate combined
    Product of SO_2 ppm, 24-hour average, and COH's equal to 1.2; or product of SO_2 μg/m³, 24-hour average, and particulate μg/m³, 24-hour average, equal to 393 $\times$ 10³

CO
    46 mg/m³ (40 ppm), 8-hour average

Oxidant (O_3)
    1200 μg/m³ (0.6 ppm), 1-hour average

NO_2
    3000 μg/m³ (1.6 ppm), 1-hour average; 750 μg/m³ (0.4 ppm), 24-hour average

---

[a] Federal Register, Vol. 36, No. 206, October 23, 1971, 15593.

levels over the preceding hours, these readings give him no information as to whether the levels will rise or fall during the succeeding hours. Since it takes hours before emergency plans can be put into effect and their impact on pollution levels felt, it is possible that by the time the community had responded, the situation to which they were responding could have disappeared.

The only way to avert this dilemma is to develop and utilize our capability of forecasting the advent and persistence of the stagnation conditions during which an episode occurs and our capability of computing pollution concentration build-up under stagnation conditions. The details of how these forecasts and computations are made are discussed in Chapter 22, Section IV, but at this point, the foregoing discussion should explain the reason for the box in Fig. 5-2 marked Air Pollution Potential Forecasts. The connecting box marked Social and Political Considerations provides a place in the system for the public debates, hearing and action processes necessary to develop, well in advance of an episode, what control tactics to use, and when to call an end to the emergency. The public needs to be involved because alternatives have to be written into the scenario as to where, when, and in what order to impose restrictions on sources. This should be done in advance and should be well publicized, because during the episode there is no time for public debate. In any systems analysis, the system must form a closed loop with feedback to keep the system under control. It will be noted that the system for tactical episode control is closed by the line connecting Episode Control Tactics to Sources, which means that the episode control tactics are to severely limit sources during the episode.

IV. Control Strategy

Now let us consider the system for long-range strategy for air pollution control. Those elements in this system that have not yet been discussed include several listed under Sources and Their Control and all those listed under Strategy for Air Pollution Control. Control of sources is effected in several ways: We can (1) use devices to remove all or part of the pollutant from the gases discharged to the atmosphere; (2) change the raw materials used in the pollution producing process; or (3) change the operation of the process so as to decrease Pollutants Emitted. These are Control Methods (see Table 5-2). Such control methods have a cost associated with them, and are the Cost Functions that appear in the system. There is always the option of seeking Alternative Products or Processes which will provide the same utility to the public but with less Pollutants Emitted. Such products and processes have their own Cost Functions.

Just as it costs money to control pollution, it also costs the public money not to control pollution. All the adverse Air Pollution Effects represent economic burdens on the public for which an attempt can be made to assign dollar values, i.e., the cost to the public of damage to vegetation, materials, structures, animals, the atmosphere, and to human health. These costs are called Damage Functions. To the extent that there is knowledge of Cost Functions and Damage Functions, the Cost Effectiveness of control methods and strategies can be determined by their interrelation. Cost Effectiveness is an estimate of how many dollars of damage can

TABLE 5-2

Control Methods

I. Applicable to all emissions
   A. Decrease or eliminate production of emission
      1. Change specification of product
      2. Change design of product
      3. Change process temperature, pressure or cycle
      4. Change specification of materials
      5. Change the product
   B. Confine the emissions
      1. Enclose the source of emissions
      2. Capture the emissions in an industrial exhaust system
      3. Prevent drafts
   C. Separate the contaminant from effluent gas stream
      1. Scrub with liquid

II. Applicable specifically to particulate emissions
   A. Decrease or eliminate particulate production
      1. Change to process that does not require blasting, blending, buffing, calcining, chipping, crushing, drilling, drying, grinding, milling, polishing, pulverizing, sanding, sawing, spraying, tumbling, etc.
      2. Change from solid to liquid or gaseous material
      3. Change from dry to wet solid material
      4. Change particle size of solid material
      5. Change to process that does not require particulate material
   B. Separate the contaminant from effluent gas stream
      1. Gravity separator
      2. Centrifugal separator
      3. Filter
      4. Electrostatic precipitator

III. Applicable specifically to gaseous emissions
   A. Decrease or eliminate gas or vapor production
      1. Change to process that does not require annealing, baking, boiling, burning, casting, coating, cooking, dehydrating, dipping, distilling, expelling, galvanizing, melting, pickling, plating, quenching, reducing, rendering, roasting, smelting, etc.
      2. Change from liquid or gaseous to solid material
      3. Change to process that does not require gaseous or liquid material
   B. Burn the contaminant to CO_2 and H_2O
      1. Incinerator
      2. Catalytic burner
   C. Adsorb the contaminant
      1. Activated carbon

be averted per dollar expended for control. It gives information on how to economically optimize an attack on pollution, but it gives no information as to the reduction required in pollution to achieve acceptable public health and well-being. However, when acceptable public health and well-being can be achieved by different control alternatives, it behooves us to utilize the alternatives that show the greatest cost effectiveness.

For the determination of what pollution concentrations in air are compatible with acceptable public health and well-being, use is made of Air Quality Criteria, which are statements of the air pollution effects associated with various air-quality levels. It is inconceivable that any jurisdiction would accept levels of pollution it recognizes as damaging to health. However, the question of what constitutes damage to health is judgmental and therefore debatable. The question of what damage to well-being is acceptable is even more judgmental and debatable. Because they are debatable, the same Social and Political Considerations come into the decision-making process as in the previously discussed case of arriving at episode control tactics. Cost Effectiveness is not a factor in the acceptability of damage to health, but it is a factor in determining acceptable damage to public well-being. Some jurisdictions may opt for a pollution level that allows some damage to vegetation, animals, materials, structures, and the atmosphere as long as they are assured that there will be no damage to their constituent's health. The concentration level the jurisdiction selects by this process is called an Air Quality Standard. This is the level the jurisdiction says it wishes to maintain.

Adoption of air-quality standards by a jurisdiction effects no air pollution control. Control is effected by the limitation of emission from sources, which, in turn, is achieved by the adoption and enforcement of Emission Standards. However, before emission standards are adopted, the jurisdiction must make some social and political decisions as to which of several philosophies of emission-standard development are to be utilized, and as to which of the several responsible groups in the jurisdiction should bear the brunt of the control effort—its homeowners, its landlords, its industries, or its institutions. This latter type of decision-making is called Emission Allocation. It will be seen in Fig. 5-2 that the system for strategic control is closed by the line connecting Emission Standards and Sources, which means that long-range pollution-control strategy is to apply emission limitations to sources.

V. Economic Considerations

The situation with regard to economic considerations has been so well stated in the First Report of the British Royal Commission on Environmental Pollution (1), that this chapter shall end with a quotation from that report.

Our survey of the activities of the Government, industry and voluntary bodies in the control of pollution discloses several issues which need further enquiry. The first and most difficult of these is how to balance the considerations which determine the levels of public and private expenditure on pollution control. Some forms of pollution bear more heavily on society than others; some forms are cheaper than others to control; and the public are more willing to pay for some forms of pollution control than for others. There are also short and long-term considerations: in the short-term the incidence of pollution control on individual industries or categories of labor may be heavy; but . . . what may appear to be the cheapest policy in the short-term may prove in the long-term to have been a false economy.

While the broad outlines of a general policy for protecting the environment are not difficult to discern, the economic information needed to make a proper assessment of the considerations referred to in the preceding paragraph . . . seems to us to be seriously deficient. This is in striking contrast with the position regarding the scientific and technical data where, as our survey has shown, a considerable amount of information is already available and various bodies are trying to fill in the main gaps. The scientific and technical information is invaluable, and in many cases may be adequate for reaching satisfactory decisions, but much of it could be wasted if it were not supported by some economic indication of priorities and of the best means of dealing with specific kinds of pollution.

So, where possible, we need an economic framework to aid decisionmaking about pollution, which would match the scientific and technical framework we already have. This economic framework should include estimates of the way in which the costs of pollution, including disamenity costs, vary with levels of pollution; the extent to which different elements contribute to the costs; how variations in production and consumption affect the costs; and what it would cost to abate pollution in different ways and by different amounts. There may well be cases where most of the costs and benefits of abatement can be assessed in terms of money. Many of the estimates are likely to be speculative, but this is no reason for not making a start. There are other cases where most of the costs and benefits cannot be given a monetary value. In these cases decisions about pollution abatement must not await the results of a full economic calculation: they will have to be based largely on subjective judgments anyway. Even so, these subjective judgments should be supported by as much quantitative information as possible, just as decisions about health and education are supported by extensive statistical data. Further, even if decisions to abate pollution are not based on rigorous economic criteria, it is still desirable to find the most economic way of achieving the abatement.

References

1. Royal Commission on Environmental Pollution. First Report, Command 4585, H. M. Stationery Office, London, 1971.

Suggested Reading

Cleaning Our Environment, The Chemical Basis for Action. Report of the Subcommittee on Environmental Improvement, Committee on Chemistry and Public Affairs. 249 pp. American Chemical Society, Washington, D. C., 1969.
Wolozin, H. Ed., "The Economics of Air Pollution," 318 pp. Norton, New York, 1966.

Questions

1. Give some examples of chemical reactions among primary pollutants to form secondary pollutants.
2. Trace the history of our knowledge of the photochemical production of secondary pollutants in the atmosphere.
3. Explain why certain important long-range air pollution control strategies will not suffice for short-term episode control and vice versa.
4. Develop an episode control scenario for a single large coal-fired steam electric generating station.
5. In Fig. 5-2, the words Social and Political Considerations appear several times. Discuss these considerations for the various contexts involved.
6. Discuss the relative importance of Air Quality Criteria and Cost Effectiveness in the setting of Air Quality Standards.
7. The quotation in Section V contains "what may appear to be the cheapest policy in the short-term may prove in the long-term to have been a false economy." Give some examples of this.
8. Draw a simplified version of Fig. 5-2 with less than ten boxes.
9. How would one go about developing an air pollution Damage Function for human health?

Chapter 6

HISTORY OF THE PROBLEM

I. Before the Industrial Revolution

One of the reasons the tribes of early history were nomadic was to periodically move away from the stench of the animal, vegetable, and human wastes they generated. When the tribesmen learned to use fire, they used it for millenia in a way that filled the air inside their living quarters with the products of incomplete combustion. Examples of this can still be seen today in some of the more primitive parts of the world. After its invention, the chimney removed the combustion products and cooking smells from the living quarters, but for centuries the open fire in the fireplace caused its emission to be smoky. In 61 A.D. the Roman philosopher Seneca reported thus on conditions in Rome.

> As soon as I had gotten out of the heavy air of Rome and from the stink of the smoky chimneys thereof, which, being stirred, poured forth whatever pestilential vapors and soot they had enclosed in them, I felt an alteration of my disposition.

Air pollution, associated with burning wood in Tutbury Castle in Nottingham, was considered "unendurable" by Eleanor of Aquitaine, the wife of King Henry II of England and caused her to move in the year 1157. One hundred-sixteen years later, coal burning was prohibited in London; and in 1306, Edward I issued a royal proclamation enjoining the use of "sea-coal" in furnaces. Elizabeth I barred the

53

burning of coal in London when Parliament was in session. The repeated necessity for such royal action would seem to indicate that coal continued to be burned despite these edicts. By 1661 the pollution of London had become bad enough to prompt John Evelyn to submit a brochure "Fumifugium, or the Inconvenience of the Aer, and Smoake of London Dissipated (together with some remedies humbly proposed)" to King Charles II and the Parliament. This brochure has recently been reprinted and is recommended to students of air pollution (1). It proposes means of air pollution control that are still viable in the twentieth century.

The principal industries associated with the production of air pollution in the centuries preceding the Industrial Revolution were metallurgy, ceramics, and preservation of animal products. In the bronze and iron ages, villages were exposed to dusts and fumes from many sources. Native copper and gold were forged, and clay was baked and glazed to form pottery and bricks before 4000 B.C. Iron was in common use and leather was tanned before 1000 B.C. Most of the methods of modern metallurgy were known before 1 A.D. They relied on charcoal rather than coal or

FIG. 6-1. Lead smelting furnace. From G. Agricola, "De Re Metallica," Book X, p. 481, Basel, Switzerland, 1556. Translated by H. C. Hoover and L. H. Hoover, *Mining Magazine*, London, 1912. Reprinted by Dover Publications, Inc., New York, 1950.

FIG. 6-2. A pottery kiln. From Cipriano Piccolpasso, "The Three Books of the Potter's Art," fol. 35C, 1550. Translated by B. Rackham and A. Van de Put, Victoria and Albert Museum, London, 1934.

coke. However, coal was mined and used for fuel before 1000 A.D., although it was not made into coke until about 1600; and coke did not importantly enter metallurgical practice until about 1700. These industries and their effluents as they existed before the year 1556 are best described in a book "De Re Metallica" published in that year by Georg Bauer, known as Georgius Agricola (Fig. 6-1). This book was translated into English and published in 1912 by Herbert Clark Hoover and his wife (2).

Examples of the air pollution associated with the ceramic and animal product preservation industries are shown in Figs. 6-2 and 6-3, respectively.

FIG. 6-3. A kiln for smoking red herring [From H. L. Duhamel du Monceau, "Traité général des pêches," Vol. 2, Sect. III, Plate XV, Fig. 1, Paris 1772].

II. The Industrial Revolution

The Industrial Revolution was the consequence of the harnessing of steam to provide power to pump water and move machinery. This occurred during the period between the early years of the eighteenth century when Savery, Papin, and Newcomen designed their pumping engines, and culminated in 1784 in Watt's reciprocating engine. The reciprocating steam engine reigned supreme until displaced by the steam turbine in the twentieth century.

Steam engines and steam turbines require steam boilers, which, until the advent of the nuclear reactor, were fired by vegetable or fossil fuels. During most of the nineteenth century, coal was the principal fuel, although some oil was used for steam generation late in the century.

The predominant air pollution problem of the nineteenth century was smoke and ash from the burning of coal or oil in the boiler furnaces of stationary power plants, locomotives, and marine vessels, and in home heating fireplaces and furnaces. Great Britain took the lead in addressing this problem, and, in the words of Sir

Hugh Beaver (3):

> By 1819, there was sufficient pressure for Parliament to appoint the first of a whole dynasty of committees 'to consider how far persons using steam engines and furnaces could work them in a manner less prejudicial to public health and comfort.' This committee confirmed the practicability of smoke prevention, as so many succeeding committees were to do, but as was often again to be experienced, nothing was done.
>
> In 1843, there was another Parliamentary Select Committee, and in 1845, a third. In that same year, during the height of the great railway boom, an act of Parliament disposed of trouble from locomotives once and for all (!) by laying down the dictum that they must consume their own smoke. The Town Improvement Clauses Act two years later applied the same panacea to factory furnaces. Then 1853 and 1856 witnessed two acts of Parliament dealing specifically with London and empowering the police to enforce provisions against smoke from furnaces, public baths, and washhouses and furnaces used in the working of steam vessels on the Thames.

Smoke and ash abatement in Great Britain was considered to be a health agency responsibility and was so confirmed by the first Public Health Act of 1848 and the later ones of 1866 and 1875. Air pollution from the emerging chemical industry was considered a separate matter and was made the responsibility of the Alkali Inspectorate created by the Alkali Act of 1863.

In the United States, smoke abatement (as air pollution control was then known) was considered a municipal responsibility. There were no federal or state smoke abatement laws or regulations. The first municipal ordinances and regulations limiting the emission of black smoke and ash appeared in the 1880's and were directed toward industrial, locomotive, and marine rather than domestic sources. As the nineteenth century drew to a close, the pollution of the air of mill towns the world over had risen to a peak (Fig. 6-4); vegetation damage from the smelting of sulfide ores was recognized as a problem everywhere it was practiced.

III. The Twentieth Century

A. 1900–1925

During the period 1900–1925 there were great changes in the technology of both the production of air pollution and its engineering control, but no significant changes in legislation, regulations, understanding of the problem or public attitudes toward the problem. As cities and factories grew in size, the severity of the pollution problem increased.

One of the principal technological changes in the production of pollution was the replacement of the steam engine by the electric motor as the means for operating machinery and pumping water. This change transferred the smoke and ash emission from the boiler house of the factory to the boiler house of the electric generating station. At the start of this period coal was hand fired in the boiler house. By mid-period it was mechanically fired by stokers; by the end of the period, pulverized coal firing had begun to take over. Each form of firing had its own characteristic emissions to the atmosphere (Table 6-1).

At the start of this period steam locomotives came into the heart of the larger cities. By the end of the period the urban terminii of many railroads had been

Fig. 6-4. Engraving (1876) of a metal foundry refining department in the industrial Saar region of West Germany. From The Bettmann Archive, Inc.

TABLE 6-1

Characteristic Emissions of Different Forms of Coal Firing[a]

		Emission per ton of coal burned (lb)		
Pollutant	Notes	Electric generating plants	Industrial plants	Domestic and commercial plants
Particulate matter	Hand-fired	N.A.[b]	N.A.	20
	Cyclone furnace	2A[c]	N.A.	N.A.
	Wet bottom P.C.	13–24A[d]	N.A.	N.A.
	Dry bottom P.C.	17A	17A	N.A.
	Spreader stokers	13–20A[d]	13–20A	N.A.
	All other stokers	5A	5A	5A
Sulfur oxides	—	38 S[e]	38 S	38 S
Nitrogen oxides	(as NO_2)	20	20	8
Carbon monoxide	—	0.5	3	50
Hydrocarbons	(as methane)	0.2	1	10
Aldehydes	(as formaldehyde)	0.005	0.005	0.005

[a] Adapted from Smith and Gruber (21).

[b] N.A. = Not Applicable.

[c] A: Ash content in percent, i.e., when A = 10%, 2A = 2 × 10 = 20 lb particulate/ton of coal.

[d] Lower number is without fly-ash reinjection; higher number is with reinjection. (Values with reinjection are loadings reaching the fly-ash control equipment.)

[e] S: Sulfur content in percent, i.e., when S = 2%, 38 S = 2 × 38 = 76 lb sulfur oxides per ton of coal.

electrified, thereby transferring much air pollution from the railroad right-of-way to the electric generating station. The replacement of coal by oil in many applications decreased ash emissions from those sources. There was rapid technological change in industry. However, the most significant change was the rapid increase in the number of automobiles from almost none at the turn of the century to millions by 1925 (Table 6-2).

The principal technological changes in the engineering control of air pollution were the perfection of the motor-driven fan, which allowed large-scale gas-treating systems to be built; the invention of the electrical precipitator, which made particulate control in many processes feasible; and the development of a chemical engineering capability for the design of process equipment which made the control of vapor phase effluent feasible.

B. 1925–1950

This was the period during which present-day air pollution problems and solutions emerged. The Meuse Valley (Belgium) episode (4) occurred in 1930; the Donora, Pennsylvania (5) episode occurred in 1948; and the Poza Rica, Mexico episode (6) in 1950. Los Angeles smog first appeared in the 1940's (Fig. 6-5). The Trail, British Columbia, smelter arbitration (7) was completed in 1941. The first

TABLE 6-2

Annual Motor Vehicle Factory Sales in the United States[a]

Year	Total	Year	Total
1900	4,192	1940	4,472,286
1905	25,000	1945	725,215
1910	187,000	1950	8,003,056
1915	969,930	1955	9,169,292
1920	2,227,347	1960	7,869,221
1925	4,265,830	1965	11,057,366
1930	3,362,820	1970	8,239,257
1935	3,971,241		

[a] Data include factory sales for trucks, buses, and automobiles.

National Air Pollution Symposium in the United States was held in Pasadena, California in 1949 (8), and the first United States Technical Conference on Air Pollution was held in Washington, D.C. in 1950 (9). The first large-scale surveys of air pollution were undertaken—Salt Lake City, Utah (1926) (10); New York, New York (1937) (11); and Leicester, England (1939) (12), and the episodes noted above were intensively investigated (4–6). Air pollution research got a start in California. The technical foundation for air pollution meteorology was established in the search for means of disseminating, and protecting against, chemical, biological, and nuclear warfare agents. Toxicology came of age. The stage was set for the air pollution scientific explosion of the last half of the twentieth century.

A major technological change was the building of natural gas pipelines, and, where this occurred, there was rapid displacement of coal and oil as home heating fuel, with dramatic improvement of air quality. Witness the much-publicized decrease in black smoke in Pittsburgh, Pennsylvania (Fig. 6-6) and St. Louis, Missouri. The diesel locomotive began to displace the steam locomotive, thereby slowing the pace of railroad electrification. The internal combustion engine bus started its displacement of the electrified street car. The automobile continued to proliferate (Table 6-2).

During this period, no significant national air pollution legislation or regulations were adopted anywhere in the world. However, the first state air pollution law in the United States was adopted by California in 1947.

C. 1950–1970

In Great Britain, a major air pollution disaster hit London in 1952 (13), resulting in the passage of the Clean Air Act in 1956 and an expansion of the authority of the Alkali Inspectorate.

The principal changes that resulted were in the means of heating homes. Pre-

FIG. 6-5. Los Angeles smog. From Los Angeles County, California.

viously, most heating was by burning soft coal on grates in separate fireplaces in each room. A successful effort was made to substitute smokeless fuels for the soft coal used in this manner; and central or electrical heating for fireplace heating. The outcome was a decrease in "smoke" concentration, as measured by the blackness of paper filters through which British air was passed from 175 μg/m^3 in 1958 to 75 μg/m^3 in 1968 (14). In 1959, an International Air Pollution Conference was held in London (15).

During these two decades almost every country in Europe as well as Japan, Australia, and New Zealand, experienced serious air pollution in its larger cities. As a result, in these decades the several countries first enacted national air pollution control legislation. By 1970, major national air pollution research centers had been set up at the Warren Springs Laboratory, Stevenage, England; at the Institut National de la Santé et de la Recherche Medicale at Le Visinet, France; the Rijksinstituut Voor de Volksgezondheid, Bilthoven and the Instituut voor Gezondheidstechniek-TNO, Delft, The Netherlands; the Statens Naturvardsverk, Solna, Sweden; the Institut für Wasser-Boden-und Luft-hygiene, Berlin and the Landesanstalt für Immissions und Bodennutzungsshutz, Essen, Germany. The important air pollution research centers in Japan are too numerous to mention.

In the United States, the smog problem continued to worsen in Los Angeles and appeared in large cities all over the nation (Fig. 6-7). In 1955 the first Federal air pollution legislation was enacted providing federal support for air pollution research, training, and technical assistance. Responsibility for the administration of the federal program was given to the Public Health Service (PHS) of the U.S. Depart-

FIG. 6.6 (a) Pittsburgh before decrease in black smoke. From Allegheny County, Pennsylvania.

ment of Health, Education, and Welfare, and remained there until 1970, when it was transferred to the new U.S. Environmental Protection Agency (EPA). The initial Federal legislation was amended and extended several times between 1955 and 1970, greatly increasing Federal authority, particularly in the area of control (see Chapter 27).

As in Europe, air pollution research activity expanded tremendously in the

Fɪɢ. 6-6. (b) Pittsburgh after decrease in black smoke. From Allegheny County, Pennsylvania.

United States during these two decades. The headquarters of the Federal research activity was at the Robert A. Taft Sanitary Engineering Center of the PHS in Cincinnati, Ohio during the early years of the period and at the Research Center of EPA in Research Triangle Park, North Carolina at the end of the period. An International Air Pollution Congress was held in New York City in 1955 (16) and a Second International Clean Air Congress was held in Washington, D.C. in 1970

FIG. 6-7. Smog in New York City. From Wide World Photos.

(17). Three National Air Pollution Conferences were held in Washington, D.C. in 1958 (18), 1962 (19), and 1966 (20).

Technological interest during these twenty years focused upon automotive air pollution and its control, and upon sulfur oxide pollution and its control by sulfur oxide removal from flue gases and fuel desulfurization. Air Pollution Meteorology came of age and, by 1970, mathematical models of the pollution of the atmosphere were being energetically developed. A start was made in elucidating the photochemistry of air pollution. Air-quality monitoring systems became operational all over the world. A wide variety of measuring instruments became available.

IV. The 1970's

The highlight of the 1970's is the emergence of the ecological or total environmental approach. Organizationally this has taken the form of departments or ministries of the environment in governments at all levels through the world. In the United States there is a federal Environmental Protection Agency and, in states like New York and cities like New York City, there are counterpart organizations charged with responsibility for air and water quality, solid waste sanitation, noise abatement, and the control of the hazards associated with radiation and the use of pesticides. This is paralleled in industry, where formerly diffuse responsibility for these areas is the increasing responsibility of an environmental protection coordinator. Similar changes are evident in research and education.

The 1970's marks the decade of pollution controls being built into pollution sources—automobiles, power plants, factories—at the time of original construction rather than later on.

V. The Future

The air pollution problems of the future are predicated on the burning of more and more fuel as the population of the world increases. During the lifetime of the students using this book, there will be some respite offered by nuclear fuels. Some of the agonizing environmental decisions of the next thirty years will be to choose between fossil fuel and nuclear power sources and the extent to allow depletion of future reserves of both fossil and nuclear fuels for present needs. Serious questions will arise as to whether to conserve or to use these reserves—whether to allow unlimited growth or to curb it.

References

1. The Smoake of London–Two Prophecies [Selected by James P. Lodge, Jr.]. Maxwell Reprint, Elmsford, New York, 1969.
2. Agricola, G. "De Re Metallica," Basel, 1556. [English translation and commentary by H. C. Hoover and L. H. Hoover, Mining Magazine, London 1912]. Dover, New York, 1950.
3. Beaver, Sir Hugh E. C., The growth of public opinion, in "Problems and Control of Air Pollution" (F. S. Mallette, ed.). Reinhold, New York, 1955.
4. Firket, J. Bull. Acad. Roy. Med. Belg. 11, 683 (1931).
5. Schrenk, H. H., Heimann, H., Clayton, G. D., Gafefer, W. M., and Wexler, H. U.S. Pub. Health Serv. Bull. 306 (1949), 173 pp.
6. McCabe, L. C., and Clayton, G. D. Arch. Ind. Hyg. Occup. Med. 6, 199 (1952).
7. Dean, R. S., and Swain, R. E. Report submitted to the Trail Smelter Arbitral Tribunal, U.S. Bur. Mines Bull. 453, U.S. Government Printing Office, Washington, D.C. (1944).
8. Proceedings of the First National Air Pollution Symposium, Pasadena, Calif. Sponsored by Stanford Research Inst. in cooperation with California Institute of Technology, University of California, and University of Southern California, 1949.
9. McCabe, L. C. (ed.), "Air Pollution" (Proc. U.S. Tech. Conf. Air Pollution, 1950), 847 pp. McGraw-Hill, New York, 1952.
10. Monett, O., Perrott, G. St. J., and Clark, H. W., U.S. Bur. Mines Bull. 254 (1926), 98 pp.
11. Stern, A. C., Buchbinder, L., and Siegel, J., Heat. Piping Air Cond. 17, 7–10 (1945).

12. Atmospheric Pollution in Leicester—A Scientific Survey, D.S.I.R. Atmospheric Research Technical Paper No. 1. H.M. Stationery Office, London, 1945.

13. Ministry of Health, Mortality and Morbidity during the London Fog of December 1952, Rep. of Public Health and Related Subject No. 95. H.M. Stationery Office, London, 1954.

14. Royal Commission on Environmental Pollution, First Report (Feb. 1971). H.M. Stationery Office, London, Cmnd 4585, 52 pp.

15. *Proc. Diamond Jubilee Int. Clean Air Conf. 1959* (National Society for Clean Air, London (1960).

16. Mallette, F. S. (Ed.), "Problems and Control of Air Pollution," *op cit.*, 272 pp.

17. Englund, H. and Beery, W. T., (eds.), Proceedings of the Second International Clean Air Congress, Washington, D.C., 1970. Academic Press, New York, 1971. 1354 pp.

18. Proceedings of the National Conference on Air Pollution, Washington, D.C., 1958. *U.S. Pub. Health Serv. Pub.* 654 (1959).

19. Proceedings of the National Conference on Air Pollution, Washington, D.C., 1962. *U.S. Pub. Health Serv. Pub.* 1022 (1963).

20. Proceedings of the National Conference on Air Pollution, Washington, D.C., 1966. *U.S. Pub. Health Serv. Pub.* 1669 (1967).

21. Smith, W. S., and Gruber, C. W., "Atmospheric Emissions from Coal Combustion—An Inventory Guide," 999-AP-24, Environmental Health Series, Air Pollution, Public Health Service, U.S. Dept. of Health, Education, and Welfare, Cincinnati, Ohio, 1966.

Suggested Reading

Singer, C., Holmyard, E. J., Hall, A. R., and Williams, T. I. (Ed.), "A History of Technology," 5 Vols. Oxford Univ. Press, London and New York, 1954–1958.

"Air Pollution." Monograph No. 46, World Health Organization, Geneva, Switzerland, 1961, 442 pp.

Questions

1. Describe the physical principles by which a chimney operates. What is "chimney draft"? How is a chimney designed to produce a certain amount of draft? How is chimney draft regulated?

2. How are charcoal and coke made? How do they differ from the parent substances from which they are made? What are the air pollution problems associated with their production?

3. Draw and appropriately label cross sections of the types of steam boilers and boiler furnaces used for (a) large pulverized coal-fired stationary power plants, (b) small gas-fired stationary power plants, (c) oil-fired marine steam generation, and (d) hand-fired coal-burning steam locomotives.

4. Draw cross sections through the grates and furnace used for burning coal by (a) hand-firing, (b) underfeed stoker, (c) spreader stoker, and (d) traveling grate stoker.

5. Draw a flow chart of a typical sulfide ore smelting process showing where in the process what air pollutants are released to the atmosphere.

6. Discuss the influence of the perfection of the motor driven fan on the engineering control of air pollution.

7. Prepare a table showing the growth of the diesel engine as means for power for electric generation and vehicular propulsion.

8. Discuss the likely future role of gas turbines as means for power for electric generation and vehicular propulsion.

9. What have been the most important developments in the "History of the Problem" since the publication of this book?

Chapter 7

AIR QUALITY

The terms ambient air, ambient air pollution, ambient levels, ambient concentrations, ambient air monitoring, ambient air quality, etc., occur frequently in air pollution parlance. The intent is to distinguish pollution in the outdoors subject to transport and diffusion by the wind (i.e. ambient), from contamination of the air indoors by the same substances.

The air inside a factory building can be contaminated by release of contaminants from industrial processes to the air of the workroom. This is a major cause of occupational disease. Prevention and control of such contamination is part of the practice of industrial hygiene. Among the procedures that industrial hygienists use to prevent exposure of workers to such contamination are industrial ventilation systems that remove the contaminated air from the workroom and discharge it, either with or without treatment to remove the contaminants, to the ambient air outside the factory building.

The air inside a home, office or public building may be contaminated by such sources as fuel-fired cooking or space-heating ranges, ovens, or stoves that discharge their combustion products to the room; by solvents evaporated from inks, paints, adhesives, cleaners, or other products; and by other pollutant sources indoors. If some of these sources exist inside a building, the pollution level of the air inside the building might be higher than that outside the building. However, if there are none of these sources inside the building, the pollution level inside would be expected to be lower than the ambient concentration outside because of the ability of the surfaces inside the building—walls, floors, ceilings, furniture, and fixtures—to adsorb or react with gaseous pollutants and attract and retain particulate pollutants, thereby partially removing them from the air breathed by occupants of the building. This adsorption and retention would occur even if doors and windows were open, but the difference between outdoor and indoor concentrations would be even greater if they were closed, in which case air could enter the building only by infiltration through cracks and walls.

I. Averaging Time

The variability inherent in the transport and diffusion process, the time variability of source strengths and the scavenging and conversion mechanisms in the at-

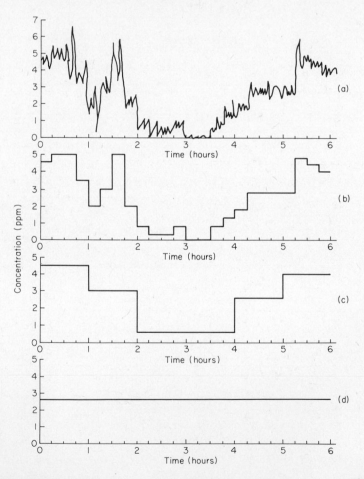

FIG. 7-1. The same atmosphere measured by (a) a rapid response instrument and by sampling and analytical procedures that integrate the concentration arriving at the receptor over a time period of (b) 15 minutes, (c) 1 hour, and (d) 6 hours.

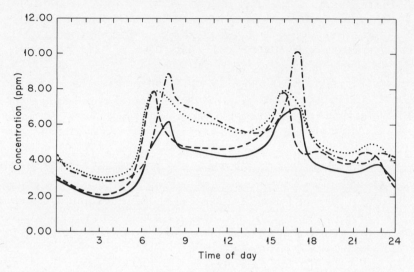

Fig. 7-2. Diurnal variation of carbon monoxide concentration, St. Louis, Missouri, 1967–1968 (1). Spring (———); summer (- - -); fall (• • • • •); winter (- • -).

mosphere, which cause pollutants to have an effective half-life, result in variability in the concentration of a pollutant arriving at a receptor. Thus a continuous record of the concentration of a pollutant at a receptor, as measured by an instrument with rapid response time, might look like Fig. 7-1(a). If, however, instead of measuring with a rapid response instrument, the measurement at the receptor site was made with sampling and analytical procedures that integrated the concentration arriving at the receptor over various time periods, e. g., 15 minutes, 1 hour, or 6 hours, the resulting information would look variously like Figs. 7-1(b), (c), (d), respectively. It should be noted that from the information in Fig. 7-1(a), it is possible to mathematically derive the information in Figs. 7-1(b), (c), and (d), and it is possible to derive the information in Figs. 7-1(c) and (d) from that in Fig. 7-1(b). The converse is not true. With only the information from Fig. 7-1(d) available, Figs. 7-1(a), (b), and (c) could never be constructed, nor could Figs. 7-1(a) and (b) be constructed from Fig. 7-1(c); nor Fig. 7-1(a) from Fig. 7-1(b). In these examples the time intervals involved in Figs. 7-1(b), (c), and (d), i.e., 15 minutes, 1 hour, and 6 hours respectively are *averaging times* of the measurement of pollutant exposure at the receptor. The averaging time of the rapid response record [Fig. 7-1(a)] is an inherent characteristic of the instrument itself, its recorder, and its sampling system. It can closely approach becoming an instantaneous record of concentration at the receptor. However, in most cases it is not desirable for it to do so because such an instantaneous record cannot be put to any practical air pollution control usage. What such a record tells is something of the turbulent structure of the atmosphere, and as such has some utility in meteorological research. In communications science parlance, an instantaneous recording has too much noise. It is therefore necessary to filter or damp out the noise in order to extract the useful information about pollu-

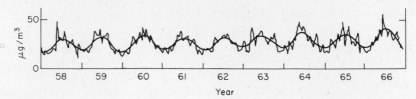

Fɪɢ. 7-3. Seasonal variation of suspended particulate matter concentration. Composite of 20 nonurban sites, United States (2).

tion concentration at the receptor that the signal is trying to tell. This damping is achieved by building time lags into the response of the sampling, analysis,and recording systems (or into all three); by interrogating the instantaneous output of the analyzer at discrete time intervals, e. g., once every minute or once every five minutes, and only recording this extracted information; or by combination of both damping and periodic interrogation.

II. Cyclic Influences

The most significant of the principal cyclic influences on variability of pollution concentration at a receptor is the diurnal cycle (Fig. 7-2). First, there is a diurnal pattern to source strength. In general, emissions from almost all categories of sources are less at night than during the day. Factories and businesses shut down or reduce

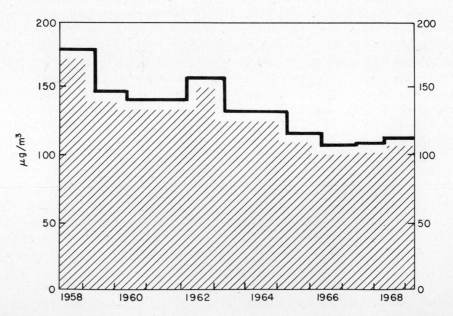

Fɪɢ. 7-4. Annual trend of sulfur dioxide concentration in the United Kingdom 1958–1968 (3).

activity at night. There is less automotive, aircraft, and railroad traffic, electrical usage, cooking, home heating, and refuse burning at night. Second, there is a diurnal pattern to transport and diffusion that will be discussed in detail later in this book.

The next significant cycle is the week-end–weekday cycle. This is entirely a source-strength cycle associated with the change in the pattern of living on weekends as compared with weekdays.

Finally, there is the seasonal cycle associated with the difference in the climate and weather over the four seasons—winter, spring, summer, and fall (Fig. 7-3). The climatic changes affect source strength, and the weather changes affect transport and diffusion.

On an annual basis, some year-to-year changes in source strength may be expected as a community, a region, a nation, or the world increases in population or changes its patterns of living. Source-strength changes will be downward if control efforts or changes in technology succeed in preventing more pollution emission than would have resulted from increase in population (Fig. 7-4). These changes are called trends. Although an annual trend in source strength is expected, none is expected in climate or weather, even though each year will have its own individuality with respect to its weather.

III. Measurement Systems

Many methods of air-quality measurement have inherent averaging times. In selecting methods for measuring air quality or assessing air pollution effects, this fact must be borne in mind (Table 7-1). Thus an appropriate way to assess the influence of air pollution on metals is to expose identical specimens at different locations and compare their annual rate of corrosion among the several locations. Since soiling is mainly due to the sedimentation of particulate matter from the air, experience has shown that this can be conveniently measured by exposing open-topped receptacles to the atmosphere for a month and weighing the settled solids. Human health seems to be related to day-to-day variation in pollutant level. It is accepted practice the world over to assess suspended particulate matter levels in the air by a 24-hour filter sample, which in the United States is acquired by a high-

TABLE 7-1

Air-Quality Measurement

Measure of averaging time	Cyclic factor measure	Measurement method with same averaging time	Effect with same averaging time
Year	Annual trend	Metal specimen	Corrosion
Month	Seasonal cycle	Dustfall	Soiling
Day	Weekly cycle	Hi-vol	Human health
Hour	Diurnal cycle	Sequential sampler	Vegetation damage
Minute	Turbulence	Continuous instrument	Irritation (odor)

volume sampler, known to workers in the field as a hi-vol. Because an agricultural crop can be irreparably damaged by an excursion of the level of several gaseous pollutants lasting just a few hours, record of such excursion requires a measuring procedure that will give hourly data. The least expensive procedure capable of doing this is the sequential sampler, which will allow a sequence of 1- or 2-hour samples day after day for as long as the bubblers in the sampler are routinely serviced and analyzed. As has already been discussed, the hourly data from sequential samplers can be combined to yield daily, monthly, and annual data.

None of the above methods will tell the frequency or duration of exposure of any receptor to irritant or odorous gases when each such exposure may exceed their irritation or odor-response threshold for only a matter of minutes or seconds. The only way that such exposure can be measured instrumentally is by an essentially continuous monitoring instrument, the record from which will yield not only this kind of information, but also all the information required to assess hourly, daily, monthly, and annual phenomena.

IV. The Arrowhead Chart (4)

Because these several cycles are all embedded in a year's data on pollution concentration at a receptor, one year is a very useful time period to use in analyzing such data. A very useful format in which to display the data for analysis is that of Fig. 7-5, which has as its abscissa, averaging time expressed in two different time

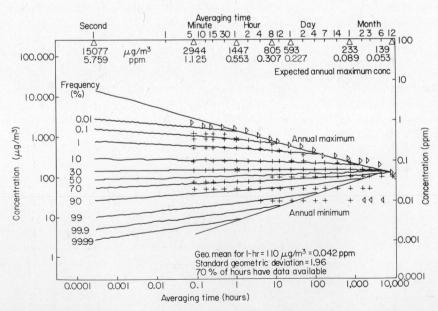

FIG. 7-5. Computer printout of concentration versus averaging time and frequency for sulfur dioxide at site 256. Washington, D.C., Dec. 1, 1961 to Dec. 1, 1968 (4).

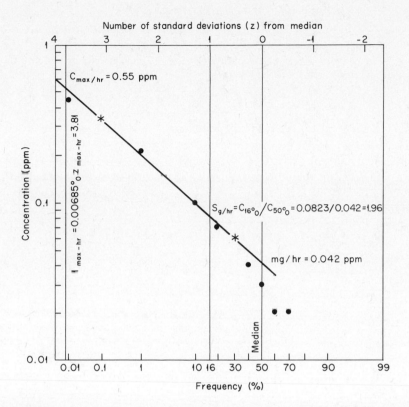

FIG. 7-6. Frequency of 1-hour average sulfur dioxide concentrations equal to or in excess of stated values. Washington, D.C. Dec. 1, 1961 to Dec. 1, 1968 (4).

units, and as its ordinate, concentration of the pollutant at the receptor. This type of chart is called an arrowhead chart and includes enough information to fully characterize the variability of concentration at the receptor.

To understand the meaning of the information given, let us concentrate on the data for 1-hour averaging time. In the course of a year there will be 8760 such values, one for each hour. If all 8760 values are arrayed in decreasing value, there will be one maximum value and one minimum value. (For some pollutants the minimum value is indefinite if it is below the minimum detectable value of the analytical method or instrument employed.) In this array of values, the value 2628 from the maximum will be the value for which 30% of all values are greater and 70% are lower. Similarly the value 876 from the maximum will be the one for which 10% of all values are greater and 90% are lower. The 1% value will be between the 87th and 88th value from the maximum; and the 0.1% value will lie between the 8th and 9th value in the array.

The 50% value, which is called the median value, is not necessarily the same as the average value, which is also called the arithmetic mean value. The arithmetic average value is obtained by adding all 8760 values and then dividing the total by

8760. The arithmetic average value obtained for other averaging times, e. g., by adding all 365 24-hour values and dividing the total by 365, will be the same and will equal the annual arithmetic average value. The median value will equal the arithmetic average value only if the distribution of all values will allow this to occur, e. g.,

$$1\ 2\ 3\ 4\ 5\ 6\ 7\ 8\ 9\ 10 \quad \text{Median} = 5.5 \quad \text{Mean} = 5.5$$

$$1\ 3\ 3\ 3\ 5\ 6\ 6\ 6\ 9\ 13 \quad \text{Median} = 5.5 \quad \text{Mean} = 5.5$$

Another value you will run into in this connection is the geometric mean which is the Nth root of the product of n values. This measure is significant because the ordinate scale of Fig. 7-5 is logarithmic (as, incidentally, is the abscissa scale). The relationship between the arithmetic and the geometric mean values is generally

$$\bar{y} = \bar{y}_g s_g^{0.5 \ln s_g} \tag{7-1}$$

where $\bar{y}$ = arithmetic mean, $\bar{y}_g$ = geometric mean, and s_g = standard geometric deviation.

Further discussion of the significance of mean values and standard deviations will be found in the discussions of sampling in Chapters 14 and 25, and in any textbook on statistics. The standard geometric deviation of any array of air-quality data at a receptor for any one averaging time may be obtained graphically by the procedure shown in Fig. 7-6, by taking advantage of the fact that most such data are lognormally distributed and will therefore plot as a straight line on log-probability graph paper.

V. Air-Quality Levels

Air-quality levels vary between concentrations so low that they are less than the minimum detectable values of the instruments we use to measure them; and maximum levels that are the highest concentrations ever measured. The best single descriptor of the air quality of a site is its arrowhead chart. The best descriptor of an arrowhead chart (short of reproducing the entire chart) is its maximum concentration line, since from it, the characteristics of the entire chart can be comprehended. The best descriptors of the maximum concentration line are any of the parameters needed to define a line, in this case, two or more points on the line. Table 7-2 gives this kind of information for the six gases measured by the Continuous Air Monitoring Program of the U.S. Environmental Protection Agency (7). It is summarized data from United States cities over a period of 4 years. There could be locations in the communities in which these measurements were made, or in other communities in the United States or elsewhere where higher values would have been obtained. However no attempt has been made to cull the world's air-quality data for this purpose. The range of concentrations obtained from analysis of 24-hour hi-vol filter samples from 247 stations of the U.S. National Air Sampling Network (5) is shown in Table 7-3.

TABLE 7-2

Ambient Air Pollutant Concentrations in the United States (1962-1968) in $\mu g/m^3$ (4)

Pollutant	Global[a] background concentration	Maximum urban concentrations for different averaging times						
		5 minutes	1 hour	8 hour	1 day	1 month	1 year	Annual geometric mean
Sulfur dioxide	5.2	5,090	4,440	2,680	2,060	920	470	370
Nitric oxide	2.5–25	2,600	2,300	1,230	650	310	123	118
Nitrogen dioxide	9.4–75	2,400	1,280	700	490	190	113	102
Carbon monoxide	115	104,000	91,000	50,500	38,000	24,200	19,500	14,900
Organic compounds[b]	980	59,000	19,600	13,700	11,100	5,900	2,600	3,300
Oxidant	20	2,740	1,020	690	350	140	78	73
Particulates	—	—	—	—	—	—	—	58–180[c]

[a] Authors' estimates.
[b] Also called Hydrocarbons.
[c] Range for 58 urban areas 1961–1965.

TABLE 7-3

Selected Particulate Constituents as Percentages of Gross Suspended Particulates (1966-1967)[a]

	Urban		Nonurban					
	(217 Stations)		Proximate[b] (5)		Intermediate (15)		Remote[c] (10)	
	$\mu g/m^3$	%	$\mu g/m^3$	%	$\mu g/m^3$	%	$\mu g/m^3$	%
Suspended particulates	102.0		45.0		40.0		21.0	
Benzene soluble organics	6.7	6.6	2.5	5.6	2.2	5.4	1.1	5.1
Ammonium ion	0.9	0.9	1.22	2.7	0.28	0.7	0.15	0.7
Nitrate ion	2.4	2.4	1.40	3.1	0.85	2.1	0.46	2.2
Sulfate ion	10.1	9.9	10.0	22.2	5.29	13.1	2.51	11.8
Copper	0.16	0.15	0.16	0.36	0.078	0.19	0.060	0.28
Iron	1.43	1.38	0.56	1.24	0.27	0.67	0.15	0.71
Manganese	0.073	0.07	0.026	0.06	0.012	0.03	0.005	0.02
Nickel	0.017	0.02	0.008	0.02	0.004	0.01	0.002	0.01
Lead	1.11	1.07	0.21	0.47	0.096	0.24	0.022	0.10

[a] United States National Air Sampling Network (5).
[b] Near urban areas.
[c] Far from urban areas.

References

1. 1968 Data Tabulations and Summaries. St. Louis, National Air Pollution Control Administration Publication No. APTD 69-19, U.S. Dept. of Health, Education, and Welfare, Public Health Service, Raleigh, North Carolina (1969).
2. Spirtas, R., and Levin, H. J. Patterns and Trends in Levels of Suspended Particulate Matter. *J. Air Poll. Contr. Ass.* **21** (6), 329–333.
3. Royal Commission on Environmental Pollution—First Report. Cmnd 4585, H. M. Stationery Office, London, 1971.
4. Larsen, R. I., "A Mathematical Model for Relating Air Quality Measurements to Air Quality Standards." Publication AP-89 (Preliminary Publication) Office of Air Programs, Environmental Protection Agency, Research Triangle Park, North Carolina Sept. 1971—58 pp.
5. Air Quality Data for 1967 (Revised 1971), Office of Air Programs Publication No. APTD 0741, Environmental Protection Agency, Research Triangle Park, North Carolina (1971).

Suggested Reading

Munn, R. E. "Biometeorological Methods," 336 pp. Academic Press, New York, 1970.
Lowry, W. P. "Weather and Life—An Introduction to Biometeorology," 305 pp. Academic Press, New York, 1969.

Questions

1. How does the range of concentrations of contaminants in air of concern to the industrialist hygienist differ from those of concern to the air pollution specialist? To what extent are air sampling and analytical methods in factories or in the ambient air the same or different?

2. Using the data of Fig. 7-1, draw the variation in concentration over the 6-hour period as it would appear using sampling and analytical procedures which integrate the concentration arriving at the receptor over time periods of 30 minutes and 2 hours, respectively.

3. Sketch the appearance of a strip chart record measuring one pollutant for a week to show the weekday–weekend cycle.

4. Draw a chart showing the most probable trend of the concentration of air-borne particles of horse manure in the air of a large midwestern United States city from 1850 to 1950.

5. Describe an air-quality measurement system to assess the levels and types of aeroallergens.

6. Using the data of Fig. 7-5, determine the frequency with which a 30-minute average SO_2 concentration of 700 $\mu g/m^3$ would be likely to be exceeded at site 256 in Washington, D.C.

7. What is the arithmetic mean concentration in ppm of an array of hourly average sulfur dioxide concentration data having a geometric mean of 0.042 ppm and a standard geometric deviation of 1.96 (cf. Fig. 7-6)

8. If the data record on which Fig. 7-6 is based was continuous and no data had to be discarded because of instrumental or recorder maintenance or malfunction, how many individual data values went into the development of the figure? Can you explain why the higher values fit the straight line distribution better than the lower values.

9. Prepare a table describing air-quality levels in your community, or in the nearest community to you that has such data available.

Chapter 8

THE CHEMISTRY OF AIR POLLUTION

Chemistry is fundamental to air pollution effects and control. Pollutants are discharged to the atmosphere as a result of chemical reactions involving combustion and industrial processes. Once in the air, the chemical pollutants may react with one another to form additional compounds. Chemistry is also involved in the abatement or control of pollutants prior to discharge into the air. In brief, chemistry may be considered the servant science of air pollution. It is important to delineate the chemicals of major interest to air pollution as well as the principles involved in the chemical reactions of pollutants with each other and with the environment.

I. The Periodic Table

Natural science is a study of the behavior of atoms and/or particles that comprise the atom. The number of atoms in the universe is almost beyond human comprehension—1 mole (22.4 liters at standard conditions: 0°C and 760 mm Hg) contains 6.023×10^{23} atoms or molecules. The enormous number of molecules in the universe allows a major simplification—properties of matter depend upon a statistical or average behavior of molecules. The periodic table organizes the chemical behavior of elements as a function of atomic number.

Arrangement of elements of the periodic table into eight groups (families of elements) and seven periods (energy levels) emphasizes pronounced similarities of physical and chemical properties. The properties of any given element do not differ greatly from those of its neighbors; property changes differ in a regular manner across (periods) or down (groups) the periodic table. The chemical and physical properties of elements are a periodic function of their atomic numbers. Chemistry is thus considered to be "organized," rather than a series of random un-related facts. Just as there is "order" to elements, so too there is order to chemical compounds and reactions. By deduction, based upon a knowledge of the periodic table, one may proceed from the chemistry of a known element (or compound) to that of another element (or compound).

In the periodic table, elements increase in atomic number from left to right—each period ending with atomic numbers 2, 10, 18, 36, 54, and 86; the atomic number indicates the number of protons or electrons in the nucleus and periphery, respectively. Neutrons (atomic mass number minus atomic number) in the nucleus may differ for the same atomic number; an element having the same atomic number but different atomic mass is known as an isotope. Metallic elements are located on the left-hand side of the periodic table and nonmetals and gases on the right; metalloids, showing characteristics of both metals and nonmetals, are in the center. Table 8-1 shows transitions in properties of periods and groups.

Proceeding from left to right within a period, the boiling point and density in-crease, go through a maximum, and decrease. The boiling point decreases when going down a group for metals (vice versa for nonmetals), whereas the density in-creases. In going down a group, the atom size increases; the atom size decreases across a row as the electrons are in the same energy level causing a contraction be-tween nucleus and electrons. In forming positive ions (loss of an electron—oxida-tion) or negative ions (gain of an electron—reduction), size increases with gain of an electron. Metallic elements lose electrons readily (low ionization energy), while the nonmetals gain electrons readily (high electronegativity). Cesium is the most active metal (best reducing agent), whereas fluorine is the most active gas or non-metal (best oxidizing agent).

To form compounds, atoms must come sufficiently close (2–3 Å) so that valence electrons may be shared or exchanged. Electronegativity is a measure of the electron

TABLE 8-1
Variation of Properties of Groups and Periods in Periodic Table

Period ————————————————————————————————→

Group						
Li	Be	B	N	O	F
Active metal	Metal	Nonmetallic solid	Inactive gas	Moderately active gas	Most active gas
Na	Mg	Al	P	S	Cl
Very active metal	Active metal	Metal	Nonmetallic solid	Nonmetallic solid	Active gas
Rb	Sr	In	Sb	Te	I
More active metal	Active metal	Active metal	Moderately active metal	Metal	Nonmetallic solid

pulling ability of atoms. Elements with low electronegativity (low ionization energy) react readily with elements of high electronegativity to form strong ionic bonds. Two elements with about the same electronegativity react to form weaker covalent bonds. There is no sharp distinction between ionic and covalent bonds; gradations of both bond types within a molecule depend upon the electronegativity of the species.

Chemical reactions thus take place between electrons in outer energy levels of separate atoms. The number and arrangement of electrons in the outer levels determine which atoms will react with other atoms. The outermost energy level cannot contain more than eight electrons, a stable configuration. In compound formation the reaction may be either endothermic (energy added) or exothermic (energy liberated).

II. Chemical Bonds

Chemists categorize molecules or compounds by two basic types of bonding—ionic or covalent. Electropositive elements (metals) lose an electron to electronegative elements (nonmetals) to form an ionic or electrovalent bond. Compounds of carbon with hydrogen, oxygen, nitrogen, and other elements share electrons to form covalent or coordinate covalent bonds. The hydrogen bond involves molecules containing one or more hydrogen atoms covalently bound to atoms of high electron affinity that have one or more unshared electron pairs. More than one bond type is possible in compounds.

Ionic or electrovalent bond

$$Na^+ + \;Cl\; \longrightarrow Na^{\oplus} + \;Cl\;^{\ominus} \tag{8-1}$$

$$Mg + \;O\; \longrightarrow Mg^{2+} + \;O\;^{2-} \tag{8-2}$$

Covalent bond

$$H:C:H \qquad H:H \qquad H:C:C:O:H \tag{8-3}$$

Coordinate covalent bond

$$H:N:H + B:Cl \longrightarrow H:N:B:Cl \tag{8-4}$$

(Once formed, the coordinate covalent bond is indistinguishable from the covalent bond.)

Hydrogen bonding

$$H \overset{x}{\bullet} \overset{\bullet\bullet}{\underset{\bullet\bullet}{F}} \bullet - - - HF - - - -HF \tag{8-5}$$

$$H \overset{x}{\bullet} \underset{\underset{H}{\bullet\bullet}}{\overset{\bullet\bullet}{O}} \bullet - - - \underset{H}{HO} - - - -\underset{H}{HO} \tag{8-6}$$

The chemical bond is not a physical structure; it is an energy relationship between atoms which holds them together in a molecule. To form compounds, atoms must be close enough for the electron cloud to overlap—forming the chemical bond.

A. Ionic Bond

The ionic bond is an electron-transfer process in which electrons are transferred from atoms of one reactant to one or more other reactants:

$$2\ Na \rightarrow 2\ Na^+ + 2\ e^- \tag{8-7}$$

$$2\ e^- + F_2 \rightarrow 2\ F^- \tag{8-8}$$

$$2\ Na + F_2 \rightarrow 2\ Na^+ + 2\ F^- \tag{8-9}$$

Ionic bonds are formed by reaction between one element with a low ionization potential (the ease with which an atom will lose an electron) and another element with a high electron affinity (the ease with which an element will gain an electron). The loss of one or more electrons by the metallic element to the nonmetallic element allows each formed ion to stabilize the outside electron shell with "noble-gas structures." Elements with fewer than four electrons in the outer energy level tend to lose electrons, whereas those with more than four tend to gain electrons.

In the formation of an ionic or electrovalent compound, positive or negative ions are formed by transfer of electrons. The resultant ionic species will be an electrolyte, i. e., it will conduct electric current. When an ionic solid (Na^+Cl^-) melts, the oppositely charged ions acquire a degree of independence; energy is required to separate ions, accounting for the fact that ionic compounds have high melting points.

B. Covalent Bond

Carbon or organic chemistry is structured by sharing electrons of functional groups. By sharing electrons, none of the atoms involved in compound formation attains a noble gas structure, but they do approach such a configuration closely. Multiple covalent bonds between atoms form double and triple bonds.

When a covalent bond solid melts, the atoms move as a unit or molecule. Because electrostatic forces are not involved, it takes less energy to melt a covalent than an electrovalent compound; covalent bonds are weaker than ionic bonds. Covalent compounds are poor conductors of electricity.

Organic compounds built upon covalent bond formation have molecular weights ranging from sixteen for methane to several million for complicated proteins.

C. Resonance

Many compounds have properties not compatible with any single electronic formula written for them. In such compounds, the formula may be represented by two or more electronic configurations (i.e., hybrid); this phenomenon is known as resonance. Resonance does not imply that there are two electronic forms that switch back and forth—there is one real electron structure—the difficulty is in describing it.

Four resonance configurations are postulated for sulfur dioxide—no single structure conforms to observed properties and octet rule.

$$(8\text{-}10)$$

D. Geometry of Formed Molecules

When atoms combine to form a molecule, there is a definite and predictable geometric relationship established between the atoms involved. As diagrammed in Fig. 8-1, there is a bond angle of 104.5° between the two hydrogen atoms and the oxygen atom of water. The bond angle is 109° 28′ between the four hydrogen atoms of methane (tetrahedron). Hydrogen and hydrogen chloride molecules are linear.

The way in which one molecule reacts with another depends, in part, on the shape of molecules. Most organic compounds, i. e., butane (C_4H_{10}) are represented as if they were flat, two-dimensional models; in reality, compounds have depth in addition to length and breadth (i. e., they are three-dimensional).

One consequence of the shape of molecules is the physical separation of electric charges. Highly electronegative compounds pull the electron cloud toward them, leaving the remaining portion of the molecule positively charged. Thus the molecule or compound has a positive and negative end and is said to exhibit polarity. Water is a polar molecule with a negative charge (dipole) on oxygen and positive charges on hydrogen atoms. As positive and negative charges exist because of uneven electron distribution, the hydrogen of one water molecule forms a hydrogen bond with the oxygen atom of an adjacent molecule.

III. Gases

A gas consists of widely separated molecules traveling in a straight line at high rates of speed; gas molecules frequently collide. A gas has little order and no definite shape. A liquid, like a gas, has no definite shape but is more ordered and denser. Solids differ from gases and liquids in that solids have a definite shape and are almost completely ordered. Solids and liquids, unlike gases, are not appreciably affected by changes in pressure or temperature.

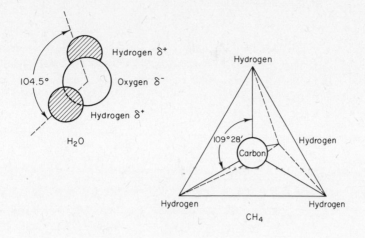

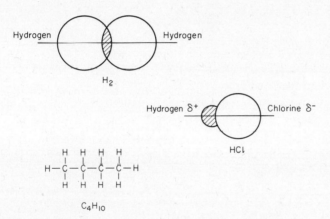

FIG. 8-1. Molecule geometry.

A. Ideal Gas Equation

The ideal gas law equation, $PV = nRT$, is a combination of the laws of Boyle, Charles, and Gay-Lussac. Value of the constant R depends upon the units of pressure (P), volume (V), and temperature (T).

The ideal gas equation is used to convert a gas from one set of conditions (P, V, T) to standard conditions (STP) or to any desired conditions. The atmosphere is generally considered as an ideal gas; deviations from the perfect gas law may be expected at high temperatures and pressures.

B. Dalton's Law of Partial Pressure

Dalton's law of partial pressure states that at constant temperature, the total pressure exerted by a mixture of gases is equal to the sum of the partial pressures of the individual components of the mixture:

$$p_{total} = p_1 + p_2 + p_3 + \cdots + p_n \tag{8-11}$$

Thus the composition of a gaseous component in the atmosphere (or gas sample) can be determined if its partial pressure is known.

C. Graham's Law of Diffusion

Graham's law states that the rate of diffusion of a gas is inversely proportional to the square root of its density, ρ:

$$\frac{Y_1}{Y_2} = \left(\frac{\rho_2}{\rho_1}\right)^{1/2} \tag{8-12}$$

where Y is the diffusion rate. As density of a gas is proportional to its molecular weight

$$\frac{Y_1}{Y_2} = (MW_2/MW_1)^{1/2} \tag{8-13}$$

Graham's law implies that lighter gases diffuse more rapidly from a container than do heavier gases. Experimentally, permeation tubes have been developed to allow a gas to diffuse from the container at a fixed rate into a known air stream to prepare accurate concentrations for research and instrument calibration purposes.

D. Velocity of Gas Molecules

The root-mean-square velocity of gas molecules, u, may be calculated from simple parameters:

$$u = (3RT/MW)^{1/2} \tag{8-14}$$

or

$$u = (3p/\rho)^{1/2} \tag{8-15}$$

The total number of colliding gas molecules per cubic centimeter per second (Z_g) is represented by

$$Z_g = 1.41^{-1}\, \pi u\, (n_v \sigma_m)^2 \tag{8-16}$$

where n_v is the number of molecules per cubic centimeter and σ_m is the molecular diameter.

The average distance a molecule travels before colliding (mean free path, λ_m) is

$$\lambda_m = (1.41\, \pi n_v \sigma_m{}^2)^{-1} \tag{8-17}$$

An understanding of velocity of molecules and frequency of collisions helps explain why atmospheric concentrations of strong reducing agents (SO_2) can exist in the presence of strong oxidizing agents (O_3).

Not all molecules of a gas have the same average speed. Because of collisions, some molecules acquire a higher energy while others lose energy. The distribution of

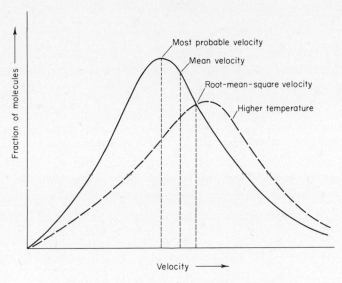

Fig. 8-2. Distribution of molecular velocities.

velocities of a gas with temperature is shown in Fig. 8-2. As noted, molecular velocities range from zero to very high values. As the temperature is raised, the maximum velocity is shifted to the right, corresponding to an increase of molecular velocity with increasing temperature. Temperature is a measure of average kinetic energy of molecules. For gases to react in the open atmosphere, they must first collide; if sufficient energy is available, reaction may occur.

E. Adsorption of Gas Molecules

Gases may be adsorbed on the surface of a solid or absorbed beneath the surface of the solid; the term sorption applies to both phenomena. As a general rule, an increase in pressure or a decrease in temperature increases sorption of a gas.

Two types of adsorption are generally recognized—physical adsorption (Van der Waals) and chemical or activated adsorption (Langmuir). Van der Waals adsorption is relatively weak and is roughly related to the ease of liquefaction of the gas, i. e., sulfur dioxide would be more strongly adsorbed on a solid than carbon monoxide.

Langmuir based his adsorption isotherm on the assumption that chemisorbed gas layers are normally only one molecule in thickness:

$$a_g = K_1 p / (K_2 p + 1) \qquad (8\text{-}18)$$

where a_g is the amount of gas adsorbed/unit mass, p is the partial pressure of gas, and K_1, K_2 are constants.

Adsorption of atmospheric gases on solids exerts a strong influence in air pollution effects. A corrosive gas may be sorbed on metal, building stone, or even fabrics; or the corrosive gas may be sorbed on particulate matter which, in turn, might affect the material. The sorption of corrosive gases on particulate matter subsequently retained in the lung is suggested as a prime reason for air-pollution-caused respiratory illnesses and deaths.

IV. Liquids and Solids

From kinetic theory, a liquid may be considered as a continuation of the gas phase into regions of small volumes and very high molecular attraction. In a liquid, cohesive forces are sufficiently strong to form a condensed state, but are not strong enough to prevent a considerable translational energy of individual molecules.

Solids may be classified as either crystalline or amorphous. Crystalline solids have a definite molecular or ionic configuration characteristic of the substance; amorphous solids do not have a definite configurational arrangement.

As a liquid is cooled, the translational energy of molecules decreases until, at the freezing point, the molecules are forced to arrange themselves in a geometric pattern characteristic of each substance. When crystallization starts, heat is either absorbed or evolved. Crystal forms are grouped into six crystal systems characterized by the number of geometric axes (three or four), the angles between them, and the intercepts among them. From measurements of these parameters, crystal identification is possible.

Interest in solid matter in air pollution is mainly concerned with the physical and chemical properties of the material. Particles sorb gases and scatter light; particles act as catalysts in promoting atmospheric reactions. Particle hardness erodes metals by impingement.

A. Vapor Pressure of Liquids

When the rate of escape of molecules from a liquid becomes equal to the rate of return of the molecules to the liquid, the established equilibrium is the saturated vapor pressure of the liquid. Vapor pressure of a liquid increases with temperature and decreases with increasing pressure.

In the air environment liquids with a high vapor pressure and low boiling point exist as gases. Liquid droplets exist in ambient atmospheres, but vaporize at a rate depending upon vapor pressure. Water droplets will vaporize until saturation pressure is reached at a specific temperature; fog is a case in point.

B. Surface Tension

A molecule in the interior of a liquid is completely surrounded by other molecules, whereas a molecule on the surface is not so surrounded. A surface molecule has a resultant inward attraction, contracting the surface of the liquid to the smallest possible area. Surface tension (s) is defined as the force in dynes acting in the surface at right angles to any line of 1 cm in length:

$$s = r'h_c\rho g/2 \qquad (8\text{-}19)$$

where r' is the radius of curvature of the surface, h_c is the height of capillary column, g is the acceleration due to gravity, and ρ is the density of liquid.

The greater the surface tension, the greater the hydrostatic pressure and the faster the spread of the liquid over a surface. The concept applies to a liquid droplet on a fabric, paint, metal, or stone surface.

C. Viscosity of Liquids

Liquids exhibit a greater resistance to flow than gases. The coefficient of viscosity η through a capillary tube of radius r is defined as

$$\eta = \pi p r^4 t / 8 L V_L \tag{8-20}$$

where p is the pressure head, t is the flow time, L is the length of flow, and V_L is the volume of liquid. Most important to air chemistry is the fall of a liquid or solid body through any fluid media as expressed by Stoke's law (also see Section 19-II):

$$V_t = (gD^2/18\eta)(\rho - \rho_m) \tag{8-21}$$

where V_t is the velocity of body, D is the body diameter, ρ is the body density, and ρ_m is the medium density.

In air the particle radius is the critical parameter; each size particle falling in air has a specific terminal velocity. For very small particles, where r is of the order of the distance between molecules of the fluid, an additional correction factor is applied.

D. Scattering and Absorption of Light

Liquid and solid particles scatter and absorb light. Light attenuation by scattering and absorption is related to wavelength of light as well as particle size, concentration, and physical characteristics.

A simplified formula may be used to estimate meteorological range, L_m of particulate matter in the atmosphere (1) (see also Section 21-I):

$$L_m = \frac{5.2\,\rho r}{X\kappa} \tag{8-22}$$

where ρ is the particle density, r is the particle radius, X is the particle concentration and κ is the scattering area ratio. From the above equation, a doubling of particle concentration in the air reduces the visibility (related to L_m) to half of its former value.

E. Solutions of Nonelectrolytes

The colligative properties of solutions depend upon the number of particles in solution but not upon the nature of the particles. Colligative properties include
(i) vapor pressure lowering (see also Section 19-I)

$$e_s = e_{sa}N_L \tag{8-23}$$

where e_{sa} is the vapor pressure above a solution surface, e_s is the vapor pressure above a pure solvent, and N_L is the mole fraction of solvent;

ii. freezing point lowering (ΔT_f),

$$\Delta T_f = -1.8°C \ (O_m) \tag{8-24}$$

where O_m is molality;

iii. boiling point elevation (ΔT_b),

$$\Delta T_b = 0.5°C \ (O_m) \tag{8-25}$$

and   iv. osmotic pressure (Π)

$$\Pi = nRT/V_L \tag{8-26}$$

Each of these properties effectively lowers the vapor pressure of solution droplets and reduces the rate of evaporation. Solution droplets remain in the atmosphere for relatively long time periods (depending upon humidity and size).

F. Solutions of Electrolytes

Substances which when dissolved in water or other appropriate solvents yield solutions that conduct electric current are called electrolytes. Although electrolytes exhibit enhanced colligative properties over nonelectrolytes, the electrolyte dissociation is of more interest to air chemistry. Strong electrolytes have a high degree of dissociation in solution (good conductors of current); weak electrolytes exhibit poor conductance and a low degree of dissociation.

In an electrolyte solution, free ions exist as well as undissociated molecules, i. e.,

$$MA \rightleftharpoons M^+ + A^- \tag{8-27}$$

At equilibrium the dissociation constant K_d depends upon the activity a of the species:

$$K_d = [(a_{M^+})(a_{M^-})]/(a_{MA}) \tag{8-28}$$

The mean activity coefficients of strong electrolytes increase with decreasing molality. The variation of activity coefficient with concentration is called the ionic strength μ_s:

$$\mu_s = \tfrac{1}{2}(O_{m_1}, z_{a_1}^2 + O_{m_2}, z_{a_2}^2 + O_{m_3}, z_{a_3}^2 \ldots) \tag{8-29}$$

where O_m represents molalities of ionic species and z_a the corresponding valencies (charge).

For any acid HA dissolved in water,

$$HA + H_2O \rightleftharpoons H_3O^+ + A^- \tag{8-30}$$

the equilibrium constant in terms of activities is

$$K_e = (a_{H_3O^+} \times a_{A^-})/(a_{HA} \times a_{H_2O}) \tag{8-31}$$

The activity of the water molecules is approximately equal to that of pure water and by convention is taken as unity; the hydronium ion is written as H^+. The simplified version of the above formula is

$$K_e = (a_{H^+}a_{A^-})/a_{HA} \tag{8-32}$$

For strong or relatively strong acids, the equilibrium constant ranges from almost unity to 10^{-4}; weak acids have equilibrium constants ranging from 10^{-5} to 10^{-36}.

Interest in the ionic strength of acids and bases is pertinent to air pollution because of corrosive effects. Corrosive effects may be direct as in a reaction of an acid on a metal; the effect may be indirect in the oxidation of SO_2 in human or vegetative tissues to H_2SO_4.

V. Chemical Reactions

A. Thermochemistry

Two questions must be asked concerning a chemical reaction—Will it proceed and how fast will it proceed? Both questions are valid for reactions involving a single state of matter (homogeneous) or more than one state (heterogeneous). Thermochemistry involves energy changes that accompany chemical reactions.

The three laws of thermodynamics consider energy transformations. The first law states that although energy can be converted from one form to another, it cannot be created or destroyed; the second law concludes that all processes in nature tend to occur only with an increase in entropy, and that the direction of change

TABLE 8-2

Heats of Formation of Compounds at 25 °C

	ΔH kcal/mole		ΔH kcal/mole
Gases		Solids	
HCl	−22.0	KCl	−104.4
H_2O	−57.8	NaCl	−98.3
CO	−26.6	AgCl	−30.3
CO_2	−94.4	$HgCl_2$	−53.4
NO	+21.5	$MgCl_2$	−153.3
NO_2	+7.9	NaBr	−86.7
SO_2	−70.9	KI	−78.9
H_2S	−4.8	CuO	−38.5
NH_3	−11.1	FeO	−64.3
CH_4	−18.1	Fe_2O_3	−198.5
C_2H_2	+53.9	BaO	−126.1
CH_3OH	−48.4	NaOH	−102.0
C_2H_5OH	−56.9	KNO_3	−118.1
		$CuNO_3$	−73.1
Liquids		Na_2SO_4	−332.2
H_2O	−68.4	$CuSO_4$	−184.3
HNO_3	−42.0	$MgSO_4$	−313.0
H_2SO_4	−189.8	$CaCO_3$	−283.5
C_6H_6	+11.6	BaO	−126.1
CH_3OH	−57.5	KOH	−102.0
C_2H_5OH	−67.1		

always leads to an entropy increase; the third law notes that the entropy of a pure crystalline solid may be taken as zero at the absolute zero of temperature.

Knowledge of the energy relationships of chemical equations applies when atmospheric reactions are considered. Molecules in the atmosphere obey the same laws as in "classic" chemistry. When an atmospheric reaction is postulated, the energy requirements must be met. As in classic chemistry, reactions in the atmosphere involve, among others, the heats of reaction, of formation, of combustion as well as solution and dilution. When sulfur reacts with oxygen to form sulfur dioxide energy is liberated.

$$S_{solid} + O_{2gas} \rightarrow SO_{2gas} \qquad \Delta H_{25°C} = -70.9 \text{ kcal} \qquad (8\text{-}33)$$

As shown in Table 8-2 heats of formation of oxides are generally negative, indicating that energy input is not required; with regard to iron oxides, the formation of

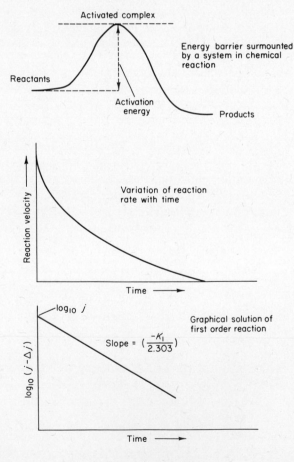

FIG. 8-3. Chemical kinetics.

Fe_2O_3 is more likely than FeO because greater heat is liberated during formation of Fe_2O_3. Catalysis must be considered in atmospheric reactions, as well as reversibility of the reaction.

Formation of acetylene from its components in the atmosphere requires an energy input of 53.9 kcal/mole. This amount of energy must be available for the reaction to proceed.

B. Chemical Kinetics

Chemical kinetics attempts to resolve the important question: "How rapidly and by what mechanism does a reaction take place?" Although thermochemically favored for a reaction to occur, molecules involved must be raised to a state of higher energy. As diagrammed in Fig. 8-3, if energy input (activation energy) is sufficient for reactants to form an activated complex, the reaction will continue. The presence of a catalyst may reduce the activation energy. The variation of the reaction rate with time is also shown in Fig. 8-3, i. e., $-dC/dt$, where C is the reactant concentration.

The rate of a chemical reaction, based in part on the law of mass action, is directly related to the order of reaction. The order of a chemical reaction is the sum of all exponents to which concentrations in the rate equation are raised. Reaction orders of zero, one, two, and higher are known; fractional reaction order is possible. The reaction rate is determined only on the basis of experimental data; order determination based on chemical equation alone is not sufficient.

A unimolecular reaction ($J \rightarrow$ products) must obey the equation,

$$\ln j/(j - \Delta j) = Kt \qquad (8\text{-}34)$$

where j is the initial concentration of J and Δj is the decrease in concentration to time t. If the calculated K's are constant, the reaction is first order.

Solution to a first-order reaction by rearrangement of the above equation to the form,

$$\log_{10} (j - \Delta j) = (Kt/2.303) + \log_{10} j \qquad (8\text{-}35)$$

is graphed in Fig. 8-3.

The fractional life method may also be used to test order of reaction. When one-half of the reactant has undergone decomposition.

$$j - \Delta j = j/2, \quad \text{while} \quad t = t_{1/2}, \text{ or } t_{1/2} = (\ln 2)/K \qquad (8\text{-}36)$$

Thus for a first-order reaction, the half-life period is a constant independent of the initial concentration.

Kinetic research has demonstrated that a reaction mechanism for even a simple atmospheric reaction may be extremely complex. Chain reactions, opposing reactions, radical formation, catalysis, consecutive reactions, inhibitors, and temperature effects must be considered in kinetic studies. When a reaction occurs in stages, the reaction rate (order) is determined by the slowest step or stage.

C. Photochemistry

Photochemistry is concerned with characteristics of chemical reactions from exposure of a system to radiation. The subject is important to air chemistry in that certain atmospheric reactions will proceed even though sun radiation reaching the surface of the earth is not less than about 3000 Å wavelength. Synthesis, decomposition, polymerization, isometric change, oxidation and reduction reactions can occur by exposure to radiant energy.

Only those radiations that are absorbed are effective in producing chemical change. Furthermore, according to Lambert-Beer's law, equal fractions of the incident radiation are absorbed by successive layers of equal thickness of the light-absorbing substance (see also Section 17-IV)

$$R = R_o e^{-k'CL} \tag{8-37}$$

where R_o is the intensity of the incident light, R is the intensity after passing through material, k' is the adsorption coefficient constant, L is the distance of travel, and C is the concentration of solution.

Each molecule taking part in a photochemical reaction takes up one quantum of the radiation causing the reaction, or

$$E = (2.859/\lambda) \times 10^5 \text{ kcal/mole} \tag{8-38}$$

where E is the energy absorbed per mole and λ is the wavelength, Å.

Table 8-3 lists energy equivalents at various wavelengths; energy equivalents decrease with increasing wavelengths.

VI. Reactions in the Atmosphere

Reactions in the atmosphere are directly related to classic concepts of science, including thermodynamics and reaction kinetics. Factors that make atmospheric chemistry more complex and more interesting scientifically include

i. Low concentrations of the order of 100 $\mu g/m^3$

TABLE 8-3

Energy Values at Different Wavelengths

Wavelength (Å)	Color region	Energy (kcal/mole)
2000	Ultraviolet	142.95
3000	Ultraviolet	95.30
4000	Violet	71.48
5000	Blue-green	57.18
6000	Yellow-orange	47.65
7000	Red	40.84
8000	Near infrared	35.74

ii. Changes in concentrations of pollutants depending upon emissions and meteorology

iii. Changes in pollutant phase depending upon concentration, pressure, and temperature

iv. Changes in the type of pollutant dependent upon emission and atmospheric reactions

v. Removal of pollutants from the atmosphere dependent upon a host of factors

A. Photochemical Air Pollution

Photochemical air pollution was first recognized as a serious phenomenon in Los Angeles in the mid-1940's. Despite tremendous research (2–4) the problem has not been resolved either scientifically or politically.

The overall dependency of photochemical air pollution on various factors may be written

$$\text{Photochemical}\atop\text{Air pollution} = \int \frac{(\text{NO}_x \text{ conc}) (\text{organic conc}) (\text{sunlight intensity}) (\text{temperature})}{(\text{wind speed}) (\text{inversion height})} \quad (8\text{-}39)$$

Sunlight (light energy) in the range 3000–4600 Å initially decomposes NO_2

$$NO_2 + h\nu \rightarrow NO + O \qquad\qquad (8\text{-}40)$$

$$O + O_2 + M \rightarrow O_3 + M \text{ (third body)} \qquad\qquad (8\text{-}41)$$

$$O_3 + NO \rightarrow NO_2 + O_2 \qquad\qquad (8\text{-}42)$$

until a dynamic equilibrium

$$NO_2 + O_2 \overset{h\nu}{\rightleftharpoons} NO + O_3 \qquad\qquad (8\text{-}43)$$

is established. After the initial decomposition step, many speculative mechanisms have been proposed; one such mechanism is presented (5):

$$RH + O \rightarrow R\cdot + \text{products} \qquad\qquad (8\text{-}44)$$

$$RH + O_3 \rightarrow \text{products (including } R\cdot) \qquad\qquad (8\text{-}45)$$

$$NO + R\cdot \rightarrow NO_2 + R\cdot \qquad\qquad (8\text{-}46)$$

Radicals are lost by chain-terminating steps but in reactions of a free radical with NO or NO_2, one major product formed is peroxyacetyl nitrate (PAN):

$$NO_2 + R\cdot \rightarrow CH_3COOONO_2 \text{ (PAN)} + \text{products} \qquad\qquad (8\text{-}47)$$

As noted in Fig. 8-4, the NO_2 concentration increases as the NO concentration decreases. The O_3 formation starts at about the time the NO_2 level peaks. PAN and aldehydes are continuously formed. The formation of products depends upon the particular type of organic compound present; olefinic compounds are most reactive, alkyl benzenes are moderately reactive, and compounds such as benzene, acetylene, and C_1–C_3 paraffins are not reactive. During daylight, the photochemical formation of ozone follows the prevailing winds; once the sun sets the reaction mechanism stops. Formulation of a precise mechanism for photochemical air pollution is virtually impossible considering the large number of organic compounds in the air and

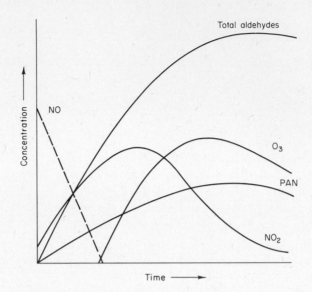

FIG. 8-4. Variation of concentration of reactants and products in photochemical air pollution.

the multitude of steps in the chain reaction; each postulated step must be confirmed on a thermodynamic and kinetic basis.

The question of reactivity of organic compounds is of interest in that differences in reactivity and chemical structure of organic compounds are related to different effects of photochemical air pollution, i. e., eye irritation, visibility reduction, rubber cracking, and vegetation damage. In eye irritation, for example, monoalkylbenzenes and 1,3-butadiene are highly reactive while in ozone yield (rubber cracking) 1,3-butadiene is highly reactive and monoalkylbenzenes have low activity. Although the mechanism for the formation of photochemical aerosols (visibility reduction) is not completely clear, SO_2 is involved. The blue haze over mountainous areas is attributed to a photochemical reaction involving terpenes emanating from vegetation.

VII. Radioactivity

Atomic transformation, natural and man-made, involves three major rays—alpha (α), beta (β), and gamma (γ). Alpha rays or particles are doubly charged helium atoms; when an alpha particle is emitted from an element, the product has the properties of an element two places to the left of the parent in the periodic table. Beta rays are negatively charged particles; when an element releases a beta particle, the product has the property of an element one place to the right of the parent in periodic table. Gamma rays are true electromagnetic radiations and result from nuclear transformations or shifts of orbital electrons. Examples of atomic transformations are as follows:

$$^{14}_{7}N + {}^{4}_{2}He \longrightarrow {}^{17}_{8}O + {}^{1}_{1}H \tag{8-48}$$

$$^{238}_{92}U \longrightarrow {}^{4}_{2}He + {}^{234}_{90}Th \tag{8-49}$$

$$^{2}_{1}H + {}^{3}_{1}H \longrightarrow {}^{4}_{2}H + {}^{1}_{0}n \tag{8-50}$$

Energy involved in nuclear transformations may be followed through Einstein's energy-mass equivalent formula:

$$E = Mc^2 \tag{8-51}$$

where E = total particle energy, M is the particle mass and c is the speed of light Energy liberated per pound of fissionable material such as uranium or plutonium is 0.9×10^{13} cal or 3.6×10^{10} BTU.

Although natural radioactivity was recognized at the turn of the century, air concentrations were not considered a serious danger to human health, vegetation, or animals. With the development of the atomic bomb and nuclear power industry, air concentrations of radioactive material increased throughout the world; the disposal of radioactive wastes from the nuclear power industry and the long half-life of radioactive elements (29 years for ^{90}Sr) are of major concern to mankind.

References

1. Robinson, E., Effect on physical properties of the atmosphere, *in* "Air Pollution" (A. C. Stern, ed.) Vol. I, pp. 349–400. Academic Press, New York, 1968.
2. Katz, M., Photochemical reactions of atmospheric pollutants. *Can. J. Chem. Eng.* **48**, 3–11 (1970).
3. Altshuller, A. P., and Bufalini, J. J., *Environ. Sci. Technol.* **5**, 39–64 (1971).
4. Schuck, E. A., and Stephens, E. R., Oxides of nitrogen, *in* "Advances in Environmental Sciences and Technology" (J. N. Pitts, Jr. and R. L. Metcalf, eds). Wiley (Interscience), New York, 1969.
5. Friedlander, S. K., and Seinfeld, J. H., *Environ. Sci. Technol.* **3**, 1175–1181 (1969).

Suggested Reading

Changnon, Jr., S. A., *Bull. Amer. Meteorol. Soc.* **50**, 411–421 (1969).
Dyer, A. J., and Hicks, B. B., *Quart. J. Roy. Meteorol. Soc.* **94**, 545–554 (1968).
Larsen, R. I., Zimmer, C. E., Lynn, D. A., and Blemel, K. G., *J. Air Pollut. Contr. Ass.* **17**, 85–93 (1967).
Peterson, E. K., *Environ. Sci. Technol.* **3**, 1162–1169 (1969).
Sargent, F., *Bioscience* **17**, 691–697 (1967).
Altshuller, A. P., *Bull. WHO* **40**, 616–623 (1969).
Goldsmith, J. R., *Sci. J.* **5**, 44–49 (1969).
McKay, H. A. C., *Chem. Industry*, 1162–1165, Aug. 1969.
Seaborg, G., *J. Chem. Ed.* **46**, 626–634 (1969).
Stephens, E. R., *J. Air Pollut. Cont. Ass.* **19**, 181–185 (1969).

Questions

1. (a) List the number of protons, neutrons and electrons contained in the following elements: H, C, O, S, F, Na, Kr, Pb, and U.
   (b) What is the valence of the following ions: chloride, aluminum, ferric, sulfate, phosphate, sodium, and sulfide?
2. (a) What are ionic and covalent bonds?
   (b) Describe general characteristics of ionic and covalent bonds.
3. Derive the ideal gas law based upon the laws of Boyle, Charles, and Guy-Lussac.
4. (a) Define pH mathematically.
   (b) Why is a strong acid more destructive in corrosion than a weak acid?
5. (a) What are factors which affect the concentration of pollutants in the atmosphere?
   (b) What is the effect of low pollutant concentrations on atmospheric chemical reactions?
6. (a) Describe the meaning of Eq. (8-39).
   (b) What is the lower limit wavelength light reaching the surface of earth? Why?
   (c) What is the energy of 3000 and 4200Å radiation in kcal/mole?
   (d) Why does photochemical air pollution cease when the sun sets?
7. (a) Postulate chemical reactions which may be involved between nitrogen oxides emitted by supersonic transports and ozone, at 16 to 32 km above earth's surface.
   (b) If the ozone layer in the stratosphere is destroyed, what effects will this have at the earth's surface?
8. (a) Water covers about 75% of the earth's surface. Why might one anticipate that waters (oceans) are "sinks" for air pollution?
   (b) If oceans are "sinks" for pollution, might they also be sources of pollution? Why?

Part II

THE EFFECTS OF AIR POLLUTION

Chapter 9

EFFECTS OF AIR POLLUTION ON INERT MATERIALS

A recent report claimed that the annual economic loss of inert materials from air pollution in the United States approximated $104 billion (1). A partial listing of reported losses from soiling and deterioration of inert materials is shown in Table 9-1; the authors estimated that the loss ranged from $32 billion to $320 billion.

An annual air pollution consequence (loss) of $104 billion on inert materials is almost beyond the realm of reason. Yet a trend agreement with time exists between the Midwest Institute results and those of other investigators. In 1949 Gibson (2) estimated that direct economic loss from air pollution, excluding all social and indirect costs, was $1.5 billion annually for the United States. Gustavson (3) in 1958 felt that $7.5 billion was the cost of air pollution to society. In the early 1960's, Department of Health, Education, and Welfare estimates for air pollution effects increased from $7 billion to 12 billion (4). On the basis of data reported by Michelson and Tourin (5), a nationwide cost of $16 billion was calculated. Data from the New York–New Jersey area, extended on a population ratio basis to that of the United States, resulted in a value of $30 billion (6).

High economic losses were reported in other countries throughout the world. The Beaver Report (7), modified by Scorer (8), concluded that air pollution costs for Great Britain were $1.4 billion. Air pollution costs for France were put at $0.6 billion (9).

TABLE 9-1

Ranking of Economic Losses on Inert Materials by Soiling and Deterioration

Material[a]	Annual economic loss (10^9)	
	Soiling	Deterioration
Paint	35.0	1.2
Zinc	24.0	0.8
Cement, concrete	5.4	0.3
Nickel	1.0	0.3
Cotton fiber	0.3	0.2
Synthetic rubber	0.07	0.14
Tin	0.05	0.14
Nylon fiber	0.01	0.04
Paper	1.12	0.02
Leather	2.5	0.02
Building stone, brick	0.18	0.04
Brass, bronze	0.17	0.01
Acrylic plastics	0.03	0.01
Acetate fiber	0.02	0.007
Cellulose plastics	0.01	0.004
Gray iron	No effect	0.002
Malleable iron	No effect	0.001
Gold	<0.001	0.0006
Total	100.0	3.8

[a] See reference 1 for complete list of materials.

Admittedly costs for air pollution consequences to inert materials are difficult to determine with sufficient accuracy to satisfy scientists, economists, and politicians. Yet these "estimates," when plotted, indicate a distinct trend with time (Fig. 9-1); the trend correlates with increases in both population and pollution. And reported results over a span of 20 years indicate that the cost ratio of control costs to effects costs ranged from 0.1 to 0.3 (2, 6, 10).

Considering future control costs Jackson and Wohlers (11) projected that for the Delaware Valley (5–6 million people) the costs in 1980 to reduce emissions 25% below the 1960 level would be $187 million; control costs for the United States in 1980 for the same reduction approximates $7 billion.

I. Corrosive Atmospheres and Effects

Shown in Fig. 9-2 are generalized types of corrosive atmospheres related to air pollution. Corrosive atmospheres range from acidic and alkaline materials to salts, biological agents, and organic compounds, to the "normal atmosphere." Too frequently we forget that even a "crystal-clear atmosphere" deteriorates inert ma-

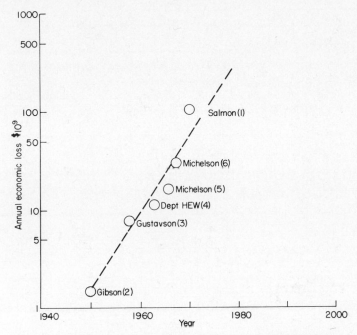

FIG. 9-1. Economic losses to inert materials in the United States

terials. Sunshine weakens fabrics, paper, and leather. Alternate freezing and thawing cracks building materials when water seeps or is absorbed into stone or concrete. Water or high humidity is a necessary ingredient for most if not all corrosive effects upon inert materials. Dust, sand, or salt particles in high winds cause pitting of painted and metallic surfaces.

TABLE 9-2

Air Pollution Effects on Inert Materials

Material	Manifestation	Measurement
Metal	Spoilage of surface; gross loss of metal	Reflectance; weight gain or loss
Building stone	Discoloration; leaching	Not usually quantitative
Fabrics, dyes	Discoloration; fading, weakening	Reflectance; loss of tensile strength
Leather	Weakening; embrittlement	Loss of tensile strength
Paper	Embrittlement	Decreased folding resistance
Paint	Discoloration	Reflectance; not usually quantitative
Rubber	Cracking	Loss of elasticity; cracks

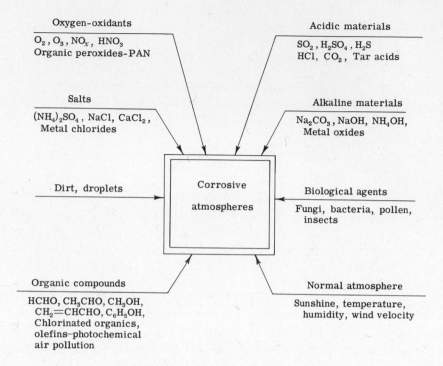

Oxygen-oxidants
O_2, O_3, NO_x, HNO_3
Organic peroxides-PAN

Acidic materials
SO_2, H_2SO_4, H_2S
HCl, CO_2, Tar acids

Salts
$(NH_4)_2SO_4$, NaCl, $CaCl_2$,
Metal chlorides

Alkaline materials
Na_2CO_3, NaOH, NH_4OH,
Metal oxides

Dirt, droplets

Corrosive
atmospheres

Biological agents
Fungi, bacteria, pollen,
insects

Organic compounds
HCHO, CH_3CHO, CH_3OH,
CH_2=CHCHO, C_6H_5OH,
Chlorinated organics,
olefins–photochemical
air pollution

Normal atmosphere
Sunshine, temperature,
humidity, wind velocity

FIG. 9-2. Types of corrosive atmospheres.

With inert material, soiling and deterioration are the prime air pollution effects (Table 9-2). It is important to emphasize that both effects can occur in either a short period of time with high pollutant concentrations or over long periods of time with low concentrations. Hydrogen sulfide can darken lead base paints in minutes; a sulfuric acid aerosol can cause a run in women's stretched stockings minutes after contact; ozone will crack rubber under tension instantly. Conversely Cleopatra's Needle disintegrated more in 90 years at Victoria Embankment in London than in the 3000 years it resided in Egypt. The frieze on the Parthenon in Athens showed greater effects of air pollution in the "industrial period" than in the prior 2200 years.

II. Metals

Metal corrodes either because of the direct action of an acidic and alkaline compound:

$$H_2SO_4 + Zn \rightarrow ZnSO_4 + H_2 \tag{9-1}$$

$$2\,NaOH + 2\,H_2O + Zn \rightarrow Na_2Zn(OH)_4 + H_2 \tag{9-2}$$

or as a result of a formed electrolytic cell with the pollutant as an electrolyte

$$\text{Anode:}\quad Zn \rightarrow Zn^{2+} + 2e \tag{9-3}$$

$$\text{Cathode:}\quad 2\,H^+ + 2e \rightarrow H_2 \tag{9-4}$$

In the formed cell, the metal goes into solution at the anode and hydrogen is liberated at the cathode.

Metals are protected from corrosion by painting, by addition of other elements (carbon steel, which contains 0.02% copper is less resistant than steel, which contains 0.2% copper with traces of nickel or chromium), or by cathodic protection. Coatings of metallic zinc or aluminum will protect steel (iron) even if the latter is exposed to the atmosphere at a surface discontinuity; the zinc or aluminum (anode) corrodes, but the steel (cathode) is protected. Copper or nickel (cathode) will not protect steel (anode).

The formation of a metal oxide or a basic salt at the metal surface protects or minimizes corrosion. Aluminum forms Al_2O_3. Nickel forms a basic nickel sulfate, $NiSO_4 \cdot Ni(OH)_2$. Copper forms either the basic sulfate or chloride, $CuCl_2 \cdot 3Cu(OH)_2$. Lead forms a protective sulfate, $PbSO_4$.

Moisture is critical to corrosion. Below 60% relative humidity corrosion is slow; at 80% there is a marked increase in corrosion; above 80% corrosion is very high; and finally corrosion is minimal during periods of rain (100%). Temperature directly affects rate of corrosion. A rise in temperature will increase chemical reaction, hence corrosion rate. Cool, dry climates are least corrosive.

Sulfur dioxide is the most detrimental pollutant in the corrosion of metals; the higher the ambient sulfur dioxide concentration in an urban area, the higher the corrosion rate. Wohlers *et al.* (12) rearranged sulfur dioxide corrosion data to show a relationship with urban population (Fig. 9-3) for the early and middle 1960's.

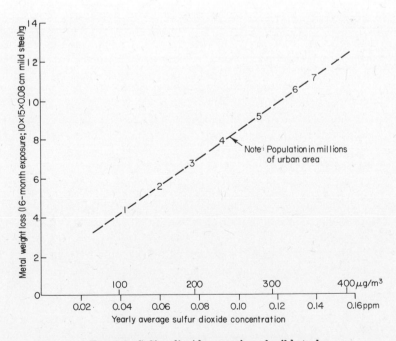

FIG. 9-3. Sulfur dioxide corrosion of mild steel.

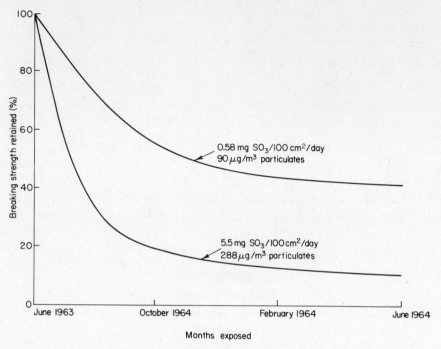

FIG. 9-4. Breaking strength of cotton duck retained after atmospheric exposure.

III. Building Stone

Pollutant action on stone is principally that of an acid dissolving a carbonate. Dolomitic stone is attacked rapidly in polluted atmospheres.

$$CaCO_3 + SO_2 \rightarrow CaSO_3 + CO_2 \tag{9-5}$$

$$CaSO_3 + \tfrac{1}{2}O_2 \rightarrow CaSO_4 \tag{9-6}$$

$$CaCO_3 + CO_2 + H_2O \rightarrow Ca(HCO_3)_2 \tag{9-7}$$

$$Ca(HCO_3)_2 + SO_2 + \tfrac{1}{2}O_2 \rightarrow CaSO_4 + 2\,CO_2 + H_2O \tag{9-8}$$

$$CaCO_3 \cdot MgCO_3 + 2\,SO_2 + O_2 \rightarrow CaSO_4 + MgSO_4 + 2\,CO_2 \tag{9-9}$$

The formed calcium sulfate is relatively soluble in water. Sulfur dioxide and carbon dioxide, to a minor extent, are the prime pollutants of concern.

Building stone that does not contain $CaCO_3$ or $MgCO_3$, e. g., granite, gneiss, or sandstone, is not as readily attacked by air pollutants. Building bricks are resistant to the action of SO_2 or H_2SO_4, but the concrete mortar between bricks is affected.

Much of the decay of building stone is attributable to formation of hard "skins" on the surface; skin formation blisters and flakes from the surface leaving new areas for atmospheric attack. Alternate freezing and thawing also helps expose additional surfaces for pollutant reaction. The action of acidic gases upon building stone is thought to be accelerated by particulate matter.

IV. Fabrics and Dyes

The soiling and fading of textiles is apparent to the general public. In addition to loss of aesthetic appeal, the functional life of the fabric is reduced because of required additional cleaning.

Fabrics and dyes are affected by particulate matter, acidic gases (SO_2), as well as oxidants (O_3, NO_2, PAN). Deterioration by pollution is additive and may accelerate fabric deterioration caused by biological organisms, sunlight, oxygen, humidity, and changes in temperature.

When textile materials are exposed to polluted air large particles may settle by gravity while smaller particles may be filtered from the air; thermal and static electrical precipitation play roles in particulate deposition. Gases may be sorbed directly by the fabric fibers or by the retained particulate matter.

Outdoor exposure tests to study the breaking strength of exposed fabrics were revealing (13). Soiling and degradation of cotton duck and print fabric, with minor variations, followed increasing sulfation rates and suspended particulate matter (Fig. 9-4). The pH determinations showed only a limited degree of correlation between strength loss and lowered pH of the exposed sample. It is believed that fabric degradation was most directly related to sulfur dioxide as measured by the sulfation rate. Fabrics that exhibit the least capacity for SO_2 absorption are least affected by acidic pollutants.

Salvin (14) found that the fading of dyes by photochemical action involved chemical reactions in which the dye molecule was activated by sunlight; the changes involved both oxidation and hydrolysis of the dye molecule. Blue disperse dyes and violet dyes redden when exposed to low concentrations of nitrogen oxides. By rearrangement of functional groups on the dye molecules, a dye resistant to nitrogen oxides was developed (Fig. 9-5). Fabrics dyed with Disperse Blue 27 faded to light

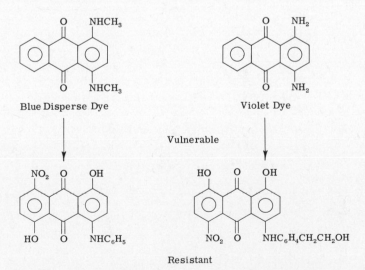

FIG. 9-5. Dyes vulnerable and resistant to nitrogen oxides.

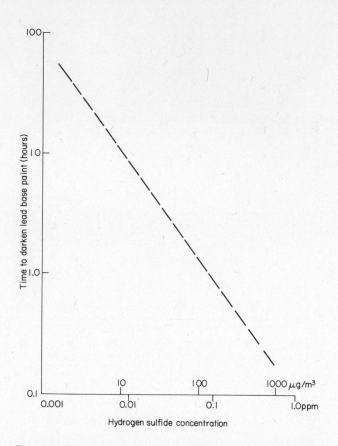

FIG. 9-6. Estimated time necessary to darken lead base paint.

yellow (or colorless) in the presence of low concentrations of ozone even when the fabrics were not exposed to sunlight. Oxidation of the dye molecule is suspect. Dyes vulnerable to nitrogen oxides reddened in Chicago and Los Angeles at 470 μg/ m^3 NO_2 (0.25 ppm), but did not redden in Sarasota or Phoenix at 19 μg/m^3 NO_2 (0.01 ppm). Ozone-sensitive dyes faded in Los Angeles at 412 μg/m^3 O_3 (0.21 ppm), but not in Chicago at 10 μg/m^3 O_3 (0.005 ppm); ozone fading occurred in rural and surburban areas of Sarasota and Phoenix (118–216 μg/m^3 O_3).

V. Paper and Leather

Effects of air pollutants are not as marked inside a building, as when material is directly exposed to weather. Both paper and leather materials sorb SO_2 and other pollutants to weaken or embrittle.

Hudson *et al.* (15) found that paper sorbed SO_2 linearly for 4 days with concentrations ranging from 2620 to 13×10^6 μg/m^3 (1–5000 ppm). It is suspected that the deterioration of modern paper is partially dependent upon the trace metal content as well as surface finish, pattern design and "sweat" deposits; as might be expected, paper is more sorptive than plastic materials.

It has been speculated that the rotting of leather upholstery in London as early as 1800 was a result of sulfur compounds (H_2SO_4) in the air. Leather safety belts exposed to urban atmospheres were found to contain 7% H_2SO_4 (16). Sheepskin valves used to pump air into large electropneumatic organs formerly wore out in 20–30 years; now they hold up for only five years.

VI. Paint

Paints are designed to decorate and protect surfaces. During normal wear, paint chalks moderately to continuously clean the surface. A hardened paint surface will resist sorption by gases although the presence of relatively high concentrations of 2620 to 5240 $\mu g/m^3$ SO_2 (1–2 ppm) will increase the drying time of newly painted surfaces.

Hydrogen sulfide reacts with lead base pigments

$$Pb^{2+} + H_2S \rightarrow PbS + 2\ H^+ \qquad\qquad (9\text{-}10)$$

to blacken white and light tinted paints. Wohlers and Feldstein (17) concluded that lead base paints could discolor surfaces in several hours at a concentration of 70 $\mu g/m^3$ H_2S (0.05 ppm) (Fig. 9-6). In time the black lead sulfide will oxidize to the original color. Paints pigmented with titanium or zinc will not form a black

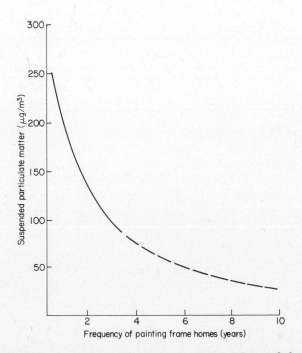

FIG. 9-7. Frequency of exterior painting as a function of suspended particulate matter.

precipitate. Alkyd or vinyl vehicles and pigments contain no heavy metal salts for reaction with H_2S.

Painted surfaces are also dirtied by particulate matter. Contaminating dirt can readily become attached to wet or tacky paint where it is held tenaciously and forms focal points for gaseous sorption for further attack. Dirt that collects on roofs, in gutters, blinds, screens, window sills, or other protuberances is eventually washed over external surfaces to mar decorative effects. The effect of suspended particulate concentration on the frequency of painting homes is shown in Fig. 9-7 (18).

VII. Rubber

Although it was known for some time that ozone cracks rubber products under tension, the problem was not related to air pollution. During the early 1940's it was

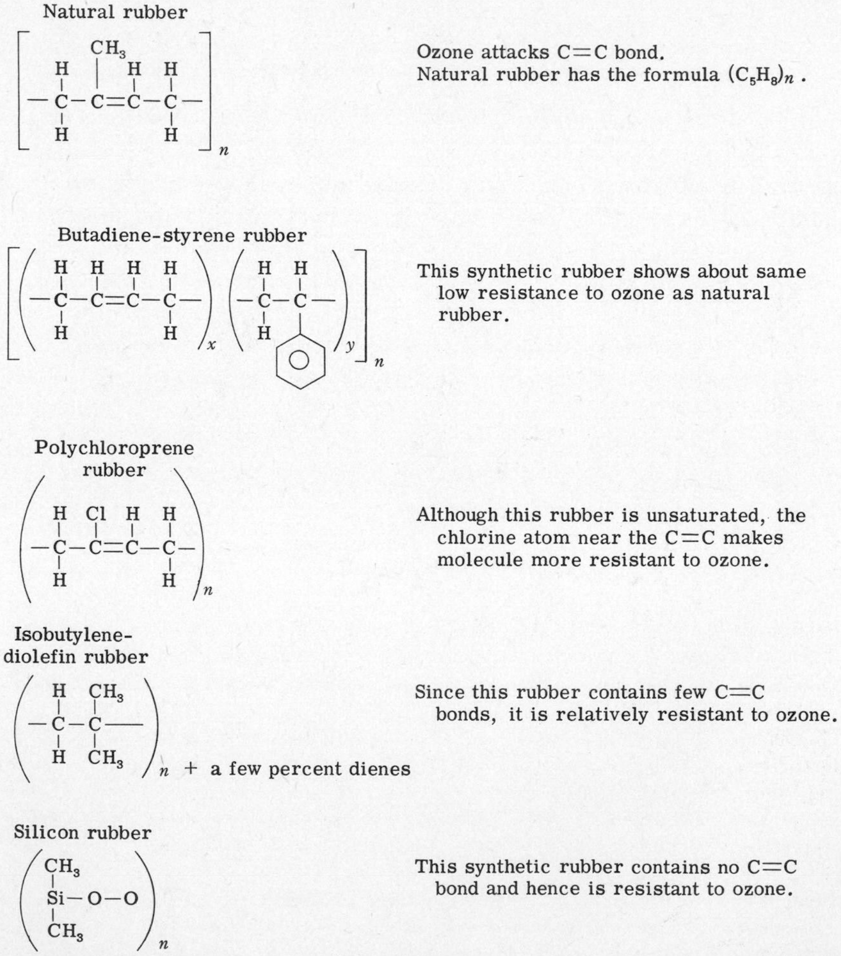

Natural rubber

Ozone attacks C=C bond.
Natural rubber has the formula $(C_5H_8)_n$.

Butadiene-styrene rubber

This synthetic rubber shows about same low resistance to ozone as natural rubber.

Polychloroprene rubber

Although this rubber is unsaturated, the chlorine atom near the C=C makes molecule more resistant to ozone.

Isobutylene-diolefin rubber

Since this rubber contains few C=C bonds, it is relatively resistant to ozone.

Silicon rubber

This synthetic rubber contains no C=C bond and hence is resistant to ozone.

FIG. 9-8. Susceptibility of natural and synthetic rubbers to attack by ozone.

discovered that rubber tires stored in warehouses in Los Angeles developed serious cracks. Intensified research soon identified ozone as the causative agent that resulted from atmospheric reaction between sunlight (3000–4600 Å), oxides of nitrogen and specific type organic compounds, i. e., photochemical air pollution.

Natural rubber is comprised of polymerized isoprene units. When rubber is under tension ozone attacks the carbon–carbon double bond, breaking the bond. The broken bond leaves adjacent C=C bonds under additional stress, eventually breaking and placing still additional stress on surrounding C=C bonds. This "domino" effect can be discerned from the structural formulas in Fig. 9-8. The number of cracks as well as the depth of the cracks in rubber under tension is related to ambient concentrations of ozone.

Rubber products may be protected against ozone attack by the use of a highly saturated rubber molecule, the use of a wax inhibitor which will "bloom" to the surface, and the use of paper or plastic wrappings to protect the surface. Despite these efforts, rubber products still "crack" more on the United States West Coast than on the East Coast.

VIII. Miscellaneous Materials

The consequences of air pollution to inert materials are undoubtedly limited by imagination or the time required to ferret out technical details. Newspaper articles reported that organic vapors from wax used for floor polishing abnormally eroded relay contact points; a carpenter was awarded a $300,000 settlement when ropes holding a scaffold were allegedly weakened by air pollutants. Ozone in the air was reported responsible for off-flavors in the manufacture of dried powdered milk. Los Angeles smog was claimed responsible for the discoloration of women's dyed hair. Valuable oil paintings in non-air-conditioned museums were found damaged by air pollutants. Windows have been etched by air pollutants. Air pollution has been suggested as the cause for shortened life of television tubes.

References

1. Salmon, R. L., Systems Analysis of the Effects of Air Pollution On Materials, Project N. 3323-D, Midwest Research Institute, Kansas City, Missouri, January 1970.
2. Gibson, W. B., The economics of air pollution. Proceedings First National Air Pollution Symposium, Pasadena, California, 1949.
3. Gustavson, R. G., What are air pollution's costs to society? Proceedings National Conference on Air Pollution. Public Health Service Publication No. 654, U.S. Government Printing Office, Washington, D.C., 1958.
4. Terry, L. T., Let's clear the air. Proceedings National Conference on Air Pollution, Public Health Service Publication No. 1022, U.S. Government Printing Office, Washington, D.C., 1962.
5. Michelson, I., and Tourin, B., *Public Health Service Reports* **81**, 505–510 (1966).
6. Michelson, I., "The Costs of Living in Polluted Air Versus the Costs of Controlling Air Pollution." A Report to the United States Public Health Service Conference on Air Pollution Abatement in the New York–New Jersey Area, January 11, 1967.

7. Beaver, H., "Committee on Air Pollution Report." Her Majesty's Stationery Office, London, 1954.
8. Scorer, R. S., *J. Inst. Fuel* **30,** 110–123 (1957).
9. Leclerc, P., Economic effects of air pollution. European Conference on Air Pollution, Council of Europe, Strasbourg, 1964.
10. Hanks, J. J., and Kube, H. D., *Harvard Business Rev.* Harvard University, Cambridge, Massachusetts, 1966.
11. Jackson, W. E., and Wohlers, H. C., "Determination of Air Pollution Costs and the Cost of Air Pollution Reduction in the Delaware Valley." Drexel University, Philadelphia, Pennsylvania, June, 1970.
12. Wohlers, H. C., Katz, E., Malfara, L., and Plater-Zyberk, J., "Air Pollution as a Function of Population." Proceedings Inst. of Environmental Sciences, Anaheim, California, April 1969.
13. Brysson, R. J., Trask, B. J., and Cooper, A. S., Jr., *Amer. Dyest. Rep.* **57,** 15–19 (1968).
14. Salvin, V. S., *Amer. Dyest. Rep.* **52,** 33–41 (1964).
15. Hudson, F. L., Grant, R. L., and Hockey, J. A., *J. Appl. Chem.* **14,** 441–447 (1967).
16. Yocom, J. E., and McCaldin, R. O., Effects of air pollution on materials and the economy, *in* "Air Pollution" (A. C. Stern, ed.), Vol. I, pp. 617–654. Academic Press, New York, 1968.
17. Wohlers, H. C., and Felstein, M., *J. Air Pollut. Contr. Ass.* **16,** 19–21 (1966).
18. Cheaper to clean up than not to clean up. *Environ. Eng. Sci.* **1,** 968 (1967).

Suggested Reading

"Air Quality Criteria for Carbon Monoxide." National Air Pollution Control Administration, Department of Health, Education, and Welfare, AP-62, Washington, D.C., March 1970.
"Air Quality Criteria for Hydrocarbons," National Air Pollution Control Administration, Department of Health, Education, and Welfare, AP-64, Washington, D.C., March 1970.
"Air Quality Criteria for Nitrogen Oxides." National Air Pollution Control Administration, Department of Health, Education, and Welfare, AP-84, Washington, D.C., January 1971.
"Air Quality Criteria for Particulates." National Air Pollution Control Administration, Department of Health, Education, and Welfare, AP-49, 801 N. Randolph Street, Arlington, Virginia, February 1969.
"Air Quality Criteria for Photochemical Oxidants." National Air Pollution Control Administration, Department of Health, Education, and Welfare, AP-63, Washington, D.C., March 1970.
"Air Quality Criteria for Sulfur Oxides." National Air Pollution Control Administration, Department of Health, Education, and Welfare, AP-50, 801 N. Randolph Street, Arlington, Virginia, Feb. 1969.
Antler, M., and Gilbert, J., *J. Air Pollut. Contr. Ass.* **13,** 405–416 (1963).
Haagen-Smit, A. J., Brunella, M. F., and Haagen-Smit, J. W., *Rubber Age (London)* **85,** 615–620 (1959).
Hudson, J. D., *J. Iron Steel Inst.* **148,** 161–215 (1943).
Langwell, W. H., "Conservation of Books and Documents." Pitman, London, 1957.
Tice, A. E., *J. Air Pollut. Contr. Ass.* **12,** 1–7 (1962).
Upham, J. B., *J. Air Pollut. Contr. Ass.* **17,** 398–402 (1967).
Williams, J. D., Maddox, F. D., Harris, T. O., Copely, C. M., and Van Dokkenbury, W., Effects of Air Pollution, Interstate Study, Phase II Project Report. National Center for Air Pollution Control, Cincinnati, Dec. 1966.

Questions

1. (a) What is the estimated per capita cost of air pollution effects on inert materials for the United States in 1950, 1960, and 1970?

(b) Estimate the per capita cost of air pollution control for the same years.
2. Why should the soiling cost of air pollution on inert materials be about 25 times greater than the deterioration costs?
3. (a) List anodic metals that will protect iron.
   (b) What chemical principle is involved in cathodic protection of metals?
4. Assuming a relationship among corrosion, population, and sulfur dioxide, why might one expect this interdependence?
5. Compare the solubilities in water of calcium carbonate, calcium sulfite, calcium sulfate, magnesium sulfate, and dolomite.
6. Write the chemical reactions involved in the darkening and subsequent oxidation of lead base paints.
7. What is the quantitative relationship between cracking of rubber under tension and ambient concentrations of ozone?
8. Ozone is a strong oxidizing agent and sulfur dioxide is a strong reducing agent. Do these reactive chemicals exist together in the atmosphere? Explain your answer.
9. Develop the formula necessary to convert sulfur dioxide concentration from parts per million (ppm) by volume to micrograms per cubic meter ($\mu g/m^3$) air at 25°C and 1 atm.

Chapter 10

AIR POLLUTION EFFECTS ON VEGETATION AND ANIMALS

The living cell undergoes many chemical reactions and physical processes for growth; life processes exist within relatively narrow ranges of temperature, light, water, and nutrient. Life can be considered a stress phenomenon; adverse factors including air pollution increase the normal stress on cells that leads to various stages of impairment—and in the extreme—death. Biological systems have defense mechanisms to minimize applied stress; the defense mechanism is able to compensate for the presence of toxic pollutants in low concentrations and in certain cases rebuild "damaged" tissues. Or if only a few cells have been "damaged" billions of other cells remain functional. There is an optimum "set point" for each biological function, which if not exceeded allows the inner dynamic cell balance to be maintained.

The concept of stress to the living cell opens Pandora's box. If biological systems continually compensate for low air pollutant concentrations in addition to innumerable normal environmental and biological effects—known and unknown—how are we to determine the net effect of pollutant stress over the entire life of the cell?

I. Air Pollution Effects on Vegetation

Vegetation "breathes in" ambient air and in the presence of sunlight builds complex organic molecules from simple inorganic compounds from air, soil, and water.

Because the leaf contains the building mechanisms, we will emphasize this organ. Obviously leaf function depends upon the condition of the entire plant—including seed, stem, root, and flower.

A. Leaf Structure and Function

The general leaf structure is shown in Fig. 10-1. A typical leaf may be divided into three regions—the epidermis (and cutin) forms a protective layer covering the entire surface of the leaf and is continuous with the epidermis of the stem; the mesophyll, differentiated into palisade and spongy parenchyma; and the veins, which consist of vascular tissues for the transfer of water and food substances synthesized by the leaf. Gases, vapors, and small particulates pass into and out of air spaces of the mesophyll through the stomata.

Cell size commonly ranges from 0.01 to 0.1 mm in diameter. Cells consist of the wall and protoplast. The wall (and inclusions) maintain the integrity of the cell. The protoplast contains the cytoplasm in which the nucleus and chloroplasts are found. Chloroplasts contain the chlorophyll, which absorbs light energy for photosynthesis.

Two major functions of the leaf are photosynthesis and respiration (Table 10-1); a third, transpiration, moves fluids through the plant system. Photosynthesis is an energy-producing process that converts carbon dioxide and water into carbohydrates, the energy source for the subsequent synthesis of compounds such as fats, proteins, and nucleic acids. The plant uses a portion of the produced and stored energy for its life processes. Another portion of plant energy may be used as food for animal and human life. Remaining portions decay eventually into CO_2 and H_2O. Inherent biochemical reactions of plant and animal metabolism are remarkably similar.

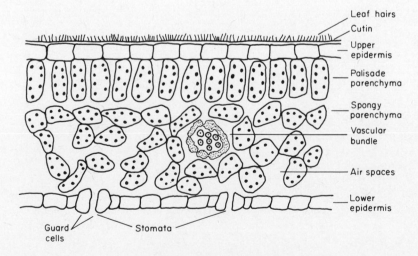

FIG. 10-1. Anatomy of the leaf.

TABLE 10-1

Photosynthesis and Respiration

$$CO_2 + H_2O + energy \xrightarrow[\text{Respiration}]{\text{Photosynthesis}} (CH_2O)_n + O_2$$

1. Oxygen given off	1. Carbon dioxide given off[a]
2. H_2O and CO_2 are raw materials	2. Sugars and O_2 are raw materials
3. Carbohydrates are synthesized	3. Carbohydrates are consumed
4. Energy is stored	4. Energy is transformed
5. Weight is increased	5. Weight is decreased
6. Goes on only in cells containing chlorophyll in presence of light	6. Goes on at all times in all living cells

[a] Note: CO_2 is also a product of vegetation decay.

B. Pollutant Stress Factors Affecting Plant Physiology

For optimum growth, vegetation requires adequate light, temperature, moisture, nutrients, and soil conditions. Under field growing conditions these requirements are rarely optimal; hence physiology of the plant may be considered to be under stress conditions. Effects of stress may or may not be observed on subsequent growth depending upon the magnitude of the stress factors.

Air pollution presents an additional factor to otherwise stressed physiological processes. Furthermore air pollutants affect vegetation differently depending upon the cells stressed (Fig. 10-2). In general, vegetation growing under optimum conditions is most susceptible to air pollution. The toxic effect of a pollutant may thus be almost directly related to the functioning of the stomata. Under active growing conditions stomata are open, allowing air to enter the leaf. Stomata openings are related to the physiological activity of the plant in that they regulate gas exchange; correlation exists between the extent of air pollution effects and the degree of opening of the stomata.

C. Pollutant-Induced Foliar Markings

Foliar markings characterize air pollution effects on vegetation. Treshow (1) listed four generalized foliar markings symptomatic of air pollutants:

1. Necrosis and bleaching of intercostal areas or leaf margins
2. Glazing or silvering of leaf surface, particularly the undersurface
3. Chlorosis or loss of chlorophyll
4. Flecking or stippling on upper leaf surface

These generalized foliar markings may also result from abnormal temperatures, poor water relations, mineral deficiencies, high winds, insects, or plant diseases. Foliar markings may be used by the experts for initial diagnosis. Table 10-2 lists leaf symptoms associated with air pollutants of concern throughout the world (1, 2, 3, 4). Care must be exercised in the diagnosis of pollutant injury from foliar markings because of the complicated natural factors noted above.

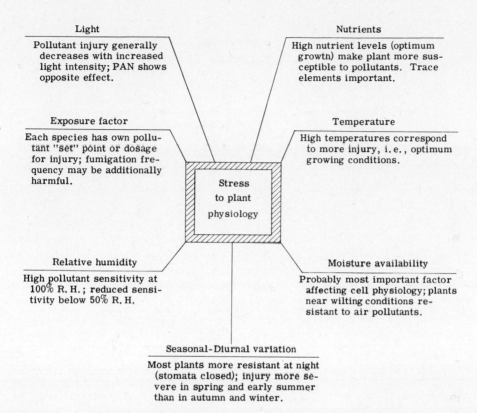

Light

Pollutant injury generally decreases with increased light intensity; PAN shows opposite effect.

Nutrients

High nutrient levels (optimum growth) make plant more susceptible to pollutants. Trace elements important.

Exposure factor

Each species has own pollutant "set" point or dosage for injury; fumigation frequency may be additionally harmful.

Temperature

High temperatures correspond to more injury, i.e., optimum growing conditions.

Stress to plant physiology

Relative humidity

High pollutant sensitivity at 100% R. H.; reduced sensitivity below 50% R. H.

Moisture availability

Probably most important factor affecting cell physiology; plants near wilting conditions resistant to air pollutants.

Seasonal-Diurnal variation

Most plants more resistant at night (stomata closed); injury more severe in spring and early summer than in autumn and winter.

FIG. 10-2. Pollutant stress factors affecting plant physiology.

The precise effect of a pollutant within the leaf is not known. With sulfur dioxide the leaf is apparently able to oxidize low concentrations to the sulfate ion without apparent harm. When a threshold is exceeded, the effects appear to result either from an excessive accumulation of sulfate ion or an imbalance in the $SH:SO_4$ ratio, i. e., a "salt effect." With excessive SO_2, the mesophyll cells (palisade parenchyma) collapse, forming necrotic or dead areas between the veins. Initially the appearance is that of a dull green color or water-soaked look. On drying, the affected areas bleach to an ivory color. It is believed that the chloroplast membrane breaks down and the protoplast becomes plasmolyzed or bleached. With grasses or pines, the point of attack is near the tip in contrast to effects to the broad-leaved plants described.

In contrast to effects of the reducing chemical sulfur dioxide, the oxidant ozone is believed to cause more damage to vegetation. Ozone primarily injures the palisade parenchyma or the spongy parenchyma in those leaves that lack palisade cells; the effect produced is localized cellular collapse and pigmentation of the cell walls. Small pale buff to reddish-brown lesions appear on the upper surface; if the stress is particularly severe, the markings may extend to the lower surface and/or widen in

TABLE 10-2

Pollutant Effects on Vegetation

Pollutant	Symptoms	Maturity of leaf affected	Part of leaf affected	Injury threshold		
				ppm (vol)	µg/m³	Sustained exposure
Sulfur dioxide	Bleached spots, bleached areas between veins, chlorosis; insect injury, winter and drought conditions may show similar markings	Middle-aged most sensitive; oldest least sensitive	Mesophyll cells	0.3	785	8 hours
Ozone	Fleck, stipple, bleached spotting, pigmentation; conifer needle tips become brown and necrotic	Oldest most sensitive; youngest least sensitive	Palisade or spongy parenchyma in leaves with no palisade	0.03	59	4 hours
Peroxyacetyl-nitrate (PAN)	Glazing, silvering, or bronzing on lower surface of leaves	Youngest most sensitive	Spongy cells	0.01	50	6 hours
Nitrogen dioxide	Irregular, white or brown collapsed lesions on intercostal tissue and near leaf margin	Middle-aged leaves most sensitive	Mesophyll cells	2.5	4700	4 hours
Hydrogen fluoride	Tip and margin burn, dwarfing, leaf abscission; narrow brown–red band separates necrotic from green tissue; fungal disease, cold and high temperatures, drought, and wind may show similar markings; suture red spot on peach fruit	Youngest leaves most sensitive	Epidermis and mesophyll cells	0.1 (ppb)	0.08	5 weeks

Pollutant	Symptoms	Leaves affected	Tissue			Time
Ethylene	Sepal withering, leaf abnormalities; flower dropping, and failure of leaf to open properly; abscission; water stress may produce similar markings	Young leaves recover; older leaves do not recover fully	All	0.05	58	6 hours
Chlorine	Bleaching between veins, tip and margin burn, leaf abscission; marking often similar to that of ozone	Mature leaves most sensitive	Epidermis and mesophyll cells	0.10	290	2 hours
Ammonia	"Cooked" green appearance becoming brown or green on drying; over-all blackening on some species	Mature leaves most sensitive	Complete tissue	~ 20	~ 14,000	4 hours
Hydrogen chloride	Acid-type necrotic lesion; tipburn on fir needles; leaf margin necrosis on broad leaves	Oldest leaves most sensitive	Epidermis and mesophyll cells	~ 5–10	~ 11,200	2 hours
Mercury	Chlorosis and abscission; brown spotting; yellowing of veins	Oldest leaves most sensitive	Epidermis and mesophyll cells	< 1	< 8,200	1–2 days
Hydrogen sulfide	Basal and marginal scorching	Youngest leaves most affected		20	28,000	5 hours
2,4-Dichlorophenoxyacetic acid (2-4D)	Scalloped margins, swollen stems, yellow-green mottling or stippling, suture red spot (2,4,5-T); epinasty	Youngest leaves most affected	Epidermis	< 1	< 9,050	2 hours
Sulfuric acid	Necrotic spots on upper surface similar to caustic or acidic compounds; high humidity needed	All	All	—	—	—

area as more cells become affected. As with sulfur dioxide, the foliar effect of ozone varies with different type leaves.

D. Field Surveys for Foliar Markings

Foliar markings, as noted in Table 10-2, are excellent indicators of past and present air pollution fumigations. Under active growing conditions pollutant markings will develop within hours. As each species has a specific "set point" or sensitivity, an estimate of the pollutant concentration may be attempted. With plants whose stomata close at night, the appearance of foliar markings is indicative of daylight fumigation. A chemical analysis for the pollutant in plant tissue is used frequently to confirm the type marking. Furthermore the observed markings remain as telltale evidence of high pollutant levels long after the episode is over. It is possible in many fumigations to plot severity of markings and locations on a map; with constant wind the apex of the markings is the pollutant source.

To conduct a vegetation foliar marking survey plants in the affected area should be carefully compared to similar plants in a control area. Consideration must be given to the relative susceptibility of the vegetation to a pollutant, the markings must correlate with wind direction and topography, a pollutant source must be in the area, chemical analysis of the vegetation should be undertaken, and consideration should be given to local conditions such as weather, insects, and disease.

A "dictionary" of the sensitivity of 13 pollutants on plants is presented in Appendix A; the list is incomplete in that each plant has not been evaluated for all the pollutants and for many plants there may be more than one species. Pollutant effects on plants have been rated as sensitive (S), intermediate (I), or resistant (R). The "dictionary" can be of great help in examination and evaluation of foliar markings in the field. Corn has been rated as sensitive to ozone, fluoride, and chlorine; resistant to sulfur dioxide and peroxyacetyl nitrate; and intermediate to 2,4-dichlorophenoxyacetic acid. Thus if the foliar markings demonstrated the pollutant to be ozone or fluoride, the information correlates with the data in the table. If, however, sulfur dioxide or PAN foliar markings were observed, the information would be contrary to what might be anticipated. In the latter case, other possible reasons for the observed markings (insect injury, winter, or drought conditions) must be investigated. If sulfur dioxide were suspected, a potential source of sulfur dioxide would have to be located in the area. Also included in Appendix A are plants native to many parts of the world. During an air pollution investigation these native plants, as well as garden flowers and commercial crops, should be examined for foliar markings.

II. Injury versus Damage

Vegetation may be either injured or damaged by air pollutants. Injury to a plant by air pollution does not necessarily mean that it is damaged. This distinction is made because too frequently the words are used interchangeably. Damage connotes economic or aesthetic loss from the intended use of a plant.

If a droplet of sulfuric acid is placed upon a leaf, the acid will dehydrate and kill the tissue it contacts. With a black spot on an otherwise green leaf (injury), the growth of the plant would not be affected—millions of other cells maintain the plant and the few hundred dead cells would not affect total development or growth. However, if a reasonable percentage of the plant surface is covered with sulfuric acid aerosol, damage to the plant will occur.

Air pollution losses to agriculture are difficult to assess. Where visible foliage marking occurs a dollar loss may be estimated. For sulfur dioxide a formula was reported years ago by O'Gara (5) for acute damage to alfalfa:

$$t(C - a) = b \qquad (10\text{-}1)$$

where C is the SO_2 concentration, ppm (vol), t is the exposure time in hours, and a, b are constants, such that

	a	b
Incipient injury	0.24	0.94
50% leaf destruction	1.4	2.1
100% leaf destruction	2.6	3.2

Guderian *et al.* (6) revised O'Gara's equation in an exponential form to better fit their data:

$$t = ke^{-a}(C - r) \qquad (10\text{-}2)$$

where k, a, and r are parameters varying with species and degree of injury.

Monetary losses to agriculture in the United States have ranged from $12 million annually in Pennsylvania (7) to $132 million for California and $500 million for the United States (3). It is believed that these values are low.

III. Pollutant Effects on Experimental Animals

For research purposes, small animals replace humans in air pollution studies. Small animals are abundant and because of a relatively short maturity span, studies can be completed over many generations. Conclusions reached on the basis of animal research may be extrapolated to humans. Although the question of health effects resulting from air pollution is still beyond grasp, results of animal studies do help clarify the way in which respiratory irritants affect humans and the manner in which the body defends against external stress (8).

A. Ozone

Research has indicated that ozone behavior on cell protoplasm is similar to that of ionization (radiomimicity) in that harmful reactions involve free radicals. Important to this line of reasoning is that ozone—like ionizing radiation—should have a "no-threshold dose-response."

Ozone (odor of Los Angeles smog) affects the olfactory nerve endings and eventually oxidizes cell sulfhydryl groups (—SH) with loss of hydrogen atoms; ionizing radiation is also known to affect the —SH groupings. Radiometric effects lead to premature ageing and chromosomal injury.

B. Nitrogen Oxides

Nitrogen dioxide, like ozone, is involved in free radical formation. Nitrogen dioxide is not as effective in this type of reaction, and, importantly, the effects of NO_2 are generally not observable until a critical concentration is reached. Lung edema resulting from NO_2 in animals and humans may be fatal; pulmonary congestion, edema, obliterative bronchiolitis, pneumonitis, and eventual death are operational sequences.

The industrial threshold for NO_2 (5 ppm) is 50 times higher than that for ozone (0.1 ppm). However, some protein changes in rabbits have been reported for exposures as low as 1880 $\mu g/m^3$ NO_2 (1 ppm) for 1 hour (reversible denaturation).

C. Sulfur Dioxide

In contrast to O_3 and NO_2, sulfur dioxide acts as an irritant gas/acid to the respiratory tract; animals differ markedly in susceptibility. The primary effect of the SO_2, a soluble gas, is on the mucous lining of the upper respiratory tract. Very little SO_2 gas is believed to penetrate deep into the lungs; the gas may reach the lungs if it is sorbed on small particulate matter ($< 1-2 \mu$).

Studies on rats have indicated that continuous exposure of 2620 $\mu g/m^3$ SO_2 (1 ppm) for periods slightly longer than a year led to increased mortality. Russian scientists reported induced reflexes in brain cortex at very short (10-second) exposures of 524 $\mu g/m^3$ (0.2 ppm).

D. Carbon Monoxide

Carbon monoxide affects the oxygen-carrying capacity of hemoglobin; hence CO levels in urban areas may be critical to animals and especially for humans with respiratory illnesses. Affinity of hemoglobin for CO is about 210-fold greater than for O_2. It has been estimated that the hazardous (toxic) levels for CO, O_3, NO_2, and SO_2 are in ratio to 20, 50, 275, and 2500, respectively.

Symptoms of CO toxicity may be related to the ratio of carboxyhemoglobin (HbCO) to hemoglobin (HbO_2); the amount HbCO formed in blood is directly related to the CO concentrations in air. Concentrations of HbCO should not exceed 5%; allowable CO dosages are based on this level.

E. Particulate Matter

Effects on animals (and humans) depend upon chemical and physical properties of the particulate matter. Corrosive materials such as acids (H_2SO_4, HCl, HF) or bases [NaOH, $Ca(OH)_2$, NH_4OH] exert direct chemical action. Inert particles induce a physiological response by slowing ciliary beat and mucus flow in the bronchial tree. Sorption of gases on small particulates increase the effect, particularly if the particles penetrate to deeper portions of the lungs. Particulates of greatest concern to urban air pollution include SO_3, H_2SO_4, fly ash, C, Fe_2O_3, asbestos, Be compounds, Pb compounds, and carcinogenic chemicals such as benzo (α) pyrene.

F. Synergism or Antagonism

Research on small animals is beginning to shed light on the mechanism of cellular response for a few air pollutants of major interest. To avoid experimental complexities most research involves a single pollutant. In urban air, there are more than one-half million known and unknown chemicals—in addition to radioactive material, bacteria, and viruses.

With the complexity of this urban "witch's brew," research involving a single pollutant cannot be considered entirely valid. Other pollutants may exert synergistic or antagonistic effects. Humidity, altitude, and temperature also exert their effects. If, at this point in time, we are beginning to understand effects of single pollutants on cellular structure and activity, it will be many decades before effects of ambient atmospheres are understood.

IV. Pollutant Effects on Large or Commercial Animals

Heavy metals on/in vegetation and water have been and continue to be toxic to animals and fish. Arsenic and lead from smelters, molybdenum from steel plants, and mercury from chlorine-caustic plants are major offenders. Poisoning by mercury of aquatic life is relatively new, while the toxic effects of the other metals have largely been eliminated by proper control of industrial emissions. Gaseous (and particulate) fluorides have caused injury and damage to a wide variety of animals—domestic and wild—as well as to fish. Accidental effects resulting from insecticides and nerve gas have been reported.

Autopsies of animals in the Meuse Valley, Donora, and London episodes revealed evidence of pulmonary edema. Breathing toxic pollutants is not, however, the major form of pollutant intake for cattle—ingestion of pollution-contaminated feeds is the primary mode.

In the case of animals we are concerned primarily with a two-step process—accumulation of airborne contaminants on/in vegetation or forage that serves as their feed, and subsequent effects of the ingested herbage on animals. In addition to pollution-affected vegetation, carnivores (man included) consume small animals that may have ingested exotic chemicals including pesticides, herbicides, fungicides, and antibiotics. Increasing environmental concern has pointed out the importance of the complete food chain for the physical and mental well-being of man.

A. Heavy Metal Effects

One of the earliest cattle problems involved widespread poisoning of cattle by arsenic at the turn of the century. Abnormal intake of arsenic results in severe colic (salivation, thirst, vomiting), diarrhea, bloody feces, and a garliclike odor on the breath; cirrhosis of the liver and spleen as well as reproductive effects may be noted. Arsenic trioxide in the feed must approximate 10 mg As/kg body weight for these effects to occur.

Cattle feeding on herbage containing 25–50 mg/kg (ppm wt) lead develop excitable jerking of muscles, frothing at the mouth, grinding of teeth, and paralysis of the larynx muscles; a "roaring" noise is caused by the paralysis of the muscles in the throat and neck.

Molybdenum symptoms to cattle include emaciation, diarrhea, anemia, stiffness, and fading of hair color. Vegetation containing 230 mg/kg Mo affects cattle.

Mercury in fish was recently found in waters in the United States and Canada. Mercury in the waters is converted into methyl mercury by aquatic vegetation. Small fish consume such vegetation, which in turn are eaten by larger fish and eventually by man; food with more than 0.5 ppm Hg (0.5 mg/kg) cannot be sold in the United States for human consumption.

B. Gaseous and Particulate Effects

Periodically, accidental emissions of a dangerous chemical affect animal well-being. During nerve gas experimentation in a desolate area in Utah, a high-speed airplane accidentally dropped several hundred gallons of nerve gas. As a result of the discharge, 6200 sheep were killed. Considering the large number of exotic chemicals being manufactured, such unfortunate accidents may be anticipated in the future.

Fluoride emissions from industries producing phosphate fertilizers or phosphate derivatives have caused damage to cattle throughout the world; phosphate rock, the raw material, can contain up to 4% fluoride, a part of which is discharged to air

TABLE 10-3

Fluoride Tolerance of Animals (ppm wt in ration, dry)[a]

Species	Breeding or lactating animals (ppm)[b]	Finishing animals to be sold for slaughter with average feeding period (ppm)[b]
Dairy, beef heifers	30	100
Dairy cows	30	100
Beef cows	40	100
Steers	—	100
Sheep	50	160
Horse	60	—
Swine	70	—
Turkeys	—	100
Chickens	—	150

[a] Data based on soluble fluoride; increased values for insoluble fluoride compounds.

[b] 1 ppm wt = 1 mg/kg.

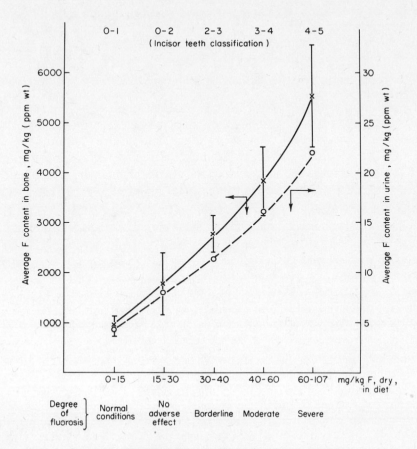

FIG. 10-3. Fluoride effects on dairy cattle (4-years old).

(and waters) during processing. In Polk and Hillsborough Counties of Florida, the cattle population decreased by 30,000 between 1953 and 1960 as a result of fluoride emissions. Since 1950, research has greatly increased our knowledge concerning the effect of fluorides on animals; standards and guides for diagnosing and evaluating fluorosis in cattle have been compiled.

Chronic fluoride toxicity (fluorosis) is the type most frequently observed in cattle. The primary effects of fluorides in cattle are in the teeth and bones. Excessive intake weakens the tooth enamel of developing teeth; the initial dulled appearance of erupted teeth can develop into soft teeth with uneven wear of molar teeth. Characteristic osteofluorotic bone lesions develop causing intermittent lameness and stiffness in the animal. Fluoride content of the bone increases with dosage despite excretion in urine and feces. Secondary symptoms include reduced lactation, nonpliable skin, and dry, rough hair coat. As shown in Fig. 10-3, the fluoride ingestion level correlates with fluoride content of bones and urine as well as incisor teeth classification (9).

Tolerance of animals for fluorides varies, dairy cattle being most sensitive and poultry least (Table 10-3). Fluorosis of animals in contaminated areas can be avoided by keeping the intake levels below those listed by incorporating clean feeds with those high in fluorides. It has also been determined that an increased consumption of aluminum and calcium salts can reduce toxicity of fluorides in animals.

V. Darwin's Missing Evidence

Examples of air pollution effects to vegetation and animals have been reviewed. Of concern to the world at large are the many factors that might change man's evolutionary path, leading to a lost race on earth; evidence leading to such a possibility exists. The incomplete facts are undoubtedly significant in that plants began to evolve over a billion years ago, animals 250 million years ago, and man one million years ago. Evidence based only on data collected over a century or so has little effect on heredity.

It is known that ionizing radiation as well as certain chemicals (pollutants) can interrupt the self-regulatory action of cells leading to odd or irregular mutations. Deformed plants, animals, and humans have resulted, but as far as we know, no heredity changes to man have been observed.

DDT is being concentrated in the fat of living cells, the final effects of which are now unknown. Cliché facts, e. g., mother's milk contains more DDT than regulations permit in milk for commercial sale, give cause for concern. Many birds face extinction partly as a result of the reproduction losses—fragile eggshell production due to pesticide intake upsetting calcium metabolism.

In Darwin's era a heredity change was started in the peppered moth *Biston betularia* in England. In the century since the Industrial Revolution many moth species have exchanged their light color and pattern for dark or even all-black coloration. These changes took place among those moths that fly at night and rest during the day on air-pollution-soiled (and darkened) tree trunks. The survival of the darkened moths against the dark background was evidently greater than that of the lighter moths. The strong correlation between industrial centers in England and a high percentage of dark forms of the moth raises questions concerning future changes in other species, including man.

It has also been noted that lichens and algae that covered trunks of deciduous trees in England up to the eighteenth century have disappeared in and around industrial areas. The frequency of the light form of the peppered moth may have depended upon the presence of the lichens.

Fortunately science has now recognized the crucial role of ecology in man's evolution. Plants supply man with oxygen and man supplies carbon dioxide for vegetation; organic decay of vegetation, animals, and man provides necessary humus. Any stress or break in the ecological chain eventually affects the physical and mental well-being of man.

References

1. Treshow, M., "Environment and Plant Response." McGraw-Hill, New York, 1970.
2. Brandt, C. C., and Heck, W. W., Effects of air pollutants on vegetation, *in* "Air Pollution" (A. C. Stern, ed.), Vol. I, pp. 401–443. Academic Press, New York, 1968.

3. Hindawi, I. J., "Air Pollution Injury to Vegetation." NAPCA Publ. No. AP-71. Superintendent of Documents, Washington, D.C., 1970.

4. Jacobson, J. S., and Hill, A. C., eds. "Recognition of Air Pollution Injury to Vegetation, A Pictorial Atlas." TR-7 Agricultural Committee, Air Pollution Control Association, Pittsburgh, Pennsylvania, 1970.

5. O'Gara, P. J., *Ind. Eng. Chem.* **14,** 744 (1922).

6. Guderian, R., von Haut, H., and Stratman, H., *Z. Pflanzenkr. (Pflanzenpathol.) Pflanzenschutz* **67,** 257 (1960).

7. Weidensaul, T. C., and Lacosse, N. L., Results of the statewide survey of air pollution damage to vegetation. Presented at Annual Meeting Air Pollution Control Association, St. Louis, Missouri, June 1970.

8. Stokinger, H. E., and Coffin, D. L., Biologic effects of air pollutants, *in* "Air Pollution" (A. C. Stern, ed.), Vol. I, pp. 445–546. Academic Press, New York, 1968.

9. Shupe, J. L., *Amer. Ind. Hyg. Ass. J.* **31,** 240–247 (1970).

Suggested Reading

"Air Quality Criteria for Carbon Monoxide." National Air Pollution Control Administration, Department of Health, Education, and Welfare, AP-62, Washington, D.C., March 1970.

"Air Quality Criteria for Hydrocarbons." National Air Pollution Control Administration, Department of Health, Education, and Welfare, AP-64, Washington, D.C., March 1970.

"Air Quality Criteria for Nitrogen Oxides." National Air Pollution Control Administration, Department of Health, Education, and Welfare, Washington, D.C., AP-84, Jan. 1971.

"Air Quality Criteria for Particulates." National Air Pollution Control Administration, Department of Health, Education, and Welfare, AP-49, 801 N. Randolph Street, Arlington, Virginia, February 1969.

"Air Quality Criteria for Photochemical Oxidants." National Air Pollution Control Administration, Department of Health, Education, and Welfare, AP-63, Washington, D.C., March 1970.

"Air Quality Criteria for Sulfur Oxides." National Air Pollution Control Administration, Department of Health, Education, and Welfare, AP-50, Arlington, Virginia, Feb. 1969.

Effect of air pollution on vegetation, *in* "Air Pollution Manual Part I," 2nd ed. (1972). American Industrial Hygiene Association, 210 Haddon Avenue, Westmont, New Jersey.

Kettlewell, H. B. D., *New Sci.* No. 385, p. 34, April 2, 1964.

Mukammal, E. I., Brandt, C. S., Neuwirth, R., Pack, D. H., and Swinbank, W. C., "Air Pollutants, Meteorology, and Plant Injury." World Meteorological Organization, Tech. Note No. 96, WMO—No. 234, TP 127, Geneva, Switzerland, 1968.

Thomas, M. D., Effects of air pollution on plants, *in* "Air Pollution." World Health Organization, Columbia Univ. Press, New York, 1961.

Questions

1. (a) What is the total weight of oxygen and carbon dioxide in the atmosphere?
   (b) Explain the reason(s) for the weight discrepancy according to the equations for photosynthesis and respiration.
2. (a) What is the size range for stomatal openings for different types of vegetation?
   (b) Name some plants that have stomata only on the upper leaf surface, the lower leaf surface, and on both upper and lower leaf surfaces.
3. Assume that the upper leaf surface of a plant is coated with a heavy layer of an inert dust. What effect(s) might be anticipated? Why?
4. Compare normal and abnormal stress factors on vegetation.
5. What steps and precautions should one consider before undertaking a vegetation survey in areas where (a) the leaves are marked, or (b) the leaves are not marked but air pollution is suspect.

6. In the vicinity of a sulfur-dioxide-emitting power plant, might one expect foliar markings on alfalfa, chrysanthemum, corn, muskmelon, potato, poplar, sweet clover, wheat, and zinnia? Explain the reason for your answers.

7. Calculate the daily fluoride intake of a dairy animal from (a) air, and (b) from food and water based upon the conditions below and assuming 100% retention of the fluoride.

    Animal breathing rate: 30 kg air/day containing 6 μg F/m^3 air (STP)

    Animal food–water intake:

    | Herbage | 10 kg containing 200 mg F/kg |
    | Water | 5 kg containing 1 mg F/kg |

Chapter 11

AIR POLLUTION EFFECTS ON HUMANS

As noted in the previous chapter, living substances are exceedingly sensitive to external stimuli; such stimuli add an undesirable stress to functioning cells. As most effects of environmental stimuli are not immediately obvious to man, we are generally unaware of physiological changes which occur to maintain internal stability (homeostasis). This biological uncertainty is the weak link in the chain of air pollution events. Despite incomplete knowledge of effects of air pollution on humans, priority of human health protection must override the lack of absolute proof. For the benefit of man and his stewardship of earth, the push toward a "zero-risk" air environment must be intensified until scientific exploration provides an optimum "set point" for each biological function.

The effects of air pollution upon the physical and mental well-being of man can be divided into two major sections—evidence that conclusively links air environment with morbidity and mortality of man, and that which only points toward the same conclusion. Initially it must be emphasized that man is a function of the total environment—not of air alone.

I. Man and the Environment

As noted in Table 11-1, Stokinger (1) prepared an etiological list of disease states of man from air, water, and food pollutants. Air pollution is considered a direct contributor to fast aging, asthma, berylliosis, emphysema, and mesothelioma, and as a contributing cause to bronchitis, cancer of the GI tract, and cancer of the respiratory tract. But the data show further that many pollutants act in multifarious ways

TABLE 11-1

Disease States for Which Evidence Points to Environmental Pollutants as Either Direct or Contributing Causes

Disease	Etiological pollutants
Accelerated ageing	Ozone and oxidant air pollutants (direct)
Allergic asthma	Airborne denatured grain protein, etc. (direct)
Cardiovascular disease	"Hard" waters, hereditary tendency, Cr deficiency, Co(?) (contributing)
Berylliosis	Airborne Be compounds (direct)
Bronchitis	Acid gases, particulates, respiratory infection, inclement weather (contributing)
Cancer—GI tract	Carcinogens in food, water, and air; hereditary tendency (contributing)
Cancer—respiratory tract	Airborne carcinogens and hereditary tendency (contributing)
Dental caries	Se (direct)
Emphysema	Airborne respiratory irritants and familial tendency (direct)
Mesotheliomas	Asbestos, associated trace metals, carcinogens in air, water, other fibers(?) (direct)
Methemoglobinomia infant death	Water-borne nitrates and nitrites (direct)
Renal hypertension	Cd in water, food, beverage in As, Se low areas(?) (contributing)

to induce, produce, or contribute to a multiplicity of disease states. Of the 12 diseases listed in Table 11-1 four diseases have a known or suspected heredity component.

Not considered in Table 11-1 are social stresses affecting humans. Calhoun (2) in a study of Norway rats, pointed out the degenerate influence of overcrowding (population density) on physical and mental well-being of the animals; all other environmental factors were optimal. In man, some data indicate that dominant individuals are most healthy and that onset of disease often results from a less dominant rank order in society (hierarchy).

Pollutants may exert their effects directly or through interaction with other agents or with some preconditioning factor(s) within the host. Environmental pollutants must be included among the multiple factors in the causality of chronic degenerative disease.

A suggested relationship between impairment and disability of human functions is shown in Fig. 11-1 (3). Low air pollutant concentrations would lead to a normal physiological adjustment, corresponding to a healthy cell. As pollutants increase in concentration, the added stress forces cells to compensate for a disturbed function. Still higher concentrations lead to cell failure resulting in disease and eventually death.

The above relationship between impairment and disability is simplified. One environmental stress imposed upon another stress does not necessarily double the effect; synergistic action may bring the total response to eleven (not two). In addition to synergism, natural antagonists exist in both environment and host. And

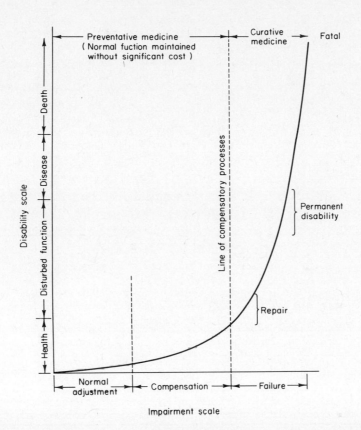

Fig. 11-1. Suggested relationship between impairment and disability. The impairment scale (normal, compensation, and failure) represents increase with aging, illness, environmental stress, etc.). The disability scale is the medical consequence of impairment.

once the homeostatic mechanism is upset, individuals have a marked different capacity to recuperate.

Normally, it may be assumed that healthy cells lead to healthy individuals and that a healthy civilization is a result of efforts of healthy individuals. Man must use his abilities to determine optimum ingredients for a homeostatic mode of life.

II. Man versus Air Pollution

The sites of air pollution effects on man are few in number. A clothed human presents a minimal number of sites for attack by air pollution. Clothing covers more than 90% of the skin area of man.

There is no general effect of pollutants on skin at normal ambient air concentrations. Exceptions do exist—salt spray and ultraviolet light at the seashore, pesticide sprays, or accidental spills of chemicals. Skin perspires and if the sweat absorbs corrosive compounds, irritation can occur.

The external coatings of the eye—conjunctivae and cornea—come into direct contact with gaseous and particulate air pollutants. Eye irritation is a major effect of photochemical air pollution, but no permanent eye injury has been attributed to photochemical air pollution. A clearing mechanism of tearing or lachrymation flushes foreign material from the eyes, avoiding damage.

Other exposed parts of the body such as hair, nails, and teeth are rarely affected by air pollution. Newspaper reports claimed that photochemical air pollution changed the color of women's tinted hair. More frequent hair washings are required for cleanliness in polluted than in clean areas. Nails and teeth are affected by arsenic, selenium, and fluorides; these chemicals are ingested through food and drink rather than by inhalation.

Internal parts of the body such as the heart, liver, and kidney are apparently not directly affected by air pollutants. If pollutants affect the circulatory system, internal organs may, in turn, become affected. Again, body defense mechanisms remove detrimental chemicals, usually via urine or feces.

Thus the obvious fact remains—the respiratory tract is the principal site of air pollution attack on the human body.

The major parts of the human respiratory system are shown in Fig. 11-2; each day 13.4 kg (31 lb) or 7570 liters (2000 gal) of air are inhaled and exhaled.

Air enters through the nose which, if in healthy condition, filters particulate matter larger than 10 μ diameter and also warms and moistens the incoming air. Inhaled air, after humidity conditioning in the nasal cavity, passes into the pharynx,

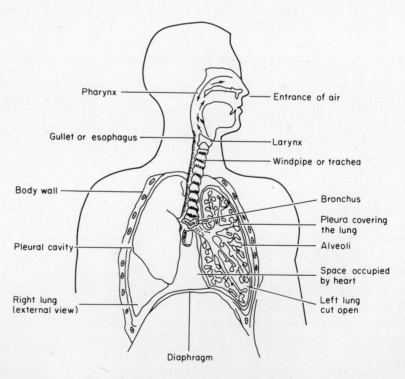

FIG. 11-2. The human respiratory system.

esophagus, larynx, and finally to the top of the trachea. The trachea branches into the right and left bronchi. Each bronchus divides and subdivides at least 20 times; the smallest units, bronchioles, are deep in the lungs. The bronchioles end in some three million air sacs, the alveoli.

During passage through the respiratory system, soluble gaseous pollutants are almost completely removed by the moist mucus membrane lining. Particulate matter, 2–10 μ diameter, settles or impinges upon the walls of the trachea, bronchi, and bronchioles; these foreign particles are escalated by ciliary action to the mouth and swallowed. Particles approximating 0.1–2 μ may reach the alveoli; once a solid particle reaches the small airways and alveoli, its residence time is measured in weeks, months, or years. Sulfur dioxide may reach the alveoli when the gas is either sorbed on small particulate matter or oxidized to a sulfuric acid aerosol of proper size.

In the lungs, oxygen is absorbed from the air into the bloodstream and carbon dioxide is removed from the blood stream. Impurities, either gaseous or particulate, initiate their major effects at this transfer point. Continuous inhalation of dusty air can slow ciliary action in the upper respiratory tract, allowing more dust to reach the lower lung; gaseous chemicals may also reduce ciliary action. Certain dusts produce pneumonconiosis, silicosis or asbestosis; other dusts seem benign. Inert particulate matter, sorbed gases on particulates, or corrosive liquid aerosols may irritate the bronchi and breathing sacs causing damage which diminishes ability of the bronchopulmonary structures to function properly and may predispose them to chronic bronchitis, emphysema, and other respiratory diseases.

III. Air Pollution Episodes

Evidence linking air pollution to human death is conclusive. As noted in Table 11-2 nine major episodes resulted in human deaths; minor episodes were probably never reported. The death-causing episodes were not isolated incidents—almost 20 other previous and subsequent episodes were described. And lesser community exposures throughout the world are now being investigated and reported.

Definitive data indicating specific pollutants or emission sources as the cause(s) for episodes do not exist. Salient facts lead to one major conclusion for the occurrence of episodes involving cities—too many people and too much industry concentrated in too small an area adverse in topographical and meteorological factors.

No single pollutant or group of pollutants responsible for the episodes has been isolated—although sulfur dioxide and particulate matter are most suspect. It cannot be said that industrial emissions were the causative agents even though steel, glass, zinc, and sulfuric acid plants were implicated in the Meuse Valley and Donora episodes; the London 1952 incident was thought to result primarily from home heating emissions. Photochemical air pollution, involving automotive emissions, may have contributed to world-wide occurrences in 1962 and subsequent years, but the earlier years did not involve necessary automotive densities. Radioactivity, pesticides, and exotic chemicals were not implicated. Unless additional research resolves the unknown pollutant culprit, we must emphasize the synergistic and an-

TABLE 11-2

Major Air Pollution Episodes

Location and date	Remarks
Meuse Valley (Dec. 1930)	6000 ill; 60 deaths; 5 previous episodes
Donora, Pennsylvania (Oct. 1948)	5910 ill; 20 deaths; 1 previous episode
Poza Rica, Mexico (Nov. 1950)	320 ill; 22 deaths; industrial accident
London, England (Dec. 1952)	Unknown number ill; 4000 deaths; 6 previous episodes and 9 subsequent episodes
New York, New York (Nov. 1953)	Unknown number ill; 165 deaths; 3 subsequent episodes
New Orleans, Louisiana (Oct. 1953)	200 ill; 2 deaths; subsequent episodes
Yokohama, Japan (1956)	Many ill; patients recovered when removed from area; subsequent episodes
Worldwide episode (Nov. 1962)	At least 1000 deaths
Eastern United States seaboard (Nov. 1966)	Many ill; data incomplete

tagonistic effect of hundreds of thousands of solid, liquid, and gaseous components in the atmosphere.

Meteorology and topography have been critical factors in each acute event. All involved areas were close to a large body of water—river or ocean. Valley trapping of pollutants was a factor for the Meuse Valley and Donora; high buildings in New York and London may have minimized dispersion of pollutants. Each episode occurred during late fall or winter when the weather was cold; fog, light winds, and atmospheric inversions persisted for three days or more (anticyclonic high-pressure systems).

The length of the episodes lasted at least three days, emphasizing the needed time for pollutant concentration build-up under inversion conditions. One might wonder whether the time necessary to develop lethal air concentrations would not shorten to perhaps one day or less as population (and pollution emission) increases.

Episode illnesses usually began after the second day of weather stagnation; during the 1952 London catastrophe illnesses began on the first day. Health effects were felt principally by the elderly and the very young. The respiratory ill suffered additional biological stress. The ratio of illnesses to deaths is imprecise because of data incompleteness; ratios of 1000:1 may be anticipated. Illness and death decreased abruptly upon cessation of stagnation weather.

Autopsies found inflammatory lesions in the lungs, including parenchyma in most cases. Coughing, shortness of breath were common complaints. The effects on people with cardiorespiratory diseases were most serious. Respiratory and cardiac illnesses were immediately obvious; the effects of episodes on the long term health prospects of the public are not known.

IV. Air Pollution Epidemiology

Numerous epidemiological investigations implicate air pollution as a major factor affecting human health in urban areas as compared to rural or less polluted areas. Admittedly, epidemiological evidence for a causative relationship between air pollution and health may never be proved to the complete satisfaction of the scientific community. Epidemiological research adds facts and incidents involving children, workers, as well as healthy and sick individuals—throughout the entire world— that implicate air pollution as an added insult to public health. Epidemiology investigates effects of all air pollutants on all people—a real world condition. Figure 11-3 lists a small fraction of epidemiological investigations showing direction of air pollution effects on humans.

Morbidity rates for chronic respiratory diseases in countries throughout the world result, in part, from differences in air pollution. Furthermore these differences in respiratory morbidity lead to differences in mortality.

Work absenteeism in England, Wales, Scotland, and the United States have been related to levels of sulfate particulates, sulfur dioxide, or smoke; absences resulted, in part, from respiratory illnesses. In some studies arthritis and rheumatism were

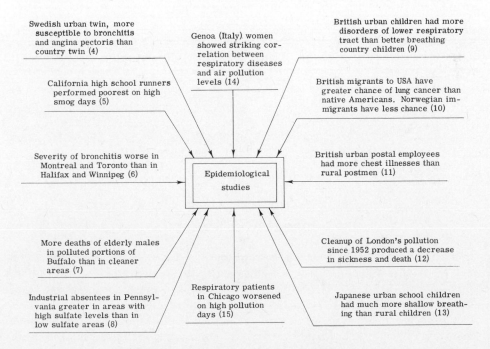

Fig. 11-3. Worldwide epidemiological air pollution studies.

TABLE 11-3

Summary of Conclusions of Epidemiological Studies on Health and Air Pollution

1. There is a wide variation in susceptibility of different persons to air pollutants.
2. Preexisting or underlying disease conditions augment stresses added by air pollution.
3. Air pollution is an additive factor aggravating effects of other substances (tobacco smoke, asbestos, etc.) in initiating disease.
4. Under some conditions, some types of air pollution can actually initiate structural and persistent disease in well persons.
5. Different types of air pollution may produce quite different physiological responses.

related to air pollution; in other studies, age distribution, conditions of work, social or climatic factors did not appear related to air pollution caused absences.

Epidemiological studies involving the effect of air pollution on children are considered most important in that children usually do not smoke, work in hazardous areas, or are not concerned with mental and physical stresses of earning a living. Research results in England, Sweden, United States, and Japan indicate enhanced respiratory problems of school children in urban areas (high pollution) compared to rural areas. More important, adverse environmental conditions early in life may detrimentally affect health in later years.

History of migrants from one country to another suggests that air pollution plays a role in the development of lung cancer. Migrants from a country with a high incidence of lung cancer retain the high mortality cancer rate when moved to a country with a lower lung cancer rate for native citizens; the reverse is also true.

In addition to smoking, air pollution has been suggested as one factor that causes increased morbidity and mortality rates for bronchitis for monozygotic and dizygotic twins in urban compared to rural areas. Athletes tend to perform better in less polluted areas than in more polluted areas.

Conclusions reached on the basis of epidemiological research point out the complexities of air pollution health effects (Table 11-3); because of these complexities no single pollutant was judged responsible for observed changes in health and it was deemed impossible to specify exact pollutant ranges that may be called "safe" for all people (16). The conclusions reached, while intensifying the concept of synergism and antagonism, emphasize the need for continued reduction in pollutant levels until a "safe" concentration can be determined from medical research.

V. Animal Experimentation

Animal research provides support for the concept that episodic events and sustained air pollution adversely affect humans. It must be reemphasized that man's complex physiology is sufficiently different from that of smaller animals so that human effects of air pollution should be studied ideally on humans.

Flexibility in animal research does have many obvious advantages over that of human test subjects. Animals may be used in much greater numbers than humans, an obvious statistical advantage. Animals have shorter life spans than man so exposures can be conducted over a number of generations under highly controlled conditions. It is possible to subject animals to high concentrations of pollutants—levels which may be toxic or dangerous to humans. Animals may be sacrificed during or upon conclusion of an experiment for pathological or analytical examination of specific organs.

Two broad aspects of animal research will be evaluated—each deals with a highly controversial human effect. Reference is made to excellent reviews on this subject (1, 17).

A. Sulfur Dioxide Synergism

A prime issue concerning the effect of sulfur dioxide on human health is the concentration (dosage) necessary to produce harmful effects. For healthy workers, a sulfur dioxide concentration limit for an eight hour industrial exposure over a lifetime has been set at 13,100 $\mu g/m^3$ (5 ppm). The threshold limit implies that a healthy worker would suffer no adverse effects from the gas during his life. Yet sulfur dioxide concentration during the 1952 London episode did not exceed 3500 $\mu g/m^3$ (1.34 ppm) daily average; in other episodes involving human deaths, the concentration of sulfur dioxide was reported as less than 2620 $\mu g/m^3$ (1 ppm) on an hourly maximum basis. Mice have been exposed to 65,500 $\mu g/m^3$ (25 ppm) sulfur dioxide for 1100 hours without obvious ill effects. The Environmental Protection Agency

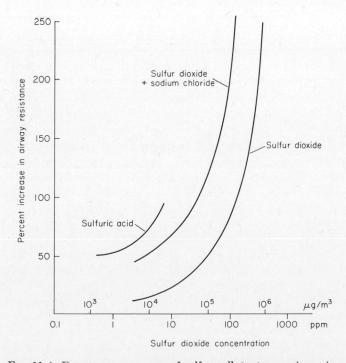

FIG. 11-4. Dose-response curves of sulfur pollutants on guinea pigs.

has set sulfur dioxide levels as low as 80 $\mu g/m^3$ (0.03 ppm) for an annual arithmetic mean and 365 $\mu g/m^3$ (0.14 ppm) for a maximum 24-hour concentration not to be exceeded more than once per year (primary standard—air-quality criteria).

The research of Amdur over many years suggests a partial explanation for the noted concentration discrepancies. As shown in Fig. 11-4, there is a synergistic effect between sulfur dioxide and sodium chloride particles (0.2 μ) affecting airway resistance of lungs; the oxidation of the sulfur dioxide to sulfuric acid on the sodium chloride particles was demonstrated by the effect of 0.8 μ sulfuric acid aerosol (18). The data showed a manyfold increased effect of sulfur dioxide when it was sorbed on a sodium chloride aerosol.

Subsequent reports (19) showed that particulate matter capable of oxidizing sulfur dioxide to sulfuric acid (ferrous iron, manganese, vanadium) caused a three- to fourfold potentiation of irritant response. Concentrations as low as 420 $\mu g/m^3$ (0.16 ppm) sulfur dioxide were used in these experiments.

The reported synergistic results on animals have not been demonstrated experimentally on man. Even though synergism of sulfur dioxide and particulate matter would explain the disparity of concentration effects on man, the data cannot be scientifically related to man until so proven by experimentation. Yet some type of synergism must be involved in the sulfur dioxide dilemma. To the present time neither the pollutants involved nor a plausible mechanism has been demonstrated for observed health effects.

B. Oxidant Susceptibility

Another puzzling question concerning human effects is whether the resistance of man to disease is decreased by air pollution. Coffin and Blommer (20) exposed laboratory mice to diluted irradiated auto exhaust (294 $\mu g/m^3$ O_3, 29,000 $\mu g/m^3$ CO) for 4 hours and then to an infectious streptococcus aerosol. Results showed a significantly increased mortality in test animals over similarly infected control animals exposed first to filtered air.

Concentrations of ozone, as determined from animal studies, indicate accelerated ageing. Thus the relevance of photochemical oxidant to aging is foreboding because of a potential sequence of damaging oxidative chain reactions in human tissue.

The decrease in disease resistance and increased rate of senescence by ozone has not been demonstrated in man. It would be understandable if human cells would act adversely to added stress of photochemical air pollution. Results of experimentation on another species cannot be extrapolated to humans.

VI. Human Health

It has been shown that episodic air pollution at high levels can kill. Epidemiological evidence reveals the fact that air pollution at "normal" concentrations has an adverse effect on human health. Animal research confirms and augments findings on humans. However, despite the abundance of research, the scientific community does not agree on "safe" levels to protect human health. Air pollution research on

TABLE 11-4

Change in the Maximum Permissible Occuptational Exposure of the Body to Ionizing Radiation[a]

Recommended rate	Comments[b]
0.1 erythema dose/year (5 rem per week)	Recommended in 1925
1 rem per week	Recommended by ICRP in 1934; used worldwide until 1950
0.5 rem per week	Recommended by NCRP in 1934; used in U.S. until 1949
0.3 rem per week	Recommended by NCRP in 1949 and ICRP in 1950; used worldwide until 1956
0.1 rem per week	Recommended by ICRP in 1956 and NCRP in 1957

[a] Values given are in addition to doses from medical and from background exposures. Values are for a 5-day work week.

[b] ICRP is the International Commission on Radiological Protection; NCRP is the National Council on Radiation Protection (U.S.).

man must determine pollutant concentrations in ambient air which do not disturb the normal homeostatic function of healthy cells. Although perhaps impossible to satisfy the entire scientific community, it is possible to define "safe" pollutant concentrations if public welfare is the prime consideration.

The question of a "safe" pollutant level for public health may be compared to changes in permissible exposure to ionizing radiation over the past half-century. Occupational exposure to ionizing radiation was initially limited to the dose that would produce erythema on the skin. As noted in Table 11-4, the permissible occupational exposure has been reduced from 5 rem per week to 0.1 rem/week (20), a factor of 50 over a half-century. It may be anticipated that "safe" air pollution levels will follow this trend as more information from medical research becomes available.

References

1. Stokinger, H. E., *Amer. Ind. Hyg. Ass. J.*, 195–217, May–June 1969.
2. Calhoun, J. B., *Sci. Amer.* **206**, 139–148 (1962).
3. Hatch, T. F., *Arch. Environ. Health* **16**, 571–578 (1968).
4. Cederlof, R., Friberg, L., and Hrubec, Z., *Arch. Env. Health* **18**, 934–940 (1969).
5. Wayne, W. S., Wehrle, P. F., and Carroll, R. E., *J. Amer. Med. Ass.* **199**, 151–154 (1967).
6. Bates, D. V., *Arch. Environ. Health* **14**, 220–227 (1967).
7. Winkelstein, W., Kantor, S., Davis, E. W., Maneri, C. S., and Mosher, W. E., *Arch. Environ. Health* **14**, 162–169 (1967) and **15**, 401–405 (1968).
8. Dohan, F. C., Everts, G. S., and Smith, B., *J. Air Pollut. Contr. Ass.* **12**, 418–422 (1962).
9. Douglas, J. W. B., and Waller, R. E., *Brit. J. Prev. Soc. Med.* **20**, 1–8 (1966).

10. Dean, G., *Proc. Roy. Soc. Med.* **57**, 984–987 (1964).
11. Holland, W. W., and Reid, D. D., *Lancet* **1**, 445–448 (1965).
12. Lawther, P. J., *Proc. Roy. Soc. Med.* **51**, 262–264 (1958).
13. Toyama, T., *Arch. Environ. Health* **8**, 153–173 (1964).
14. Petrilli, R. L., Agnese, G., and Kanitz, S., *Arch. Environ. Health* **12**, 733–740 (1966).
15. Carnow, B. W., Lepper, M. H., Shekelle, R. B., and Stamler, J., *Arch. Environ. Health* **18**, 768–776 (1969).
16. Higgins, I. T. T., and McCarrol, J. R., Types, ranges, and methods for classifying human pathophysiologic changes and responses to air pollution, *in* "Development of Air Quality Standards" (A. Atkisson and R. S. Gaines, eds.), Charles E. Merrill, Columbus, Ohio, 1970.
17. Stokinger, H. E., and Coffin, D. L., Biologic effects of air pollutants, *in* "Air Pollution" (A. C. Stern, ed.), Vol. I, pp. 445–546. Academic Press, New York, 1968.
18. Amdur, M. O., *Amer. Ind. Hyg. Ass. Quart.* **18**, 149–155 (1957).
19. Amdur, M. O., *Air Pollut. Contr. Ass. J.* **19**, 638–644 (1969).
20. Turner, J. E., and Morgan, K. Z., "Principles of Radiation Protection; A Textbook of Health Physics." Wiley, New York, 1967.

Suggested Reading

"Air Quality Criteria for Carbon Monoxide." National Air Pollution Control Administration, Department of Health, Education, and Welfare, AP-62, Washington, D.C., March 1970.

"Air Quality Criteria for Hydrocarbons." National Air Pollution Control Administration, Department of Health, Education, and Welfare, AP-64, Washington, D.C., March 1970.

"Air Quality Criteria for Nitrogen Oxides." National Air Pollution Control Administration, Department of Health, Education, and Welfare, AP-84, Washington, D.C., Jan. 1971.

"Air Quality Criteria for Particulates." National Air Pollution Control Administration, Department of Health, Education, and Welfare, AP-49, 801 N. Randolph Street, Arlington, Virginia, Feb. 1969.

"Air Quality Criteria for Photochemical Oxidants." National Air Pollution Control Administration, Department of Health, Education, and Welfare, AP-63, Washington, D.C., March 1970.

"Air Quality Criteria for Sulfur Oxides." National Air Pollution Control Administration, Department of Health, Education, and Welfare, AP-50, 801 N. Randolph Street, Arlington, Virginia, Feb. 1969.

Cassel, E. J., The health effects of air pollution and their implications for control, *In* "Law and Contemporary Problems." Duke University, Durham, North Carolina, **33**, 197–216 (1968).

Iglauer, E., *The New Yorker Magazine*, New York, April 13, 1968.

Wohlers, H. C., Air—A priceless resource, *In* "Environmental Health" (P. W. Purdom, ed.), pp. 190–289. Academic Press, New York, 1971.

Questions

1. Explain the differences in mode of attack of air pollution on humans and inert materials.
2. Elaborate on two of the direct air pollution effects and one contributing effect on humans as mentioned in Table 11-1.
3. (a) What do you believe are possible air pollution effects on humans in your community?
   (b) Briefly describe how you might gather data to support your contention.
   (c) How might the economic loss of air pollution on humans be determined?
4. Assuming a life expectancy of 70 years and a daily intake of 13.4 kg air, calculate the lifetime intake of SO_2, NO, NO_2, CO and particulate matter (100% retention) in your community.
5. Describe in some detail an air pollution episode close to your community.
6. Describe in some detail an epidemiological study close to your community.
7. Can animal air pollution research results be directly related to humans? Explain your answer.
8. Using an example, explain the effects of synergism (or antagonism) of air pollution on humans.
9. What is a "safe" air pollutant level for humans?

Chapter 12

AESTHETIC EFFECTS OF AIR POLLUTION

Considering the gross implications of air pollution effects upon the physical and mental well-being of man, one might assign a minor role to aesthetic effects. Such is not the case. Medical science has afforded us a prolonged life-span compared to our ancestors; if one is forced because of circumstances to live in a dismal urban air atmosphere, one questions the advantages of a longer life. Granted food, shelter, and clothing as prime requisites of life, man continually yearns for a fuller life with an environment not degraded by pollution.

Aesthetics pertains to a sense of the beautiful. In contrast, the effects of air pollution on man's emotions or sensations are negative. Is there a city in the world where air pollution does not adversely affect man's olfactory and visual senses? Add the senses of touch, taste, as well as hearing and the result is an irate citizen. One does not expect a city's atmosphere to be as aesthetically pleasing as the great open spaces, but one must hope for something much improved over today's blighted communities.

I. Weber-Fechner Relationship

The Weber-Fechner law states that the intensity of a sensation is proportional to the logarithm of the stimulus, or

$$\text{Sensation intensity} = K \log \text{stimulus} \qquad (12\text{-}1)$$

139

where K is an experimental constant for each sense modality. The law or relationship is not precise—psychologists now believe that sensation is a variable power function for each human sense rather than a single logarithmic relationship. The expression may be used to relate sensation and air pollution in the language of mathematics, particularly for the middle ranges of intensity.

The basic postulate of Weber's and Fechner's psychophysics was that mind and body appear to be separate entities but are actually only different sides of one reality. In more practical language, the increase of a stimulus necessary to produce an increase of sensation (intensity) in any sense is not an absolute quantity, but depends on the proportion the increase bears to the immediately preceding stimulus.

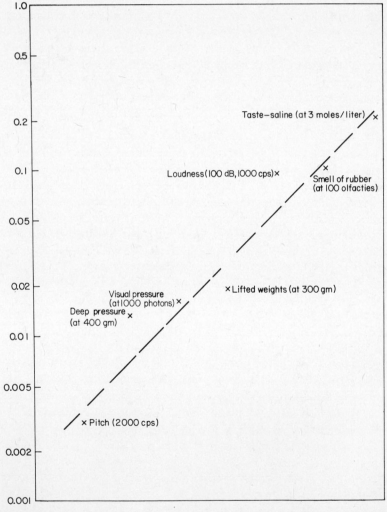

FIG. 12-1. Minimal values of the Weber fraction under optimum experimental conditions for judgment.

Through experimentation, it is possible to determine the minimum stimulus concentration for the affected sense. In like manner, it is possible to fix an upper limit beyond which no increase of stimulus produces any increase of sensation; it is necessary to depend upon a mental estimate or comparison of two or more sensations. Experimental data for the Weber fraction

$$\Delta S_t / S_t = K \tag{12-2}$$

where S_t is stimulus intensity, ΔS_t is the differential threshold, and K is the constant fraction for each type stimulus is shown in Fig. 12-1 (1). For each human sense, the minimal value for a sensation to be noted varies about seventyfold. Pitch (sound) is most sensitive at one-third of 1%, and taste, the least sensitive at 20%; sight approximates 2%.

The jump from psychophysics to air pollution is relatively simple—pollutants are external stimuli affecting human senses. Once a pollutant affects the human senses, a large increase in stimulus (concentration) is necessary for a step increase in sensation intensity. Most important from the point of view of enforcement is that the abatement of an emission source must be high to reduce sensation intensity; furthermore, if it is necessary to completely eliminate a sensation, the pollutant control must be virtually complete—"zero" emission control.

II. Smell

Odor complaints comprise the greatest number of investigations in a single category for control agencies; in the San Francisco Bay Area, the greatest number of odor complaints were lodged against metallurgical, chemical and petroleum operations. In San Francisco from 1959 through 1963, 637 complaints (34%) involved odors. For comparison, yearly odor complaints during the same period in Cincinnati approximated 18%, Philadelphia 36% and Detroit 12%. Walk through any urban area and smell vehicular exhaust, food in preparation, rendering plants, petroleum refineries, chemical operations, or burning leaves. The aroma of roasting coffee or grilling of steak and onions is pleasant. Yet consider for a moment if you lived or worked downwind of these "pleasant" odors; in a relatively short period of time one would detest the smells.

Odors in themselves are not causes of physical illness or disease, even though the average man believes the contrary; malodors are disturbing but not harmful. The psychological effects of odors are, at times, surprising. Once an odor problem is aired in a neighborhood, a series of developments occur that may verge on hysteria. Complainants have claimed damage to household content and garments; people have asserted that hair falls out; dogs bark, and hens refuse to lay; in one case, an epidemic of measles was claimed to be the result of odors.

Odors are an obvious nuisance because of the low concentration of most chemicals necessary to trigger a sensation response. Surprisingly, odor stimuli affect only a very small area of yellow brown receptor cells located in the ceiling of the inner nose. The ultimate olfactory receptors are thought to be cilia-like hairs which protrude into the mucus layer lining the entire surface of the inner nose. Olfactory nerves transmit impulses to the olfactory bulb located at the front base of the brain. At the bulb, fibers from the nose contact other nerves, which go to different parts of the brain (Fig. 12-2).

Science has not determined precisely how odors are perceived. Clearly, a material must be volatile to reach the inner nose. One theory postulates that an odorous material must be soluble in the mucus covering the olfactory receptors to stimulate a response. Another claims that the shape of the molecule must fit into special receptor sites—each receptor site has a fixed geometry for a definite odor (2). More than 30 different theories have been suggested, none entirely acceptable.

Almost all people can smell, and experts can identify thousands of distinct odors. As the nose is the instrument, subjective intensity ratings from 0 (no odor) to 4 (overpowering) have been used in research and field studies. Most odors follow the logarithmic Weber-Fechner relationship for concentration versus stimulus (intensity); Kaiser (3) plotted odor intensity of four chemicals against concentration as shown in Fig. 12-3. To reduce odor intensity rating by one unit, the odor concentration must be reduced by a factor of ten.

Odor is an intangible commodity. Odors follow air movement so that the smell may be noted one minute and not the next; eddy currents may be involved. Olfactory nerves become fatigued rapidly so that if one is surrounded by an odor, recognition is quickly lost unless the intensity increases. And the odorous material may oxidize in open atmospheres to either a more or a less noticeable odor; butanol oxidizes to highly objectionable butyric acid, while the stench of ethyl mercaptan may oxidize to less odorous sulfur dioxide.

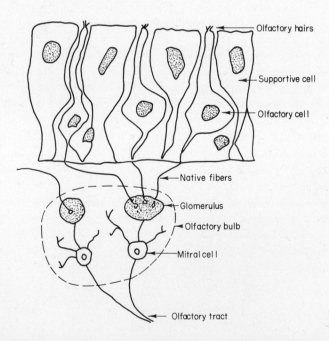

FIG. 12-2. Olfactory membrane and odor-sensitive structure.

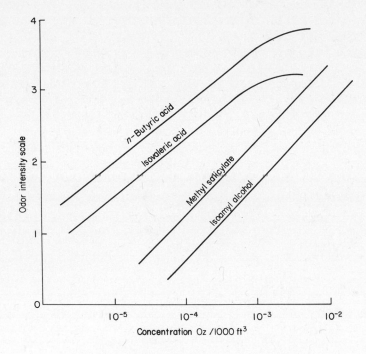

Fig. 12-3. Odor intensity versus concentration.

III. Sight

On a clear day one cannot see forever. For example, in an atmosphere containing only permanent gases and uncondensed vapors, an airplane passenger could see a mountain about 250 miles away. Visibility is reduced by particulate matter as well as gases added to the atmosphere by man and nature. Small particles scatter light and some gases (e. g., nitrogen dioxide) absorb light.

Visibility interference (scattering) by particulate matter can best be understood by considering the effects of a dirty windshield upon the driver. Dirt on the windshield is not very bothersome to the driver in shade or at night when there are no oncoming cars. However, when strong light from the sun or vehicular headlights strikes the dirty windshield, the visible range of the driver is drastically restricted. There is no light obscuration; impairment of visibility is caused by intense forward scattering of oncoming light into the eyes of the driver. The dirty windshield simulates a polluted atmosphere.

Robinson (4) presented a simplified formula for visibility reduction (also see Section 21-I):

$$L_c = 0.22 \frac{\alpha}{X_\kappa} \tag{12-3}$$

where L_c = visible range at constant lumen of 0.05, X = particulate concentra-

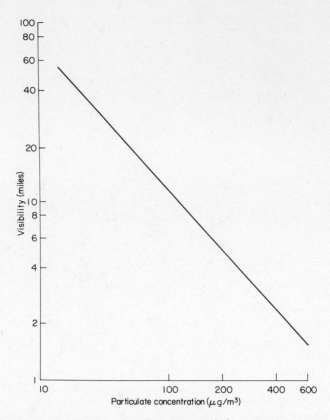

FIG. 12-4. Visibility as a function of suspended particulate concentration.

tion, mg/m³, κ = scattering area ratio, and

$$\alpha = \frac{2 \ (\text{particle radius})}{\text{wavelength of light}}$$

Visibility impairment increases as the weight of particulate matter in the atmosphere increases, as the wavelength of incident light increases, or as the particle diameter decreases. Visibility reduction is most severe when the particle size is in the size range of 0.3–0.8 μ and the relative humidity is above 70%. As an order of magnitude, one-half gram material of an optimum particle size per square meter will determine the limit of visibility.

Gaseous air pollutants have a minimal effect on visibility reduction unless they are colored, absorbing light in the visible regions. Nitrogen dioxide has a maximum light absorption at about 4500 Å. Even in low concentrations nitrogen dioxide can reduce light transmission significantly. At a concentration of 470 μg/m³ NO_2 (0.25 ppm), almost 70% of 4500 Å light is absorbed; the effect of NO_2 decreases as particulate concentration in the atmosphere is increased. The state of California has set a 1-hour ambient air-quality standard of 0.25 ppm (470 μg/m³) nitrogen dioxide based primarily on light absorption (about 20% reduction for 2 miles in "normal" atmosphere).

In open atmospheres, a correlation exists between suspended particulate concentrations and visibility (5), as noted in Fig. 12-4. The correlation implies that the particle size in a specific location is fairly uniform as the mass of one 20-μ particle is equivalent to that of one thousand 2-μ particles and one million 0.2-μ particles.

IV. Hearing

Like odor, noise is of major environmental concern. Environmental noise levels range from that produced by automobiles, buses, and trucks, to jackhammers breaking cement, to the quiet of libraries. The development of the supersonic transport (SST) tied noise and air pollution in a single ecological package. The sonic pressure of the plane, estimated to be a 50-mile swarth at 1800 mph, could be an environmental nightmare in populated areas. Unknown are the climatological effects of dumping tons of water vapor, particulate matter, and other exhaust pollutants into the stratosphere; a reduction of the stratospheric ozone has been postulated.

Sound intensity is subjectively measured by the human ear. Although the ear responds in a nonlinear fashion, experiments have demonstrated that the ear responds logarithmically in relation to the loudness of applied stimuli. Noise is commonly measured in terms of decibels (dB)—a dimensionless unit used to express the logarithm of the measured quantity to a reference quantity. The sound

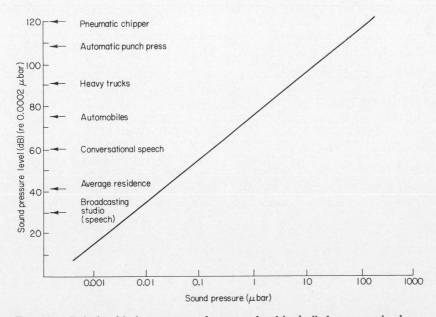

FIG. 12-5. Relationship between sound pressure level in decibels versus microbars.

pressure level is defined as

$$\text{Sound pressure level} = 20 \log \frac{\text{sound pressure (rms)}}{\text{reference sound pressure}} \qquad (12\text{-}4)$$

Figure 12-5 shows the relationship between sound pressure in microbars and in decibels.

V. Touch and Taste

Touch and taste are the human senses least affected by air pollution. Both of these senses may be used, however, as rough analytical tools for air pollution purposes.

Particulate matter settled on surfaces may be rubbed between the fingers to estimate the hardness, friability, or other physical characteristics of the material. The amount of material retained on the fingers after rubbing is a rough indication of particle size.

Odor and flavor (taste) are related; in many cases flavor enhancement depends upon odor perception. Interestingly, sulfur dioxide can be "tasted" at concentrations approximating 785 μg/m^3 (0.3 ppm), which is below the odor threshold value.

VI. Public Opinion

Air pollution affects the human senses—hence the public should be aroused to action. Presently, ecology and air environment have become a political issue throughout the world and particularly in the United States. As a measure of its interest in air environment, the Federal Government increased its spendings from less than $1 million dollars on control activities in 1956 to $100 million dollars in 1970; based on a population of 200 million, the expenditures per capita in 1970 approximated 50 cents. In the past, air was considered as a "free good" because it was available in sufficient quality and quantity—hence air supposedly had no valuable alternate markets and thus was of no concern to economists.

During the 1960's a number of public opinion questionnaire surveys were completed in the United States (Fig. 12-6). Although it is difficult to evaluate precisely the data in total because of differences in the types of questions and methodologies involved, one can conclude that a majority of people were at least "somewhat bothered" by air pollution. It was anticipated that public opinion survey findings during the 1970's would emphasize more strongly the hazards of all types of pollution.

Important socioeconomic relationships were developed from the opinion surveys concerning air pollution. Generally, people in the higher income brackets were more likely to be aware of, or perceive, hazards of air pollution than lower economic groups. More highly educated people were better able to define air pollution in terms of causal agents. As would be expected, rural populations were not as concerned as urban dwellers.

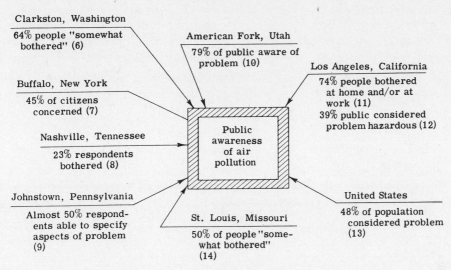

Clarkston, Washington
64% people "somewhat bothered" (6)

American Fork, Utah
79% of public aware of problem (10)

Los Angeles, California
74% people bothered at home and/or at work (11)
39% public considered problem hazardous (12)

Buffalo, New York
45% of citizens concerned (7)

Public awareness of air pollution

Nashville, Tennessee
23% respondents bothered (8)

Johnstown, Pennsylvania
Almost 50% respondents able to specify aspects of problem (9)

St. Louis, Missouri
50% of people "somewhat bothered" (14)

United States
48% of population considered problem (13)

FIG. 12-6. Public awareness of air pollution.

Studies also indicated that people with children under 18 years were more concerned than single persons or older married couples. Women were more informed on air pollution topics than were men. The majority of respondents felt the Federal Government should take the lead in air pollution control.

Perhaps the most important result of the opinion surveys was that the public has begun to understand that an erroneous trade-off was made. Formerly the public felt that air pollution had to be resolved by high-level technical and political decisions; technology ruled man. Now the populace realize that they can decide the type of air environment they desire and technicians can provide the solution; man rules technology.

VII. Sensory Perception

The use of sensory perception as a limitation to air pollution was investigated (15). Russians believe that a pollutant concentration which causes a reflexive reaction of human sense receptors should be limiting; receptors of the respiratory tract were emphasized. The threshold of the most sensitive person was used for standard setting.

Pollutant-limiting concentrations based upon sensory perception are much lower than those values considered damaging to human health. For comparison with the Russian concept, standards in the United States have been set based upon a proven adverse effect to man or his well-being. Although difficult to resolve these differences in terms of control concepts, it must be emphasized that levels of pollutants considered toxic have decreased with time as analytical techniques improved and medical research results became available. Perhaps sensory perception limitations should be used as goals for air quality.

It has been shown that the scientific approach may be used in the study of aesthetics by formulating hypotheses relevant to sensations recognized by human senses. The Weber-Fechner relationship shows that a stimulus necessary to produce a sensation is relative to the "amount" of stimulus already present. Thus a scientific method is available to confirm what we feel instinctively—dirty air must be cleaned with regard to weight to a greater extent than a clean, rural air mass.

References

1. Krech, D., and Crutchfield, R. S., "Elements of Psychology." Knopf, New York, 1960.
2. Amoore, J. E., Johnston, J. W., and Rubin, M., *Sci. Amer.* **210,** 42–49 (1964).
3. Kaiser, E. R., Odor and its measurement, *in* "Air Pollution" (A. C. Stern, ed.), Vol. I, pp. 509–526. Academic Press, New York, 1962.
4. Robinson, E., Effect on the physical properties of the atmosphere, *in* "Air Pollution" (A. C. Stern, ed.), Vol. I, pp. 349–490. Academic Press, New York, 1968.
5. Noll, K. E., Mueller, P. K., and Imada, M., *Atmos. Environ.* **2,** 465–475 (1968).
6. Medalia, N. Z., and Finker, A. L., Community Perception of Air Quality: An Opinion Survey in Clarkston, Washington. Environ. Health Service, Public Health Service Publ. No. 999—AP-10, Washington, D.C., 1965.
7. de Groot, I., Loring, W., Rihm, A., Samuels, S. W., and Winkelstein, W., *J. Air Pollut. Contr. Ass.* **16,** 245–247 (1966).
8. Smith, W. S., Schueneman, J. J., and Zeidberg, L. D., *J. Air Pollut. Contr. Ass.* **14,** 418–423 (1964).
9. Crowe, M. J., *J. Air Pollut. Contr. Ass.* **18,** 154–157 (1968).
10. Creer, R. N., Social Psychological Factors Involved in the Perception of Air Pollution as an Environmental Health Problem. Master of Science Thesis, Univ. of Utah, Salt Lake City, Utah, Aug. 1968.
11. Air Pollution—Effects Reported by California Residents. California State Department of Health, Berkeley, California, 1960.
12. Van Arsdol, M. D., Sabagh, G., and Alexander, F., *J. Health Human Behavior,* pp. 144–153, Winter, 1964.
13. Opinion Research Corporation, Survey of Public Attitudes toward Air Pollution. Princeton, New Jersey, Nov. 1966.
14. Schusky, J., *J. Air Pollut. Contr. Ass.* **16,** 72–76 (1966).
15. Ryazanov, V. A., *Arch. Environ. Health* **5,** 479–494 (1962).

Suggested Reading

"Air Quality Criteria for Carbon Monoxide." National Air Pollution Control Administration, Department of Health, Education, and Welfare, AP-62, Washington, D.C., March 1970.

"Air Quality Criteria for Hydrocarbons." National Air Pollution Control Administration, Department of Health, Education, and Welfare, AP-64, Washington, D.C., March 1970.

"Air Quality Criteria for Nitrogen Oxides." National Air Pollution Control Administration, Department of Health, Education, and Welfare, AP-84, Washington, D.C., Jan. 1971.

"Air Quality Criteria for Particulates." National Air Pollution Control Administration, Department of Health, Education, and Welfare, AP-49, 801 N. Randolph Street, Arlington, Virginia, Feb. 1969.

"Air Quality Criteria for Photochemical Oxidants." National Air Pollution Control Administration, Department of Health, Education, and Welfare, AP-63, Washington, D.C., March 1970.

"Air Quality Criteria for Sulfur Oxides." National Air Pollution Control Administration, Department of Health, Education, and Welfare, AP-50, Arlington, Virginia, Feb. 1969.

"Industrial Noise Manual" (2nd edition). American Industrial Hygiene Association, Detroit, Michigan 1966.

Leighton, P. A., "Photochemistry of Air Pollution." Academic Press, New York, 1961.

Middleton, W. E. K., "Vision through the Atmosphere." Univ. of Toronto Press, Toronto, Canada, 1952.

Questions

1. (a) What is the Weber-Fechner law and how does it relate to the human senses?
   (b) What is the difference between a law and a relationship between two factors?
2. (a) What is the definition of odor threshold?
   (b) The odor threshold for most chemicals, such as hydrogen sulfide, as reported in the literature, varies over several orders of magnitude. What factors might be involved in odor threshold discrepancies?
3. (a) Write the stepwise chemical reactions for the oxidation of n-butane to butanol, n-butanol to butanal, and n-butanal to butyric acid.
   (b) What is the odor threshold for each of these chemicals?
4. (a) Calculate the visible range in the ambient atmosphere where the mean particle size is $0.5\ \mu$ and the particulate concentration is $150\ \mu g/m^3$. Assume a wavelength of light of 5000 Å.
   (b) What is the color of light at 5000 Å and explain why this approximate wavelength light might have been chosen for the example?
5. Calculate the mass of a 1-, 10-, and 1000-μ particle, assuming a density of $2.5\ gm/cm^3$.
6. Write chemical equations that would show that emissions from supersonic transport planes (SST) could reduce the mass of stratospheric ozone. Note that these equations resemble those involved in photochemical air pollution.
7. List advantages and disadvantages of public surveys regarding air pollution.

Chapter 13

AIR-QUALITY CRITERIA AND STANDARDS

I. Air-Quality Criteria

Air-quality criteria are scientifically sound statements about effects that have been observed or inferred to have been produced by various exposures to specific pollutants. For any pollutant, air quality criteria may refer to different types of effects. For example, Figs 13-1 through 13-6 list effects on humans, experimental animals, vegetation and the atmosphere caused by various exposures to sulfur dioxide, particulates, nitrogen dioxide, carbon monoxide, photochemical oxidants and hydrocarbons, respectively. These data are from the "Air Quality Criteria" for these pollutants published by the U.S. Environmental Protection Agency (1-6). Air-quality criteria are based on analysis and critical review of available information. The criteria may stipulate conditions of exposure and may refer to sensitive population groups or to the joint effects of several pollutants with other factors. Air-quality criteria are descriptive—they describe effects that can be expected to occur wherever the ambient air level of a pollutant reaches or exceeds a specific concentration for a specific time period. Criteria will change as new information becomes available.

II. Air-Quality Standards and Goals

Air-quality standards prescribe pollutant levels that cannot legally be exceeded during a specific time in a specific geographical area. Air-quality standards should

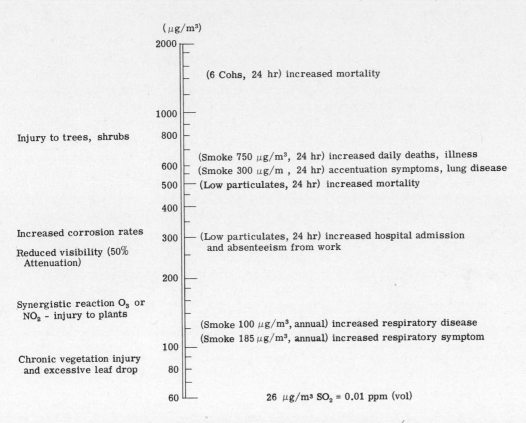

FIG. 13-1. Air-quality criteria for sulfur oxides.

be based upon air-quality criteria, with added factors of safety as desired. A relationship among air-quality criteria, air-quality standards, and other standards is shown in Fig. 13-7.

The concept of an air-quality goal requires it to be more restrictive, i. e., at lower pollutant concentration, than the air-quality standard. This requires that there be adverse effects on other than human health. For some pollutants this may be the case, but for other pollutants, a level low enough to prevent health effects will prevent all other known effects. For this latter case, the concept of an air-quality goal (based on all effects) as being more restrictive than the air-quality standard has no validity.

The U. S. Clean Air Amendments of 1970 give two special meanings to air-quality standards, primary standards—levels that will protect health, but not necessarily prevent the other adverse effects of air pollution—and secondary standards—levels that will prevent all the other adverse effects of air pollution. In this context, secondary standards are analogous to air-quality goals.

Since air pollution control is effectuated through the medium of air quality and emission standards, the principal philosophical discussions in the field of air pollution control revolve around the matter of their development and application.

III. Conversion of Effects Data to Criteria

In developing air pollution cause–effect relationships, we must be constantly on our guard lest we attribute to air pollution an effect caused by something else. Material damage due to pollution must be differentiated from that due to ultra-violet irradiation, frost, moisture, bacteria, fungi, insects, and animals. Air pollution damage to vegetation has to be differentiated from quite similar damage attributable to bacterial and fungal diseases, insects, drought, frost, soil mineral deviations, hail, and cultural practices. In the principal animal disorder associated with air pollution, i. e., fluorosis, the route of animal intake of fluorine is by ingestion, the

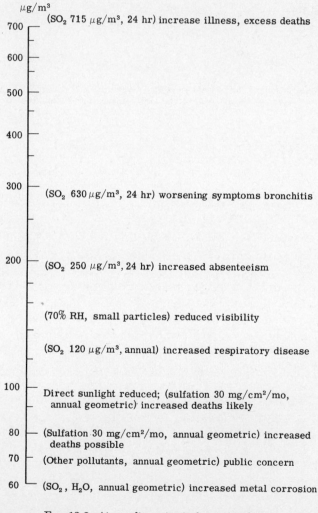

FIG. 13-2. Air-quality criteria for particulates.

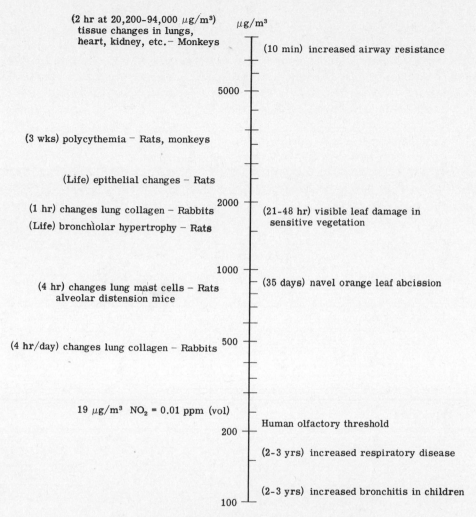

FIG. 13-3. Air-quality criteria for nitrogen dioxide.

air being the means for transporting the substance from its source to the forage or the hay used for animal feed. However, the water or feed supplements used may also have excess fluorine. Therefore these sources and disease states that may have symptoms similar to fluorosis must be ruled out before a cause–effect relationship may be established between ambient air levels of fluorine and fluorosis in animals. Similarly there are many instances of visibility reduction in the atmosphere by fog or mist for which air pollution is not a causative factor.

To study damage to materials, vegetation, and animals we can set up laboratory experiments in which most confusing variables are eliminated and a direct cause–effect relationship is established between pollutant dosage and resulting effect. We cannot do this with respect to man. Our cause–effect relationships for man are based upon (1) extrapolation from animal experimentation, (2) clinical observation of individual cases of persons exposed to the pollutant or toxicant (industrially, accidentally, suicidally, or under air pollution episode conditions), and (3) most importantly epidemiological data relating population morbidity and mortality to air pollution. There are no human diseases uniquely caused by air pollution. In all air pollution-related diseases in which there is build-up of toxic material in the blood, tissue, bone, or teeth, part or all of the build-up could be from ingestion of food or water containing the material. Those diseases which are respiratory can be caused by smoking or occupational exposure. They may be of bacterial, viral, or fungal origin quite divorced from the inhalation of man-made pollutants in the ambient air. These causes in addition to the variety of congenital, degenerative, nutritional, and psychosomatic causes of disease must all be ruled out before a disease can be attributed to air pollution. However, the common situation is that air pollution exa-

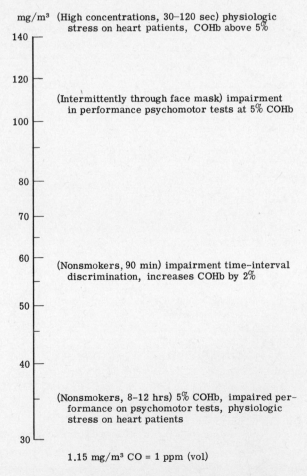

FIG. 13-4. Air-quality criteria for carbon monoxide.

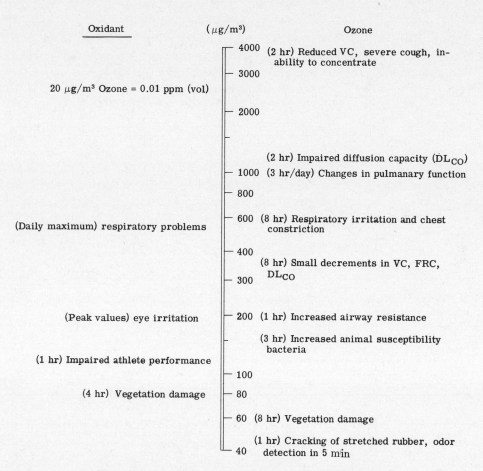

FIG. 13-5. Air-quality criteria for photochemical oxidants.

1. No demonstrable, direct health effects of gaseous hydrocarbons in ambient air

2. Injury to sensitive plants with ethylene concentrations of 1.15-575 $\mu g/m^3$ (0.001-0.5 ppm) over time period of 8-24 hr

3. Nonmethane hydrocarbon concentrations of 200 $\mu g/m^3$ (0.3 ppm C) for a 3-hr period from 6:00 to 9:00 AM can be expected to produce a maximum hourly average oxidant concentration of up to 200 $\mu g/m^3$ (0.1 ppm) near midday

FIG. 13-6. Air-quality criteria for hydrocarbons.

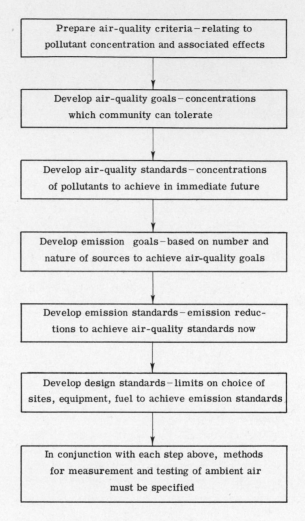

FIG. 13-7. Criteria → standards → control.

cerbates preexisting disease states. In human health, air pollution can be the "straw that breaks the camel's back."

IV. Conversion of Physical Criteria to Standards

The main philosophical question that arises with respect to air-quality standards is just what do we wish to consider an adverse effect or a cost associated with air pollution. Let us examine several categories of receptors to see the judgmental problems that arise. Most materials will deteriorate even when exposed to an unpolluted atmosphere. Iron will rust; metals will corrode, and wood will rot. To prevent deterioration protective coatings are applied. Their costs are part of the economic picture. Some materials, such as railroad rails, are used without protective coatings. There are costs associated with the decrease in their life in a polluted atmosphere as compared to an unpolluted one. One may argue that for those materials on which protective coatings are used, only those pollution levels that damage such coatings are of concern. One may further argue that some air pollution damage to protective coatings is tolerable, since by their very nature such coatings require periodic replenishment to maintain their protective integrity or appearance, and therefore that only levels that would require more frequent replenishment than would an unpolluted atmosphere should enter into the establishment of deterioration costs and of air quality standards. This argument certainly does hold with respect to the soiling of materials and structures. In fact it is frequently the protective coatings themselves that require replacement because they are dirty long before they have terminated their useful life as protectants. It can readily be shown that there are costs associated with soiling and that these include the cost of removing soil, the cost of protective coatings to facilitate the removal of soil, the premature disposal of material when it is no longer economical or practicable to remove soil, and the growth inhibition of vegetation due to leaf soiling. However, decision-making for air-quality standards related to soiling is based less on economic evaluation than on aesthetic considerations, i. e., on subjective evaluation of how much soiling the community will tolerate. This latter determination is judgmental and difficult to make. It may be facilitated by opinion surveys, but even when the limit of public tolerance for soiling is determined, it still has to be restated in terms of the pollution loading of the air that will result in this level of soiling.

An important effect of air pollution on the atmosphere is change in spectral transmission. The spectral regions of greatest concern are the ultraviolet and the visible. Changes in ultraviolet radiation have demonstrable adverse effects. To some extent people have learned to introduce dietary supplements to their food supply to compensate for the loss of naturally produced Vitamin D. For instance, ergosterol is irradiated as a food supplement for this purpose. The energy in the ultra-violet portion of the spectrum, had it not been absorbed and scattered by pollution, would have caused the human body to synthesize the required amount of the vitamin and would have produced an equivalent amount of erythema (sunburn) in the process. Given this set of facts, a jurisdiction must weigh its extra costs to provide dietary additions and achieve satisfactory sunburns and the burden of deficiency disease among those who do not use the dietary supplements, against the costs of cleaning up the air enough to let more solar ultraviolet energy penetrate its atmosphere. The same pollutants that absorb and scatter ultraviolet energy may also absorb and scatter visible light and, in so doing, decrease visibility through the atmosphere.

The fact that after a storm or the passage of a frontal system, the air becomes crystal clear and one can see for many kilometers does not give a true measure of year round visibility under unpolluted conditions. Between storms, even in unpolluted air, natural sources build up enough particulate matter in the air so that on many days of the year there would be less than ideal visibility. In many parts of the world mountains are called Smoky or Blue or some other name to designate prevalence of a natural haze, which gives them a smoky or bluish color and impedes visibility. When the Spanish first explored the area that is now Los Angeles, California, they gave it the name "Bay of the Smokes." The Los Angeles definition of the way it used to be before the advent of smog was that "You could see Catalina Island on a clear day." The part of the definition that is lacking is some indication of how many clear days there were each year before the advent of smog.

There are costs associated with loss of visibility and solar energy. These include increased need for artificial illumination and heating; delays, disruptions and accidents involving air, water, and land traffic; vegetation growth reduction associated with reduced photosynthesis; and commercial losses associated with the decreased attractiveness of a dingy community or one with restricted scenic views. However, these costs are less likely to be involved in deciding, for air-quality standard setting purposes, how much of the attainable visibility improvement to aim for than are aesthetic considerations. Just as in the previously noted case of soiling, judgment as to the limit of public tolerance for visibility reduction still has to be related in terms of the pollutant loading of the atmosphere that will yield the desired visibility. Obviously the pollutant level chosen for an air-quality standard must be the lower of the values required for soiling or visibility, otherwise one will be achieved without achieving the other. Whether the level chosen will or will not be lower than the atmospheric pollutant level required for prevention of health effects, will depend upon the aesthetic standards of the jurisdiction.

V. Conversion of Biological Criteria to Standards

There is considerable species variability with respect to damage to vegetation by any specific pollutant (see Appendix A). There is also great geographic variability with respect to where these species grow naturally or are cultivated. Because of this it is possible that in a jurisdiction none of the species particularly susceptible to damage by low levels of pollution may be among those indigenous or normally imported for local cultivation. As an example, the pollution level at which citrus trees are adversely affected, while meaningful as to air-quality standard in California and Florida, is meaningless for this purpose in Minnesota and Wisconsin. In like manner, a jurisdiction may take different viewpoints with respect to indigenous and imported species. It might set its air quality standards low enough to protect its indigenous vegetation even if this level is too high to allow the satisfactory growth of imported species. Even if a particularly susceptible species is indigenous it may be held in such low local esteem commercially or aesthetically that the jurisdiction may be unwilling to let the damage level for that species be the air-quality standard discriminator. In other words the people are willing for that species to be damaged rather than to assume the cost of cleaning up the air to prevent the damage. This same line of reasoning applies with equal logic to effects on animals both wild and domestic.

A jurisdiction may base part of its decision making regarding vegetation and animal damage on aesthetics. Its citizens may wish to grow certain ornamentals or raise certain species of pet birds or animals and allow these wishes to override the agricultural, forestry, and husbandry economics of the situation. Usually, however, the economic considerations would predominate in decision making. Costs of air pollution agricultural effects are the sum of the loss in income from sale of crops or livestock and the added cost necessary to raise the crops or livestock for sale. To these costs must be added the loss in value of agricultural land as its income potential decreases and the loss suffered by those segments of local industry and commerce that are dependent upon farm crops and the farmer for their existence. An interesting sidelight is that when such damage occurs on the periphery of an urban area, it is frequently a precursor to the breakup of such farmland into residential development with a financial gain rather than a loss to the landowner. When the crop that disappears is an orchard, grove or vineyard that took years to establish, and when useable farm buildings are torn down, society as a whole suffers a loss to the extent that it will take much time and money to establish a replacement for them at new locations. To some industries, air pollution costs include purchase of farm and ranch land to prevent litigation to recover damages; annual subsidy payments to farmers and ranchers in lieu of such litigation; and maintenance of air-quality monitoring systems to protect themselves against unwarranted litigation for this purpose.

There is a range of ambiguity in our health effects criteria data. In this range there is disagreement among experts as to its validity and interpretation. Thus from the same body of health effects data, one could adopt an air quality standard

TABLE 13-1

National Air-Quality Standards in the United States

Substance	Air-quality standard, $\mu g/m^3$	
	Primary (human health)	Secondary (all other effects)
Sulfur dioxide		
Annual arithmetic mean	80	60
24-hour max[a]	365	260
3-hour max[a]	—	1,300
Particulate matter		
Annual geometric mean	75	60
24-hour max[a]	260	150
Carbon monoxide		
8-hour max[a]	10,000	10,000
1-hour max[a]	40,000	40,000
Oxidant		
1-hour max[a]	160	160
Nitrogen oxides (e. g., NO_2)		
Annual arithmetic mean	100	100
Hydrocarbons		
3-hour max (6–9 A.M.)	160	160

[a] Not to be exceeded more than once per year.

on the high side of the range of ambiguity or one on the low side. Much soul-searching is required before accepting the results of questionable human health effects research and being accused of imposing large costs on the public by so doing; or of rejecting these results, and being accused of subjecting the public to potential damage to human health.

VI. United States National Air-Quality Standards

Although there are a large body of Air-Quality Standards adopted by national, state, provincial, and municipal jurisdictions of the world, the only one that will be reproduced here are the National Air Quality Standards of the United States (Table 13-1) (7).

References

1. "Air Quality Criteria for Sulfur Oxides." National Air Pollution Control Administration, Department of Health, Education, and Welfare, AP-50, 801 N. Randolph Street, Arlington, Virginia, Feb. 1969.
2. "Air Quality Criteria for Particulates." National Air Pollution Control Administration, Department of Health, Education, and Welfare, AP-49, 801 N. Randolph Street, Arlington, Virginia, Feb. 1969.
3. "Air Quality Criteria for Nitrogen Oxides." National Air Pollution Control Administration, Department of Health, Education, and Welfare, AP-84, Washington, D.C., Jan. 1971.
4. "Air Quality Criteria for Carbon Monoxide." National Air Pollution Control Administration, Department of Health, Education, and Welfare, AP-62, Washington, D.C., March 1970.
5. "Air Quality Criteria for Photochemical Oxidants." National Air Pollution Control Administration, Department of Health, Education, and Welfare, AP-63, Washington, D.C., March 1970.
6. "Air Quality Criteria for Hydrocarbons." National Air Pollution Control Administration, Department of Health, Education, and Welfare, AP-64, Washington, D.C., March 1970.
7. "National Primary and Secondary Ambient Air Quality Standards." Environmental Protection Agency, Federal Register, Vol. 36, No. 84, April 30, 1971, p. 8187.

Suggested Reading

Atkisson, A., and Gaines, R. S., eds. "Development of Air Quality Standards," 220 pp. Merrill, Columbus, Ohio, 1970.
Hagevik, G. H., "Decision-Making in Air Pollution Control," 217 pp. Prager, New York, 1970.

Questions

1. Why are air-quality criteria descriptive?
2. Why are air-quality standards prescriptive?
3. Evaluate the use of air-quality criteria and air-quality standards in your community.
4. Discuss the importance of the development of air-quality goals.
5. Prepare a figure similar to Fig. 13-1 through 13-6 for another pollutant.

6. Discuss the problem caused by cigarette smoking in the evaluation of epidemiological data on the effect of air pollution on respiratory disease.
7. Given the data of Fig. 13-1, how does one arrive at the air-quality standards for sulfur dioxide listed in Table 13-1.
8. Discuss the relative merits of stating air-quality standards as 1 hour, 3 hour, 8 hour, 24 hour, and annual averages.
9. Discuss the relative merits of national versus local air-quality standards.

Chapter 14

AMBIENT AIR POLLUTION SAMPLING

The sampling of pollutants and their subsequent analysis are a corner stone of air pollution control and abatement. Time and effort expended to define intent of sampling and method of analysis are amply rewarded by productive results. Inadequate sampling or incomplete analysis of pollutants produce results that may not be reliable in defining the whole problem. Cooperative information exchange by sampling and analytical personnel builds the firm foundation for a superstructure of air pollution studies.

I. Objective and Sequence of Air Pollution Sampling

While objectives of air pollution sampling are varied depending upon need, the sampling sequence is defined once objectives have been resolved; both considerations must be carefully evaluated prior to initiation of air pollution sampling.

A. Objectives of Air Pollution Sampling

Too frequently expensive continuous air sampling equipment is purchased without regard for sampling objectives. As noted in Table 14-1, fourteen air sampling objectives are listed; the list is incomplete and limited only by the need of the individual or group undertaking the program.

TABLE 14-1

*Objectives of Ambient Air
Pollution Sampling*

Air-quality determination
Alert network procedure
Analytical instrument development
Background concentration measurements
Evaluation of in-plant equipment

Government regulation requirement
Legal action support
Local source nuisance investigation
Nonair pollutant sampling
Odor study confirmation

Pollution trend monitoring
Radioactivity analysis
Regulation control and development
Research investigation

Relatively simple equipment is usually adequate to determine background levels, pollution trends, radioactivity, odor, or a local source nuisance; the use of simple equipment reduces monetary and personnel costs. More complex sampling equipment, including a telemetering system and computer data storage and retrieval, is necessary for successful and complete determination of air quality, operation of alert networks, regulation control, legal action, evaluation of in-plant equipment, government regulations, instrument development and research needs. Nonair-pollutant sampling involves indirect measures of effects of pollution, i.e., vegetation, waters, animal feeds, fruits, paints, metals, and fabrics.

An air pollution sampling program may include more than one objective. Complex sampling equipment will serve simple needs, albeit costly; simple and inexpensive equipment may be totally inadequate to serve an alert network. Trade-offs between objectives, costs, and capabilities of simple or complex equipment need not affect results if creativity and imagination are intertwined in decisions.

B. Air Pollution Sampling Sequence

Once the sampling objective has been decided, the operational sequence described in Fig. 14-1 should be followed; the initial steps are most important.

The evaluation of personnel, budget, and time limitations is critical. With limited budget and personnel, a complex air monitoring program should not be considered; if air-sampling results must be available within a short time period, it is again fruitless to consider a complex monitoring system because at least 2 years are required for site selection and preparation, for instrument delivery and shakedown, for training of station operators, and for writing and debugging of computer programming.

Integration of equipment must be related to precision (reproducibility) and ac-

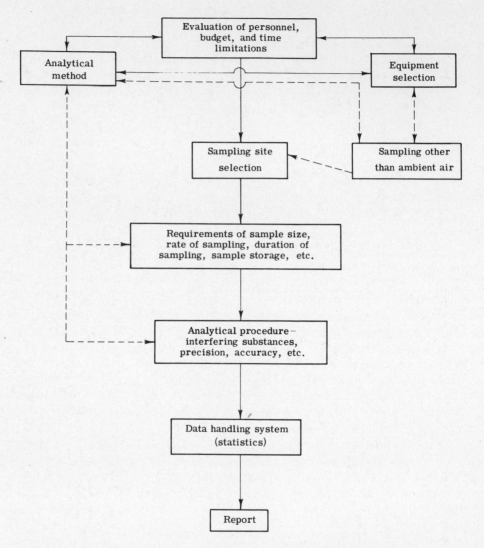

Fig. 14-1. Air pollution sampling sequence.

curacy of sampling and analytical methods. Once these decisions have been reached, items in the remaining boxes of Fig. 14-1 follow sequentially.

It is not always necessary to sample air for air pollution investigations; in many cases sampling and analysis of other materials are more appropriate. In fluorosis of cattle, vegetation sampling for fluoride content is preferable to air sampling. Olfactory evaluation of obnoxious odors is a more direct approach than instrument sampling for nuisance investigations. In many instances, air sampling results may be correlated with metal corrosion, vegetation markings, and paint discoloration.

II. Sampling System for Air Pollution Concentrations

The sampling system for atmospheric gases and particulate matter is conveniently divided into a determination of either a concentration or a level. Concentration refers to a weight of pollutant per volume of air (micrograms material per cubic meter air), or to the definition of pollutant characteristics (number of particles of a certain size per cubic meter air). Level implies a pollutant measurement on a basis other than concentration. Dustfall may be measured in terms of grams deposited per square meter per month; sulfur dioxide measured by the lead peroxide candle method is reported as milligrams of sulfur trioxide per day per 100 cm² lead peroxide surface.

The determination of a pollutant concentration involves four steps as noted in Fig. 14-2. The inlet system brings air to a collection device where the pollutant is

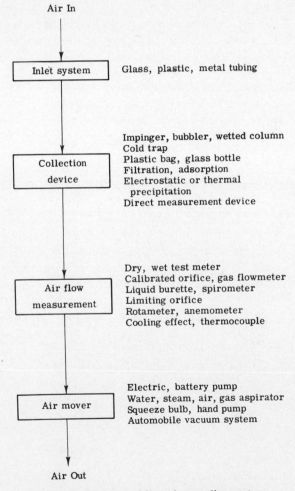

FIG. 14-2. Typical ambient air-sampling system.

either analyzed, concentrated, or fixed (stabilized) for subsequent analysis. The volume of air sampled, corrected to a fixed temperature and pressure, is determined with a flow measurement device while an air mover draws air through the system.

A. Inlet System

The inlet system should be given more attention in air sampling. Reaction or sorption of pollutants with inlet tubing is probable even with clean inlet material; dirt accumulates on the inner surface of the inlet system, additional reaction or sorption of pollutants occur. Reaction and sorption of pollutants with the inlet system are minimized with short inlet tubes.

Glass is the preferred inlet material but, because of breakage, metals or plastics have been substituted. When using plastic or metal tubings, testing must confirm that there is no loss or change in pollutant concentration with flow through the inlet system. The inlet tubing must be cleaned periodically to remove accumulated dirt along with condensed water vapor.

Wohlers *et al.* (1, 2) showed that sorption of sulfur dioxide took place when 100 ft of tygon, polyvinyl chloride, and aluminum were used as inlet material; glass, Teflon, polypropylene, and stainless steel were not affected; the nonreactive gas, carbon monoxide, was not affected by any of the above inlet systems.

With long inlet systems, the time required for a volume of ambient air to reach the sampling device must be considered; a separate high flow system may be used to reduce or minimize this time lag. The volume of the inlet system must also match the sensor response time of continuous monitoring equipment. The time constant of a sensing device corresponds to the time required for a reading to attain 63% of its true value. Peak concentrations over short intervals are markedly reduced with short sampling periods and a large volume inlet or plenum system.

B. Collection Device

The collection device must be tailored to the collection method (3). Collection methods are few in principle; devices are infinite in variety. Methods of collection of gases and particulate matter include: adsorption, absorption, freeze-out, impingement, thermal or electrostatic precipitation, direct measurement, mechanical filtration, and displacement. Any collection method could be adapted for a pollutant depending upon its chemical and physical properties. The collection device may be intermittent, semicontinuous or continuous coupled with a telemetering and computer network. Efficiency of collection varies but normally ranges between 90–100%; two units in series may be used to approach 100% efficiency. Pollutant collection and efficiency are limited only by the imagination and application of sound chemical and engineering principles on the part of the investigator.

Work horses of collection devices are the impinger, the bubbler, or the wetted column; a chemical solution is used to fix or stabilize the pollutant for subsequent analysis with minimum interference by other pollutants. Air flow through the system is limited by the reaction rate of the pollutant with the solution; too high a flow rate carries solution droplets from the system. Some interfering substances must be removed prior to absorption of the pollutant.

Cold traps, plastic bags or glass bottles may be used to obtain a sufficiently large air sample for analysis (4, 5). Cold traps concentrate the volatile pollutant; coolants vary from dry ice–isopropyl alcohol to liquid nitrogen. Samples collected in plastic bags or glass bottles are analyzed directly in the laboratory.

Filtration with paper or glass filters removes particulate matter and gases as well if the filtering medium is chemically treated. Sorption is a convenient method for the collection of gases. Activated carbon is the most common adsorbent although any solid may be used depending on chemical reactivity. A specific reactive chemical is used to absorb pollutants.

Electrostatic or thermal precipitators are used for special sampling of particulate matter. Flow rate for a thermal precipitator is low owing to the time required for particles to move from a high to a low temperature zone ($\Delta T = 750°C$).

Direct measurement devices involving optical methods (in series with the collection device) are used for both gases and particulate matter. Combustion may be used for organic material.

The collection method may be designed to be stationary or mobile for use with an automobile, ship, airplane, or rocket (6–9).

C. Air Flow Measurement

To determine pollutant concentration, it is necessary to measure accurately the air volume sampled. The total gas volume sampled may be determined using dry or wet test meters, liquid burettes, or containers of known volumes such as the spirometer. The volume of air sampled may also be determined using gas-velocity instruments that average or integrate flow over the sampling period. Gas velocity instruments include: calibrated orifices, gas flowmeters, rotameters, and thermal anemometers. Limiting orifices (hypodermic needle may be one type) are used for gas streams with a minimum of particulate matter to control volume rate of flow at constant pressure. Temperature and pressure corrections convert the volume collected to standard conditions.

D. Air Mover

Any compatible air mover may be incorporated into the sampling system. A simple electric pump may be used for flow rates as low as 1 liter per minute to as high as 2000 liters/min; centrifugal, vane, diaphragm and positive displacement units are available. Volume rate of air flow must be constant for accurate results.

For sampling locations remote from a power source, battery-powered pumps may be used. Handpumps or squeeze bulbs have been operated for short time periods when accuracy was not essential. Water, steam, air, or compressed gas aspirators have been used as air movers.

The air mover may be designed for specific sampling programs. The intake manifold of an internal combustion engine may be used as a vacuum source. Windmills and water wheels have been utilized to power air movers.

A final point—the discharged air from the system should not be close to or directly upwind from the inlet.

III. Static Samplers for Air Pollution Levels

In contrast to the concentration sampling system, it is possible to determine a pollutant level with a great saving in manpower and costs using static samplers. Surprisingly, static sampler determinations for sulfur dioxide, ozone, fluoride, and other pollutants have been related to component air concentrations. Although the level sampling system denotes the pollutant trend in an area, results found in one area may not be directly comparable with results in another area; temperature, humidity, sunshine, wind speed, topography, and building configuration for both areas may not be comparable—hence results of static sampling will differ.

Table 14-2 lists a limited number of pollutant level sampling systems. Static systems operate on the principle that when air passes over the sampler surface, the gas is sorbed; no moving parts or electric power are required. The sorbed gas on the surface is fixed by chemical reaction or by strong physical bonds to surface and subsurface molecules to form a stable chemical for subsequent analysis. Simplified pollutant static sampling systems could be devised for other reactive atmospheric gases and particulate matter.

Thomas and Davidson (10) found that lead peroxide candle results were directly related to sulfur dioxide air concentrations from a power plant; a lead peroxide planchet has been developed to reduce exposure time of the candle from a month to a week (11).

A practical sampler for atmospheric ozone (oxidant) is the cracking of rubber strips under tension. Rubber strip measurements of ozone were comparable to air concentrations determined by the colorimetric neutral potassium iodide method (12).

TABLE 14-2

Static Sampling Devices

Pollutant	Device
Fluoride	Lime paper, Spanish moss, glass etching
Hydrogen sulfide	Lead acetate paper or tiles, silver tarnishing, paint discoloration
Miscellaneous gases	Silica gel tubes (high concentrations)
Odors	Nose, tongue
Ozone	Rubber, cracking, fabric dye discoloration, eye irritation
Particulate matter	Dustfall jars, sticky paper, grease plates
Sulfur dioxide	Lead peroxide candle or planchet
Nonspecific and nonair sampling	Metal and stone corrosion, reflectance loss of paint, breaking strength of fabrics, visibility reduction, fungi growth, bioassay (weeds, vegetation, water, food, bone, blood, hair, nails, etc.)

Limed filter papers have been demonstrated to quantitatively compare fluoride levels with air concentrations and uptake by vegetation (13–15).

The lead acetate test for hydrogen sulfide gas was available as early as 1914. Wohlers and Feldstein (16) reported effects of atmospheric hydrogen sulfide on exterior lead base paints. Duckworth (17) standardized the unglazed tile method for hydrogen sulfide.

Particulate matter ($>30\ \mu$) falling into dustfall jars is about the simplest of monitoring tools for particulate matter. Refinements in methodology include sticky papers, grease plates, and roto rods (18, 19).

In recent years, electromagnetic radiations from lasers were developed to determine both particulate and gaseous pollutants. For particulates, the laser radar technique relates the backscattering and attenuation coefficients upon the signal and return (20). Infrared lasers were developed by Hanst (21) for the sampling and analysis of both organic and inorganic gaseous pollutants.

The nose, and at times the tongue, serve as both the sampling device and means of transmitting a signal impulse for recognition to the brain (22). Odor panels statistically improve the data (23, 24).

Silica gel tubes, impregnated with chemicals to react with a pollutant to produce a color change, are simple devices to sample (and analyze) many gases (25). Although detector tubes are available for most gases, the lower range of detection is generally higher than those experienced in ambient air; age of detector tubes affects results.

Many indirect sampling methods are available for relating pollutant concentration to an effect. Mineral materials (metals and stone products), as well as human, animal, and vegetation tissues, have been used for this purpose.

Vegetation is frequently used to measure pollutant effects—by determining apparent photosynthesis (26) or by color photographs using specially sensitized films (27). Fungi and bacteria have been studied in relation to air pollution (28).

Imaginative nonair-sampling methods depend only on the individual. Tansy and Roth (29) related the lead content of pigeon tissues to urban and rural air pollution. Randerson (30) found a 23% loss of solar radiation in air masses over urban as compared to rural areas. Lieberman and Schipma (7) applied monitoring instrumentation developed in aerospace research to general air pollution studies. Snow surfaces were used to collect air pollutants in Finland (31).

IV. Statistical Approach to Air Sampling

The fundamental purpose of any air-sampling program is to obtain representative results in the most economical manner; samples must reflect the "population" evaluated. The applicability of statistics rest upon the possibility of repeated observations made under essentially the same conditions.

In practice, experimental or measurement errors always exist. The investigator must be aware of sources of errors and of their potential contribution to erroneous evaluations. Errors may be systematic or random; the concepts of accuracy and precision are involved.

A. Accuracy and Precision

Accuracy may be defined as the absence of bias. Perfect accuracy reflects a situation where there is zero difference between the population mean of repeated measurements and the actual mean (Fig. 14-3). Bias or systematic errors may include examples such as the following:

1. Two monitoring instruments of the same type may yield different pollutant concentrations when measuring the same air sample, or

2. A given monitoring instrument is used to consistently measure the sulfur dioxide concentration during a period of time when combustion processes, the prime source of SO_2, are not in full operation.

In both of the above cases, the choice of what is the actual mean (reference value) for evaluation may be selected on the basis of experience or convenience.

Precision is related to the lack of precision, i.e., imprecision. Given a controlled procedure for measuring pollutant concentration, imprecision is the random scatter of concentration. An example of a random error is the spread of concentra-

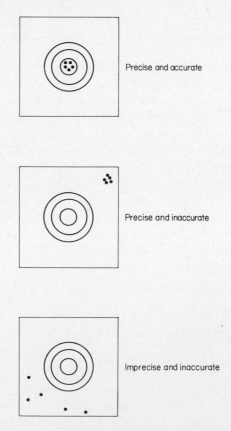

Precise and accurate

Precise and inaccurate

Imprecise and inaccurate

FIG. 14-3. Examples of precision and accuracy in target shooting.

tions attributed to instrument limitations or reading errors when other factors are controlled.

An understanding of accuracy and precision in both a qualitative and quantitative sense is required for the investigator to appreciate uncertainty effects in pollutant measurements.

B. Sampling Parameters

The investigator must make an initial evaluation of (1) instrument adequacy, (2) instrument location, and (3) site evaluation.

Each factor may influence the bias of results by a range of concentrations which may not be representative of the pollutant sampled. The instrument must be calibrated and sufficiently sensitive at anticipated pollutant concentrations. The instrument must be located at a position which will not yield consistently low or high results (i. e., representative) to preclude undesired data stratification; more than one instrument or repeated samplings at different locations could be used to obtain "average" data. Emissions from the sampling site area may vary in cyclical, seasonal, or other manner; measurements must be taken over different time frames to characterize the conditions. Furthermore pollution effects are a function of concentration and time (dosage).

Once the investigator has resolved the instrument adequacy and location, as well as site uniformity, the question of number of required samples must be determined by statistical methods. In general, the larger the number of samples, the greater the probability that the sample mean approaches the true population mean. The number of samples governs the precision of estimate of population for any statistical parameter, i. e., mean, range, standard deviation, variance, or confidence limits.

Consider the following relations (also see Section 25-III)

$$S_n{}^2 = \frac{S_N{}^2}{n} \qquad (14\text{-}1)$$

where $S_n{}^2$ is the variance of distribution of sample means, $S_N{}^2$ is the variance of population, and n is the number of samples. If $n = 1$ the variance of the sampling distribution mean equals the population variance; when $n = 2$ the variance of the sampling distribution mean is $\frac{1}{2}$ of the population variance. Obviously, if n is a very large number, the sampling distribution mean approaches zero. Thus the measurement precision can be controlled by the investigator. The investigator may, however, "trade-off" the cost of a sampling program with a reduced confidence level when less precise results are considered adequate.

C. Concentration Distribution

Shown in Fig. 14-4 is the normal frequency distribution curve where the variable x is plotted against the frequency of occurrence $f(x)$; the arithmetic mean, median and mode all coincide. In air monitoring, x may represent pollutant concentration. In actual sampling, the probability curve is not bell-shaped but skewed, i. e., there

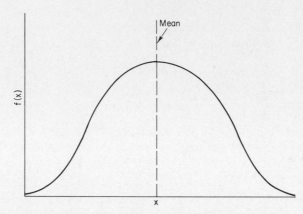

Fig. 14-4. Normal frequency distribution curve.

will be a greater number of either low or high values. In a skewed curve, the arithmetic mean (average), median (number midway between extreme values) and mode (number occurring most frequently) do not coincide.

Frequently, pollutant concentrations vary over an extremely wide range. With a wide range of values, it is cumbersome to plot results on a linear scale; also pollutant concentrations may not follow the familiar "bell-shaped" pattern. Research has shown that the frequency distribution of most air pollutant concentrations is lognormal, i.e., the logarithms of the concentrations are normally distributed (32, 33). Logarithms of concentrations may be used in statistical calculations.

V. Computer and Telemetering Systems

A remarkable achievement of the late 1960's developed computer systems for pollutant sampling, analysis, telemetering, plus data storage and retrieval. Instrument development was emphasized by Bertrand (34) who estimated a total instrument market for ambient level monitoring, stationary source emission measurement and auto exhaust measurement of nearly one-half billion dollars in the 1970's.

The air pollution monitoring system includes both sampling and analysis apparatus in a single automated instrument; the total system can include instrumentation to measure wind, temperature, turbulence, humidity, and solar radiation. All parameters that affect pollutant concentrations should be included in a complete system. Hamburg (35) estimated that a primary automated system should be based upon population—3 stations for a city under 100,000 to 11–20 for greater than 1,000,000 population; to optimize the monitoring system, a trade-off of factors must consider need, cost, and capability. A properly planned monitoring system can be built over a number of years to reduce the yearly capital outlays; remote links with high speed time sharing computers has been suggested for small organizations (36).

The automated monitoring system is but one part of a total integrated data system to support the organization in air resources management. The system should be closely interphased with data systems for emission inventory, registrations and permits, violations and complaints, fuel use, emission reduction plans, land use demographic projections, and urban planning.

References

1. Wohlers, H. C., Trieff, N. M., Newstein, H., and Stevens, W., *Atm. Environ.* **1,** 121–130 (1967).
2. Wohlers, H. C., Newstein, H., and Daunis, D., *J. Air Pollut. Contr. Ass.* **17,** 753–756 (1967).
3. Hendrickson, E. R., Air sampling and quantity measurement, *in* "Air Pollution" (A. C. Stern, ed.), Vol. II, pp. 3–52. Academic Press, New York, 1968.
4. Conner, W. D., and Nader, J. S., *Amer. Ind. Hyg. Ass. J.* **25,** 291–297 (1964).
5. Van Houten, R., and Lee, G., *Amer. Ind. Hyg. J.* pp. 465–469, Sept.–Oct. 1969.
6. Webb, J. C., Kinchen, J. C., and Scarberry, J. E., *J. Air Pollut. Contr. Ass.* **20,** 453–455 (1970).
7. Liberman, A., and Schipma, P., Air Pollution Monitoring Instrumentation, Office of Technology Utilization, Nat., Aeronautics and Space Adm., Washington, D.C., 1969.
8. Randerson, D., *J. Air Pollut. Contr. Ass.* **18,** 249–253 (1968).
9. McCaldin, R. O., and Johnson, L. W., *J. Air Pollut. Contr. Ass.* **19,** 405–409, (1969).
10. Thomas, F. W., and Davidson, C. M., *J. Air Pollut. Contr. Ass.* **11,** 24–27 (1961).
11. Huey, N. A., *J. Air Pollut. Contr. Ass.* **18,** 610–611 (1968).
12. Cherniack, I., and Bryan, R. J., *J. Air Pollut. Contr. Ass.* **15,** 351–354 (1965).
13. Adams, D. F., *Int. J. Air Water Pollut.* **4,** 247–255 (1961).
14. Robinson, E., *Amer. Ind. Hyg. Ass. Quart.* **18,** 145–148 (1957).
15. Miller, V. L., Allmendinger, D. F., Johnson, F., and Polley, D., *Agr. Food Chem.* **1,** 526–529 (1953).
16. Wohlers, H. C., and Feldstein, M., *J. Air Pollut. Contr. Ass.* **16,** 19–21 (1966).
17. Duckworth, F. S., Hydrogen Sulfide Survey, Environmental Service Bulletin No. 68-3, Metronics Assoc., Inc., Palo Alto, California, 1968.
18. Gruber, C. W., Pritchard, W. L., and Schumann, C. E., *J. Air Pollut. Contr. Ass.* **20,** 161–163 (1970).
19. Lapple, C., *Chem. Eng.* **75,** 149–156 (1968).
20. Johnson, W. B., *J. Air Pollut. Contr. Ass.* **19,** 176–180 (1969).
21. Hanst, P. L., *Appl. Spectrosc.* **24,** 161–174 (1970).
22. Wohlers, H. C., *J. Air Pollut. Contr. Ass.* **17,** 609–613 (1967).
23. Wilby, F. V., *J. Air Pollut. Contr. Ass.* **19,** 96–100 (1969).
24. Benforado, D. M., Rotella, W. J., and Horton, D. L., *J. Air Pollut. Contr. Ass.* **19,** 101–105 (1969).
25. Saltzman, B. E., *Amer. Ind. Hyg. Ass. J.* **23,** 112 (1962).
26. Taylor, O. C., Cardiff, E. A., and Mesereau, J. D., *J. Air Pollut. Contr. Ass.* **15,** 171–173 (1965).
27. Berry, P., and Markussen, K., Airborne Surveillance of Air Pollution by Spectral Photographs, Presented at Mid-Atlantic States Section APCA, Albany, N.Y., October 1968.
28. Serat, W. F., Kyono, J., and Mueller, P. K., *Atm. Environ.* **3,** 303–309 (1969).
29. Tansy, M. F., and Roth, R. P., *J. Air Pollut. Contr. Ass.* **20,** 307–309 (1970).
30. Randerson, D., *J. Air Pollut. Contr. Ass.* **20,** 546–548 (1970).
31. Laamanen, A., *Work-Environ.-Health* **5,** 42–50 (1968).
32. Zimmer, C., and Larsen, R., *J. Air Pollut. Contr. Ass.* **15,** 565–572 (1965).
33. Hunt, W. F., Jr., *J. Air Pollut. Contr. Ass.* **22,** 687–691 (1972).
34. Bertrand, R. R., *J. Air Pollut. Contr. Ass.* **20,** 801–803 (1970).
35. Hamburg, F. C., *J. Air Pollut. Contr. Ass.* **21,** 609–613 (1971).
36. Dittrich, W., Brown, W., Hallman, G., and Nelson, E. *J. Air Pollut. Contr. Ass.* **21,** 555–558 (1971).

Suggested Reading

Adams, D. F., and Koppe, R. K., Instrumenting light aircraft for air pollution research, *J. Air Pollut. Contr. Ass.* **19,** 410–415 (1969).

"Air Pollution Manual, Part I Evaluation," 2nd ed. American Industrial Hygiene Association, 210 Haddon Avenue, Westmont, New Jersey, 1972.

"Air Sampling Instruments," 4th ed. American Conference of Governmental Industrial Hygienists, P.O. Box 1937, Cincinnati, Ohio, 1972.

Bradley, C. E., and Haagen-Smit, A. J., *Rubber Chem. Technol.* **24,** 750–755 (1951).

First, M. W., *Environ. Res.* **2,** 88–92 (1969).

Hays, W. L., "Statistics." Holt, New York, 1963.

Heck, W. W., and Heagle, A. S., *J. Air Pollut. Contr. Ass.* **20,** 97–99 (1970).

Hermann, E. R., Thresholds in biophysical systems. Presented at Joint Symposium on Thresholds, American Academy of Industrial Hygiene and American Academy of Occupational Medicine, Cincinnati, Ohio, Feb. 1970.

Jacobs, M. B., "The Chemical Analysis of Air Pollutants." Wiley (Interscience), New York, 1960.

Keitz, E. L., Determining data requirements and primary performance characteristics for air quality monitoring networks. 63rd Annual Meeting, Air Pollution Control Association, St. Louis, Missouri, June 1970.

Kitagawa, T., The rapid measurements of toxic gases and vapors. The 13th International Congress on Occupational Health, New York, July 1960.

Leonardos, G., Kendall, D., and Bernard, N., *J. Air Pollut. Contr. Ass.* **19,** 91–95 (1969).

Saltzman, B. E., *Environ. Sci. Technol.* **2,** 22–32 (1968).

Stevens, R. K., and O'Keeffe, A. E., *Anal. Chem.* **42,** 143A–149A (1970).

Wohlers, H. C., Static samplers for gaseous pollutants. Proceedings Mid-Atlantic States Section, Air Pollution Control Association, Technical Conference, Albany, New York, Oct. 10, 1968.

Questions

1. Why is ambient air sampling of pollutants and subsequent analysis important in air pollution?
2. Elaborate upon five of the objectives of ambient air pollution sampling shown in Table 14-1.
3. (a) How would one determine the volume of air sampled when using either the plastic bag or glass bottle as collection devices?
   (b) Why are the flowrates markedly different when using the electrostatic or thermal precipitators?
4. (a) If the efficiency of a single bubbler as the collection device for an air pollutant was experimentally determined to be 90%, would a second bubbler in series increase the efficiency to 100%?
   (b) How would one determine the collection efficiency of a sampling system experimentally?
5. (a) What are static samplers?
   (b) What are the advantages and disadvantages of static samplers for air pollution studies?
   (c) What steps would you take to devise static samplers for CO_2, aldehydes, or pollens?
6. Define mean, median, and mode.
7. What function does an automated monitoring system have in a total integrated data system of a governmental agency?

Chapter 15

AIR POLLUTANT ANALYSIS

Once an air pollutant has been collected or "fixed" by sampling, the analyst performs his task. At first glance the task is simple—having sampled a volume of air, analyze its contents. Air pollutant analysis is complicated by two factors—low concentrations and interfering substances. In analysis of pure sodium chloride, the chemist would find 39.34% sodium; in air the sodium content approximates $1 \times 10^{-8}\%$. In air there are traces of every chemical or substance known to man; while sampling for a specific pollutant, unwanted and analytically undesirable materials are commingled.

The low pollutant concentration is overcome by sampling a sufficiently large volume of air. If the sensitivity of the analytical method is 10^{-6}g, air must be sampled until at least that weight of pollutant is "fixed." Each analytical method has an optimum concentration spread or range for precise and/or accurate results; the sample used by the analyst must be in this optimum range.

The choice of an analytical method is also dependent upon type and concentration of interfering substances present. At times it is possible to remove interfering compounds prior to final analysis. Alternatively, the method may be standardized in the presence of the interference. Or as a last resort it may be necessary to accept reduced accuracy. The degree to which data are inaccurate must be given full consideration with respect to needs of the results. A partial list of physical factors that must be considered prior to selection of an analytical method is shown in Table 15-1 (1).

TABLE 15-1

Factors Affecting Choice of Analytical Method for Air Pollutants

Stability of reagents and reagent products
Speed of chemical reactions
Type and amount of interfering substances
Temperature coefficient
Ease of calibration procedure
Degree of simplicity, specificity, sensitivity, precision, and
    accuracy

The method choice must also be based upon the capability of the laboratory in terms of budget, personnel, and equipment. Collaborative testing with governmental or industrial laboratories is suggested as a means to reconcile method accuracy and laboratory capability. Saltzman (2) tabulated data of 80 laboratories evaluating fourteen analytical methods and found a wide range of results; the suggestion was made that 5–10% of a laboratory budget should be allocated to standardization activities. Countries in Europe cooperate in the testing of analytical procedures to a greater extent than laboratories in the United States.

The number of analytical methods and instruments available for air pollutants is in the thousands. Specific details will be found in professional journals; new and improved analytical methods continually appear in the open literature, including excellent articles in many books (1, 3–8).

I. Analysis of Gaseous Pollutants

A. Wet Chemical Analysis

Wet chemical analysis, the standard procedure for classical chemistry, is a prime method for air pollutant analysis. Despite inroads of continuous monitoring instruments, wet chemical methods of analysis are still used for routine analysis of pollutants as well as instrument checks and calibrations.

In strong solutions of sulfate ions, an aliquot is acidified with hydrochloric acid and the SO_4^{2-} precipitated as barium sulfate by addition of $Ba(NO_3)_2$; the $BaSO_4$ is dried and weighed. In air pollution, this procedure could be followed but the resulting precipitate would be too small to weigh on conventional balances. In air analysis, the precipitate weight is in microgram or nanogram range in contrast to milligrams. The resultant $BaSO_4$ from an air sample, however, can be accurately determined by turbidimetric or nephelometric methods. Thus, classical methods of chemical analysis must be modified for most air pollution studies. A review of analytical methods for the important pollutant sulfur dioxide, will be used for illustration.

1. Sulfur Dioxide

The diversity of wet type methods for sulfur dioxide is noted in Table 15-2. Each method used for the analysis of sulfur dioxide must be compatible with the sampling method. The sensitivity of methods vary but are in the low range of 13–525 $\mu g/m^3$, SO_2 (0.005–0.2 ppm) found in ambient air; interfering substances vary with the method.

The West-Gaeke method is most specific for sulfur dioxide. Those methods which involve an oxidizing agent are least specific in that other oxidizable material in the air will be determined as sulfur dioxide. The formation of H_2SO_4 in the peroxide method and its determination by electroconductivity, is interfered by both gaseous and particulate matter converted to electrolytes in the absorbing solution; further, a change of temperature of the absorbing solution of 1°C will alter the conductivity by about 1%.

Particulates in the air which may interfere with gaseous analysis are normally removed by a dry filter (Millipore), a dry impactor or midget impinger placed upstream. Some SO_2 may be sorbed on the filter or trapped particulate matter.

For properly conducted monitoring and calibrated analytical procedures, the total error involved is about $\pm 10\%$ below 260 $\mu g/m^3$; accuracy increases up to 2620 $\mu g/m^3$ (1 ppm) SO_2.

In each method listed in Table 15-2 equations can be written to follow the course of the reaction. For the peroxide method:

$$SO_2 + H_2O_2 \rightarrow H_2SO_4 \tag{15-1}$$

the resulting sulfuric acid may either be titrated with sodium hydroxide

$$H_2SO_4 + 2\ NaOH \rightarrow Na_2SO_4 + 2\ H_2O \tag{15-2}$$

or determined by measuring resistance across fixed plate electrodes

$$H_2SO_4 + 2\ H_2O \rightarrow 2\ H_3O^+ + SO_4^{2-} \tag{15-3}$$

In the iodine–thiosulfate method, the sulfur dioxide reacts with iodine

$$SO_2 + I_2 + 2\ H_2O \rightarrow H_2SO_4 + 2\ HI \tag{15-4}$$

TABLE 15-2

Brief Summary of Wet Type Methods for Ambient Sulfur Dioxide

Method	Outline of procedure
Hydrogen peroxide	SO_2 in water is oxidized with H_2O_2 forming H_2SO_4, which is determined conductimetrically
Iodine	SO_2 absorbed in NaOH, resulting Na_2SO_3 titrated with I_2
Fuchsin	SO_2 absorbed in NaOH and color developed with fuchsin–formaldehyde
Iodine–thiosulfate	SO_2 absorbed in I_2–KI solution and titrated with $Na_2S_2O_3$
Disulfitomercurate (West-Gaeke)	SO_2 absorbed in $Na_2Hg\ Cl_4$ and color developed with p-rosaniline hydrochloride
Barium sulfate	SO_2 absorbed in acidified H_2O_2 and sulfate precipitated as $BaSO_4$, determined turbidimetrically

TABLE 15-3

Wet Analytical Methods for Selected Air Pollutants

Pollutant	Method and procedure
Chlorine	o-Tolidine—Yellow color in reaction of Cl_2 with o-tolidine
	Iodine–thiosulfate—Liberation of iodine from KI by chlorine
Chloride	Fajan—Titration with $AgNO_3$; dichlorofluorescein dye turns red with excess $AgNO_3$
	Volhard—Excess $AgNO_3$ added; back titrate with KCNS using Fe^{3+} as indicator
	Turbidimetric—Precipitation of AgCl with visual or spectrophotometric turbidity determination
Hydrogen sulfide	Cadmium sulfide—Precipitation of CdS in ammoniacal solution and titrate precipitate with I_2
	Methylene blue—Blue color formed with p-aminodimethyl aniline, Fe^{3+} and Cl^-; optical density determined
	Lead acetate—S^{2-} reacts with acetate solution to form PbS, which is evaluated turbidimetrically
Nitrogen dioxide	Griess-Ilosvay—NO oxidized with $KMnO_4$; NO_2 reacted with sulfanilic acid, alphanaphthylamine and acetic acid
	Saltzman—NO oxidized with $KMnO_4$; NO_2 reacted with sulfanilic acid; N-(1-naphthyl)ethylenediamine and acetic acid
	Phenoldisulfonic acid—NO_2 forms a nitrate of phenoldisulfonic acid (yellow)
Ozone (oxidant)	Alkaline iodide—O_3 absorbed in alkaline KI; I_2 liberated determined by titration or spectrophotometrically
	Neutral iodide—O_3 absorbed in neutrally buffered KI—otherwise as above or coulometrically
	Phenolphthalein—O_3 oxidizes phenolphthalein; red-purple color evaluated.
Carbon dioxide	Titrimetric—CO_2 absorbed in $Ba(OH)_2$; excess $Ba(OH)_2$ titrated with oxalic acid.
	Gravimetric—CO_2 absorbed on Ascarite (NaOH on asbestos) and weighed
Fluoride	Willard–Winter—Material converted to soluble fluorides; colored with zirconyl chloride–acid solution
	Fluorometric—Material converted to soluble fluorides; fluorescence with superchrome garnet Y or eriochrome red B
	Ion exchange—Material converted to soluble fluorides; ion exchange resin used to separate; choice of analysis
Lead	Dithizone—Diphenylthiocarbazone reacts with lead to form a brick-red complex salt

TABLE 15-3—Continued

Pollutant	Method and procedure
Aldehydes	Chromate—Lead reacts with K_2CrO_4 to form $PbCrO_4$; determined turbidimetrically MBTH—Aliphatic aldehydes react with 3-methyl-2-benzothiazolone hydrazone to form a blue dye Chromotropic acid—HCHO reacts with chromotropic acid to form a violet color Schiff-Elvove—Aldehydes react with fuchsin–sulfite reagent to form a yellow to purplish-red color

and the iodine remaining is titrated with thiosulfate

$$I_2 + 2\ Na_2S_2O_3 \rightarrow Na_2S_4O_6 + 2\ NaI \tag{15-5}$$

The West-Gaeke method involves the reaction of sulfur dioxide with the tetrachloride ion

$$Hg\ Cl_4^{2-} + 2\ SO_2 + 2\ H_2O \rightarrow [Hg(SO_3)_2]^{2-} + 4\ H^+ + 4\ Cl^- \tag{15-6}$$

the stable, nonvolatile disulfitomercurate ion forms a red-violet color when mixed with p-rosaniline hydrochloride formaldehyde mixture.

The concentration of sulfur dioxide is based upon stoichiometric relationships, and volume of air sampled corrected to a standard temperature and pressure.

2. Other Gases

Analytical methods have been developed for all gases of concern in air pollution. A description of wet methods for selected pollutants is shown in Table 15-3. Comparable analytical difficulties, as mentioned for sulfur dioxide, may be anticipated for all methods.

The analyst must carefully review the literature to select the optimum method for his objective. Each new method must be tested and calibrated in the laboratory before field monitoring is initiated.

3. Continuous Instruments

Automated instruments are available based upon almost all of the methods listed in Tables 15-2 and 15-3; indeed, SO_2-analysis instruments based upon other methods are available. Performance characteristics of instrumental methods for SO_2 analysis were reviewed by Rodes and Nelson (9); significant differences were detected with a correlation coefficient range of 0.40–0.96.

B. Physical Methods of Analysis

Inorganic and organic gases as well as particulate matter may be analyzed by spectrographic or spectroscopic methods. Two prime advantages of spectrographic analysis are the speed of analysis and, in most cases, the small air volume necessary for analysis. A portion of these methods of analysis is presented in Table 15-4.

TABLE 15-4

Physical Methods of Analysis

Method	Remarks
Flame ionization analyzer	Responds in proportion to number of carbon atoms in gas sample
Mass spectrometry	Determines charge–mass ratio of organic fragments or ions
Infrared absorption	Sample absorbs radiation in infrared region of spectrum; differences in absorption measured. Other regions of spectrum used, e. g., uv
Emission spectroscopy	Excited solid provides spectra of metals present
Flame photometry	Flame excites samples with low excitation potentials
Atomic absorption	Sample absorbs radiation; emitted radiation is a function of atoms present
Fluorescence spectroscopy	Excited sample may re-emit excess of excited energy.

In flame ionization the gaseous air sample is burned in a small hydrogen flame; ions or electrons are formed which are proportional to the number of carbon atoms present and are counted electrically. The flame detector is insensitive to water and formic acid; electronegative gases (O_2, SO_2, Cl_2) reduce the flow of ions produced. The gaseous or vaporized sample in a carrier gas may be physically separated in a packed or surface treated small diameter tubing (column). Emerging gases are analyzed with, for example, a flame ionization detector or an infrared analyzer. Sensitivities in the parts per billion range are feasible.

The heart of a chromatograph is the column maintained at a specific temperature. Packing materials consist of diatomaceous earth usually coated with liquids such as paraffin oil, silicone greases, or polyglycols. In the column, gaseous components of the air sample are sorbed and selectively removed as the stream moves through the column with a carrier gas. The gases emerge in reverse order of their retentivity and are measured by a detector. The time required for detection is related to the specific compound; as noted in Fig. 15-1, the area under the curve is the component concentration.

Ions, consisting of fragments of components of the air sample, are produced in the mass spectrometer by electron bombardment; magnetically, the fragments are separated and recorded according to their mass/charge ratio (m/e). Current density, as measured by a detector, is proportional to the number of particles in each relevant class. Monopole, quadrupole, and time of flight instruments are available for m/e in excess of 1000 for particulates as well as gases. In analysis of complex mixtures, definition of fragment m/e ratios is not sufficient to elucidate all compounds in the sample.

Many organic and inorganic compounds absorb energy in the infrared portion ($2-15 \mu$) of the spectrum. This absorption is characteristic of the compound and is related to vibration modes within the molecule. Nondispersive (one wavelength) analyzers may be used for CO; commercial dispersive infrared spectrometers usually cover the region from 0.8 to 50 μ. Other portions of the spectrum (visible and/or ultraviolet) may be used for identification purposes.

Emission spectrometers are a convenient means of analysis of metallic elements in trace concentrations in a few milligrams of sample. The sample to be analyzed is thermally excited in a flame or arc; the spectra produced (continuous, band, or line spectra) are characteristic of metals and some nonmetals (P, Si, As, C, B).

Air samples analyzed by flame photometry must be in solution and sprayed under controlled conditions into a flame. Radiation from the flame is isolated and determined by a convenient photosensitive device. The method is usually used for metal analysis; relatively large amounts of anions reduce the spectral emission of elements. Atomic absorption is based on the measurement of light absorbed at the wavelength of a resonance line by the unexcited atoms of an element; transition

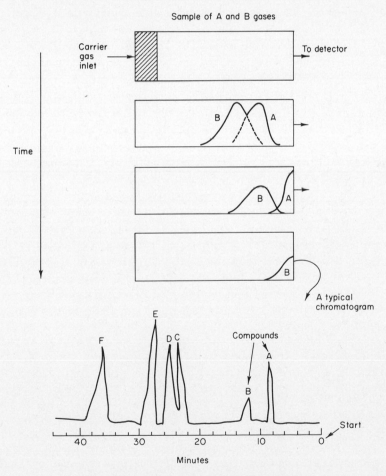

FIG. 15-1. Elution in gas chromatography.

from the ground state to a higher energy level is a measure of the number of atoms in the flame.

Fluorescence analyzes the way in which a molecule gets rid of absorbed radiation in compounds where there is a measurable interval between absorption and reemission (10^{-4}–10^{-8} seconds). Emission intensity of fluorescence at a given wavelength may be used for quantitative analysis. Fluorescence may be expected in molecules with multiple-conjugated double bonds containing at least one electron donating group ($-NH_2$ or $-OH$).

C. Thermal Analysis

Various thermal properties of air pollutants are used for gas analysis. Most instruments are based either on thermal conductivity or heat of combustion of gases.

The thermal conductivity of a gas is the quantity of heat transferred in unit time between two separated surfaces. The thermal conductivity method is usually used for a single gas but a gaseous mixture may be analyzed if one gas has a significantly different thermal conductivity than the remaining gases (He, CH_4, or CO_2—all in air). The relationship between H_2, CO, and CO_2 in automobile exhaust is measured by thermal conductivity methods. Thermal conductivity is not used for analysis of unknown mixtures of gases.

The heat of combustion evolved by a gas when it burns on a filament may be used for quantitative detection of organic air pollutants. The method is specific for a single combustible gas; as noted for thermal conductivity, the combustion method is also nonspecific. Thermal combustion cells are used to determine explosive mixtures of gases.

II. Analysis of Particulate Pollutants

The analysis of particulate pollutants must be geared to the sampling method. If the objective is to determine the weight of particulate matter in the atmosphere, filtration suffices; chemical analysis of collected particulate matter is possible. If it is desired to ascertain particle size or other physical characteristics, the inertial sampler, electrostatic or thermal collector, greased slides, and photometric methods should be used. Special experimental sampling and analytical methods may be devised for specific problems.

A. Filtration, Inertial, Electrostatic, and Thermal Approaches

Filtration, inertial, electrostatic, and thermal approaches should be considered as sampling methods as well as the start of analytical schemes. Once the particulates have been separated from an air volume, analysis begins.

The simplest analytical method for particulate matter is drying at constant temperature and weighing the material; concentration is reported as micrograms per cubic meter of air. Suspended particulate matter measured throughout the world is compared on this basis (high volume samplers). For more definitive analysis of air pollution, the particulate material is further analyzed for organic and inorganic components. A comparison may also be made between those particulates which are water-soluble and those which are benzene-soluble. The organic fraction may be analyzed for carcinogenic compounds; water soluble fraction may be analyzed for pollutants such as sulfates, nitrates, or chlorides.

Rather than weighing filtered particulate matter, light reflected from the surface or transmitted through the material may be determined. Either measurement may be related to the soiling character of the sampled air. Light-colored particles, which soil dark surfaces, are difficult to detect by this method. Identification of filtered solids may be undertaken using special microscope techniques.

Inertial devices are used to separate particles according to size ($0.5-50\ \mu$). The devices operate on the principle that large particles move in the original direction when an airstream direction is changed and impinge on prepared surfaces; if a slit or jet through which the air passes is made progressively narrower, progressively smaller particles are retained. Particles caught at various stages are counted and sized microscopically.

Electrostatic and thermal methods are used for special sampling for small quantities of dusts. With these methods particles may be examined microscopically or imbedded on an electron microscope grid for identification purposes.

B. Photometric Analysis

One disadvantage of particle-size analysis by filtration or impingement is that the dust may agglomerate or disintegrate as a result of monitoring. Photometric analysis allows measurements to be made with an absolute minimum of disturbance, i. e., *in situ* analysis.

Photometric analysis is based upon the fact that when a particle is illuminated with a beam of light, the amount of light scattered by the particle is roughly proportional to the projected area. Particles in the size range of $0.3-10\ \mu$ may be determined individually (single pulse of light scatter) or in aggregate (light scattered by all particles). Forward scattering by submicron particles is 1000-fold greater than back scattering.

Particle counting and sizing of atmospheric aerosols by photometric methods are complex and absolute interpretation of measurements is difficult; the scattered light varies in a complicated manner with the system of optics as well as with the size and physical characteristics of particles.

III. Experimental Methods

As technology progresses, improved analytical methods for pollutants become available for general or specific sampling purposes. Recent analytical methods generally have better sensitivity and specificity, thus advancing knowledge concerning air pollution and its effects.

Acoustical analytical techniques have been developed. It was determined that particles produced pressure pulses (audible clicks) when accelerated and suddenly projected into a hexacavity (10). Hofmann and Mohnen (11) described the operation of an acoustical particle counter. Single-electron charging of submicron particles, combined with current measurements, permit determination of total quantity of submicron particles (12).

Griggs *et al.* (13) described the use of a high resolution infrared spectrometer to measure polluted atmospheres. Pollutants were detected from presence of pollutant emission lines which appear at characteristic wavelengths; the analytical system may be used during day and night. Remote sensing correlation spectrometers are available for air analysis of SO_2 and NO_2 (14). Hanst (15) reviewed infrared spectroscopy and infrared lasers in air pollution research and monitoring.

Pretorius (16) used neutron activation analysis for air pollutants; sensitivity was in the nanogram range. Neutron activation analysis is an elemental analysis in which the nuclides in the sample are made radioactive for measurement with newly developed lithium-drifted germanium detectors. Brewers and Flack (17) described a quantitative determination of fluorine by means of proton bombardment. General nuclear methods in air and water pollution analysis were reviewed by Iddings (18).

IV. Radioactivity Analysis

Presently the world is primarily concerned with pollution from industry, automobiles, and power plants. As supplies of carbonaceous fuels dwindle, atomic plants are constructed to supply continued demand for electric power. Atomic powered ships and submarines exist; atomic powered automobiles, airplanes, and rockets are forecast. Radioactivity is expected to be the major pollutant of concern to future world generations.

Radioactivity in gases and particulate matter results from spontaneous disintegration of the nucleus. Energy is given off in the form of gamma rays (electromagnetic radiation), beta particles (electrons), and alpha particles (helium nuclei). Radioactive material may be present in the atmosphere as solids, gases, or liquids; sources of radioactivity are both man-made and natural.

Monitoring equipment for radioactivity measurements is identical to that previously described except for two major differences. The first is the health hazard when high ionization levels are sampled and analyzed; remote instrumentation or proper shielding eliminate this danger. The second is concerned with the life expectancy (half-life) of the sampled material. If the half-life is of the order of minutes or seconds, rapid sampling and analysis are mandatory.

Alpha and beta particles have large ionization capabilities. The ion current generated as the particles move in space is measured in a confined space under controlled conditions in an ionization chamber.

The most versatile instrument for radioactivity analysis is the Geiger-Müller detector—a modification of the ionization chamber. The G–M counter is nondiscriminating with respect to type of activity and hence determines gross activity. In the G–M counter a high potential field is produced to cause disintegrated electrons to produce other electrons in a selected gaseous medium, causing an avalanche of electrons and a high voltage discharge.

The most quantitative instrument is the scintillation detector as it is able to quantify radioactivity according to particle energy. Scintillation phosphors produce light flashes when radioactive particles hit or pass through the surface; a photomultiplier tube analyzes the number of flashes as well as the flash intensity.

V. Continuous Air-Monitoring Systems

Continuous sampling and analytical data systems are available for routine and research operations. A general flow chart for such a system is shown in Fig. 15-2. For each pollutant a continuous sampler and analyzer feeds results, usually as a voltage, to both a strip chart and tape on which the information may be translated

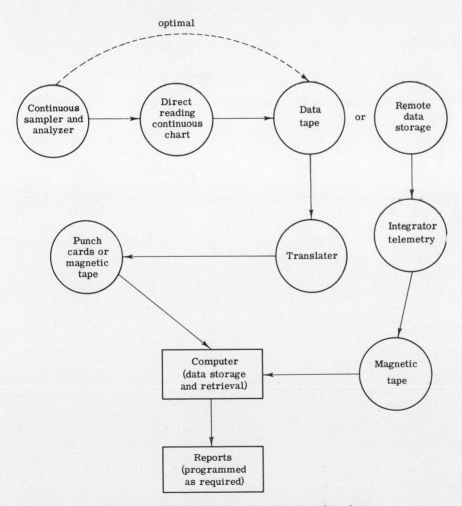

FIG. 15-2. Continuous air-monitoring data flow chart.

for computer processing. In some systems, the tape is replaced by an electronic accumulator and averager which transmits a voltage signal to the computer on demand; in other systems analog signals are telemetered to a second set of recorders at a central office. Table 15-5 lists a selection of instruments which may be incorporated into a complete air quality monitoring system (19–20). Limits of detection for commercially available instruments are presented in Table 15-6 (21).

TABLE 15-5

Continuous Automatic Gaseous Pollutant Analyzers

Method	Reaction	Comment
Sulfur dioxide		
Electrolytic conductivity (pure water)	$SO_2 + H_2O = H_2SO_3$	Good sensitivity, possible CO_2 interference, not specific
Electrolytic conductivity (H_2O_2, and H_2SO_4)	$SO_2 + H_2O_2 = H_2SO_4$	Good sensitivity, not specific, most commonly used method, most experience
Amperometric	$SO_2 + I_2 + 2\ H_2O =$ $2\ I^- + SO_4^{2-} + 4\ H^+$	Not specific
Coulometric	$SO_2 + I_2 + 2\ H_2O =$ $2\ I^- + SO_4^{2-} + 4\ H^+$ $SO_2 + Br_2 + 2\ H_2O =$ $2\ Br^- + SO_4^{2-} + 4\ H^+$	Not specific
Colorimetric	West and Gaeke	Slow response, specific
Ultraviolet	Correlation spectrometry Flame emission	Used for long path measurement; measures all sulfur gases
Oxidants		
Coulometric	$O_3 + 2\ KI + H_2O =$ $I_2 + O_2 + 2\ KOH$	Not specific, NO_2 and SO_2 interference
Colorimetric	Liberation of I_2	Not specific
Amperometric	Liberation of I_2	Not specific
Nitrogen oxides		
Colorimetric	Griess-Saltzman	Slow response, not specific, calibration problems
Coulometric	$NO_2 + Br_2 =$ $2\ Br^- + NO_2^- + H^+$	Interferences: O_3, free halogens, SO_2
Ultraviolet	Correlation spectrometry	Specific
Carbon monoxide		
Infrared	Nondispersive absorption	Specific; accuracy marginal
Gas chromatography	Retention time	Sensitivity (batch process)
Hydrocarbons		
Flame ionization	Combustion	Empirical; not specific
Gas chromatography	Retention time	Specific; information on concentration step required

TABLE 15-6

Minimum Limits of Detection (ir Parts per Million) of Selected Gaseous Pollutants with State-of-the-Art, Commercially Available Instruments

Method	NO_2	SO_2	O_3	CO	Hydro-carbons
Wet chemistry	0.01	0.01	0.02		
Chemiluminescence	0.005	0.005	0.002	100	
Electrochemical transducers	0.05	0.05		20	
Gas chromatography	10	0.005		0.1	0.02
Nondispersive infrared				0.5	
Dispersive infrared[a]	0.025	0.025	0.1	0.1	
Dispersive ultraviolet[b]	0.05	0.003	0.17		

[a] Minimum concentration to give 1% absorption in a 100-meter cell.
[b] Minimum concentration to give 1% absorption in a 20-meter cell. [Source: James Hodgeson, Environmental Protection Agency].

Various urban and statewide areas have described their monitoring systems:

Allegheny County—(22)
New York City—(23)
Washington State—(24)
St. Louis—(25)
Chicago—(26)
New York State—(27)
New Jersey—(28)

The Environmental Protection Agency has integrated aerometric data into a storage and retrieval system. A discussion of the coding structure (SAROAD) along with examples of input and output formats was presented by Nehls *et al.* (29).

References

1. Katz, M., "Measurement of Air Pollutants." World Health Organization, Geneva, 1969.
2. Saltzman, B. E., *J. Air Pollut. Contr. Ass.* **18,** 326–328 (1968).
3. Katz, M., Analysis of inorganic gaseous pollutants, *in* "Air Pollution" (A. C. Stern, ed.), Vol. II, pp. 53–114. Academic Press, New York, 1968.
4 Alshuller, A. P., Analysis of organic gaseous pollutants, *in* "Air Pollution" (A. C. Stern, ed.), Vol. II, pp. 115–145. Academic Press, New York, 1968.
5. West, P. W., Chemical analysis of inorganic particulate pollutants, *in* "Air Pollution" (A. C. Stern, ed.), Vol. II, pp. 147–185. Academic Press, New York, 1968.
6. Hoffman, D., and Wynder, E. L., Chemical analysis and carcinogenic bioassays of organic particulate pollutants, *in* "Air Pollution" (A. C. Stern, ed.), Vol. II, pp. 187–245. Academic Press, New York, 1968.
7. McCrone W. C., Morphological analysis of particulate pollutants, *in* "Air Pollution" (A. C. Stern, ed.), Vol. II, pp. 281–301. Academic Press, New York, 1968.

8. Leithe, W., "The Analysis of Air Pollutants." Ann Arbor Science Publishers, Ann Arbor, Michigan 48106 (1971).

9. Rodes, C. E., and Nelson, C. J., *J. Air Pollut. Contr. Ass.* **19**, 778–786 (1969).

10. Langer, G., *J. Colloid. Sci.* **20**, 602–609 (1965).

11. Hofmann, K. P., and Mohnen, V., *Staub* **28**, 15–19 (1968).

12. Clark, W. E. and Whitby, K. T., *Tellus* **18**, 573–586 (1966).

13. Griggs, M., Ludwig, C. B., Bartle, E. R., and Abeyta, C., Passive infrared remote sensing of air pollution, Presented at Annual Meeting, *Air Pollut. Contr. Ass.*, St. Louis, Missouri, June 1970.

14. Barringer Research Ltd., Barringer Remote Sensing Correlation Spectrometer. 304 Carlingview Drive, Rexdale, Ontario, Canada, June 1970.

15. Hanst, P. L , *Appl. Spectrosc.* **24**, 161–174 (1970).

16. Pretorius, R., *S. Afr. Med. J.* 169–171, Feb. 1970.

17. Brewers, J. M., and Flack, F. C., *Analyst* **94**, 1–14 (1969).

18. Iddings, F. A., *Environ. Sci. Technol.* **3**, 132–140 (1969).

19. Keitz, E. L., Determining data requirements and primary performance characteristics for air quality monitoring networks, Presented at Annual Meeting, *Air Pollut. Contr. Ass.*, St. Louis, Missouri, June 1970.

20. Hamburg, F. C., *J. Air Pollut. Contr. Ass.* **21**, 609–613 (1971).

21. Maugh, T. H. II., *Science* **177**, 685–687 (August 25, 1972).

22. Strange, W. J., and Popiel, W G., The development of Allegheny County's Bureau of Air Pollution Control on-line computer data processing system, Presented at Annual Meeting, *Air Pollut. Contr. Ass.*, St. Louis, Missouri, June 1970.

23. Auto Met III Data Acquisition System. Packard Bell, Newbury Park, California, Sept. 1968.

24. Dittrich, W., Brown, W., Hallman, G., and Nelson, E., *J. Air Pollut. Contr. Ass.* **21**, 555–558 (1971).

25 Copley, C. M., and Pecsok, D. A., St. Louis regional air monitoring network, Presented at Annual Meeting, *Air Pollut. Contr. Ass.*, St. Louis, Missouri, 1970.

26. Stanley, W. J., *J. Air Pollut. Contr. Ass.* **16**, 100–101 (1966).

27. Hunter, D. C., Continuous air quality monitoring in New York State. Technicon International Congress on Automated Analysis, Chicago, Illinois, June 1969.

28. Yunghaus, R., Continuous air monitoring with the auto analyzer. Technicon Symposium, New York, October 1966.

29. Nehls, G. J., Fair, D. H., and Clements, J. B., *Environ. Sci. Technol.* **4**, 902–905 (1970).

Suggested Reading

"Air Quality Criteria for Carbon Monoxide." National Air Pollution Control Administration, Department of Health, Education, and Welfare, AP-62, Washington, D.C., March 1970.

"Air Quality Criteria for Hydrocarbons." National Air Pollution Control Administration, Department of Health, Education, and Welfare, AP-64, Washington, D.C., March 1970.

"Air Quality Criteria for Nitrogen Oxides." National Air Pollution Control Administration, Department of Health, Education, and Welfare, AP-84, Washington, D.C., Jan. 1971.

"Air Quality Criteria for Particulates." National Air Pollution Control Administration, Department of Health, Education, and Welfare, AP-49, 801 N. Randolph Street, Arlington, Virginia, Feb. 1969.

"Air Quality Criteria for Photochemical Oxidants." National Air Pollution Control Administration, Department of Health, Education, and Welfare, AP-63, Washington, D.C., March 1970.

"Air Quality Criteria for Sulfur Oxides." National Air Pollution Control Administration, Department of Health, Education, and Welfare, AP-50, 801 N. Randolph Street, Arlington, Virginia, Feb. 1969.

"Air Sampling Instruments," 4th ed. Amer. Conf. of Govern. Indust. Hygienists, P.O. Box 1937, Cincinnati, Ohio, 1972.

Giever, P. M., Number and size of pollutants, *in* "Air Pollution" (A. C. Stern, ed.), Vol. II, pp. 249–280. Academic Press, New York, 1968.

Jacobs, M. B., "The Chemical Analysis of Air Pollutants" Wiley (Interscience), New York, 1960.

McCrone, W. C., Draftz, R. D., and Delly, J. G., "The Particle Atlas." Ann Arbor Science Publishers, Ann Arbor, Michigan, 1967.

Rub, F., *Wasser Luft Betr.* **13**, 455–459 (1969).

Schulte, H. F., Monitoring airborne radioactivity, *in* "Air Pollution" (A. C. Stern, ed.), Vol. II, pp. 393–424. Academic Press, New York, 1968.

Stevens, R. K., and O'Keeffe, A. E., *Anal. Chem.* **42**, 143A–149A (1970).

Willard, H. H., Merritt, L. L., and Dean, J. A., "Instrumental Methods of Analysis" Van Nostrand-Reinhold, Princeton, New Jersey, 1966.

Questions

1. Explain the difference between precision and accuracy with respect to analytical methods.
2. Elaborate upon any three factors affecting the choice of an analytical method for air pollutants.
3. (a) Explain why the iodine and conductimetric methods for ambient sulfur dioxide are not specific.
   (b) By contrast, why is the West-Gaeke method specific?
4. How would you proceed to calibrate (standardize) the conductimetric and West-Gaeke methods for sulfur dioxide?
5. Briefly describe the analytical method for an air pollutant not listed in Tables 15-2 and 15-3.
6. If the concentration of carbon atoms analyzed by the flame ionization detector were 12 ppm, what would be the equivalent concentration if each of the following chemicals were present separately: methane, ethane, ethylene, butane, butadiene, hexane, hexene, and benzene? Complete reaction is assumed.
7. Describe in greater detail one of the spectrographic methods of analysis listed in Table 15-4.
8. Describe in detail the operational sequence of any continuous air pollutant instrument.
9. What are the differences between a Geiger-Müller counter and a scintillation detector for the detection of radioactivity?
10. Describe in detail any relatively new experimental method for the analysis of an air pollutant.

Part III

THE METEOROLOGY OF AIR POLLUTION

Chapter 16

SUN, EARTH, ATMOSPHERE, WEATHER, AND CLIMATE

I. The Sun–Earth–Atmosphere System

The sun radiates energy into space. At a distance of about 1.5×10^8 km (9.3×10^7 miles) one of its planets, our Earth, intercepts a small fraction of that energy. Because earth's mass, diameter, and distance from the sun are as they are, the gravitational pull exceeds the escape velocity of most gaseous molecules, and earth has developed an atmosphere. The conjunction of these factors and the presence of an atmosphere is by no means a common occurrence in the universe. The composition of earth's atmosphere, the nature of the surface, and the geometry of the annual passage around the sun combine to produce a wide variety of conditions on earth, which we call climates.

Within the earth–atmosphere system, the sun's energy is intercepted principally at the earth's surface, and then transferred by various processes into the atmosphere. The surface, therefore, is the prime heat exchanger in the system. Because the earth rotates on its axis and because the heat exchange at the surface is not uniform in time or space, the atmosphere is set in motion. The atmosphere's motion in turn affects the manner in which solar energy is intercepted and transferred to the atmosphere. The earth–atmosphere system is dynamic and variable. Observation of the state and behavior of the system during a short time period—a sort of "snapshot"—reveals what we call weather. Integrated over longer periods of time these snapshots merge into a "time exposure" and reveal the climates of earth.

II. Duration and Intensity of Sunshine

The fundamental determinant of climate is the geometry of the earth–atmosphere system (Fig. 16-1). The fact that the earth moves in eccentric orbit around the sun once each year, turning once each day on an axis tilted about $23\frac{1}{2}$ deg. from the plane of its orbit, is the basis for the annual cycle of seasons and the diurnal cycle of night and day. The earth is closer to the sun in January (perihelion) than in July (aphelion). The eccentricity results in a 7% difference between the two times of year in the radiant flux at the outer limits of the atmosphere. With a uniform surface and no atmosphere, earth would theoretically experience a mean temperature 4°C (7°F) warmer in July than in January. Winters would be warmer in the northern hemisphere than in the southern and summers would be warmer in the southern hemisphere than in the northern. The point is that the presence of the atmosphere and the distribution of land and ocean on the surface result in circulations and patterns of heat distribution that produce outcomes different from those predicted from eccentricity and tilt alone.

Each minute, the sun radiates a total of about 56×10^{26} cal of energy. The intensity of this flux on a spherical shell at the distance of the earth is called the solar constant, S.

$$S = \frac{56 \times 10^{26} \text{ cal min}^{-1}}{4\pi \ (1.5 \times 10^{13} \text{ cm})^2} = 2.0 \text{ cal cm}^{-2} \text{ min}^{-1} \qquad (16\text{-}1)$$

S is measured on a unit area normal to the solar beam at the outer limit of the atmosphere. The flux through a unit of horizontal area is

$$S_h = S \cos Z \qquad (16\text{-}2)$$

where the zenith angle Z is measured between the local vertical and the solar beam. Clearly, the flux S_h at any moment is a function of the latitude, the date, and the hour. The solar declination δ or latitude at which the sun is directly overhead at noon, is uniquely related to the date if δ has opposite signs for north and south latitude. The hour angle η is $(15°) \times$ (number of hours before or after noon)—it is, e. g., $(15°) \times (6) = 90°$ at sunrise or sunset on the equinoxes June 21 and September 21.

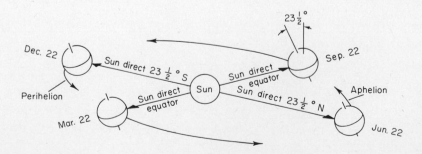

FIG. 16-1. The earth's axial tilt and orbit around the sun.

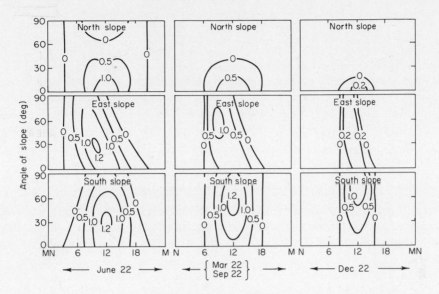

Fig. 16-2. The intensity of sunlight (measured at Trier, Germany) as a function of slope, aspect, and date for 50°N latitude. Units are langleys (ly) minute^{-1}.

Spherical trigonometry yields the result that the zenith angle, and thus through Eq. (16-2) the flux S_h, is related to the three factors by

$$\cos Z = \sin \phi \sin \delta + \cos \phi \cos \delta \cos \eta \qquad (16\text{-}3)$$

where ϕ is the latitude. Both latitude and declination are assigned positive values in the northern hemisphere, and negative in the southern. The solar azimuth ω is the angle measured between south and the direction toward the sun in a horizontal plane:

$$\sin \omega = (\cos \delta \sin \eta)/(\sin Z) \qquad (16\text{-}4)$$

Actually, of course, sunlight is generally not incident on a horizontal surface at the earth's surface. A slope inclined at an angle i from the horizontal and facing (i. e., normal to) an azimuth ω' degrees from south experiences an intensity of sunlight, in the absence of an atmosphere, of

$$S_s = S \left[\cos Z \cos i + \sin Z \sin i \cos (\omega - \omega')\right] \qquad (16\text{-}5)$$

Angles ω and ω' are negative to the east of south and positive to the west. Figure 16-2 shows, for key combinations of date and slope, the large variety of measured sunniness for slopes located at 50°N latitude. As will be discussed later, this sunniness or intensity of solar flux, is a primary determinant of local climate and of many phenomena important in problems of air pollution.

The units in Fig. 16-2 are cal cm^{-2} min^{-1} (also known as langleys min^{-1}) to be compared with the maximum equal to the solar constant, 2.0 ly min^{-1}. This measured flux on a slope naturally departs from theoretical values in very complex ways

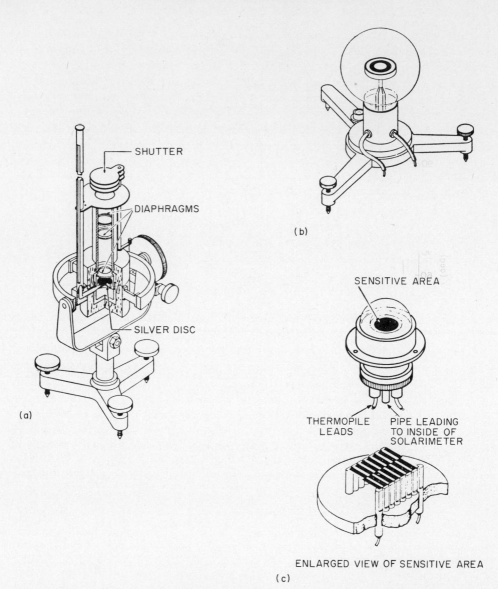

SHUTTER

DIAPHRAGMS

SILVER DISC

(a)

(b)

SENSITIVE AREA

THERMOPILE
LEADS

PIPE LEADING
TO INSIDE OF
SOLARIMETER

ENLARGED VIEW OF SENSITIVE AREA

(c)

FIG. 16-3. Various types of solarimeters. (a) The Abbot pyrheliometer collimates the direct solar beam to the blackened surface of a calibrated silver disk oriented perpendicular to the beam. The rate of temperature change of the disk is converted to a measure of intensity. (b) The Eppley pyranometer consists of two concentric silver rings, one painted white and the other black. The temperature difference between the rings, measured by a 50-junction thermopile, is a measure of the total shortwave radiation from the hemisphere seen by the rings (not necessarily the sky hemisphere if the instrument is not used horizontally). (c) The Moll-Gorcznski pyranometer exposes a series of 14 copper–constantan thermojunctions, all blackened, to the shortwave radiation from a selected hemisphere. One set of junctions is connected, for temperature compensation, to a heavy brass block. (a), (b), and (c) from "Handbook of Meteorological Instruments," Part I. British Meteorological Office, 1956. After "Physical Climatology" by W. D. Sellers, The University of Chicago Press, 1965. (b) Courtesy of the Eppley Laboratory, Inc.

because of several factors. The molecules of the air deplete the solar beam by absorption and scattering. At larger zenith angles (farther from noon) the length of the path the sunshine takes through the atmosphere is greater than at noon on the same day. Increased dirtiness (turbidity) of air further depletes the solar beam. Finally atmospheric water depletes the solar beam, slightly as vapor and greatly when condensed as cloud. The solar flux at the surface is increased by scattering from the blue part of the sky and by occasional reflection from the sides of nearby clouds—the "silver lining." On balance, the intensity of sunshine on a surface is usually reduced from the values of Fig. 16-2 because of a complex of processes. Occasionally it is momentarily larger than values of Fig. 16-2 because of cloud reflection. It is generally necessary to measure the intensity of sunshine, since calculation involves such complexity.

Figure 16-3 displays instruments that use different physical principles for measuring the intensity of sunlight and of the scattered and reflected light from sky and clouds. In practice, intensity is usually measured with respect to a horizontal surface and converted by trigonometry, if necessary, for a sloping surface.

III. Composition and Dimensions of the Atmosphere

The atmosphere is a mechanical mixture of gases. It is compressible, with a marked decrease in density at higher altitude. Figure 16-4 shows the decrease, and Table 3-1 gives the average composition of the mixture "dry air." The turbulent and convective mixing processes in the atmosphere maintain a nearly constant mixture in time and space below 15–20 km, the altitude of the tropopause, which separates the troposphere beneath from the stratosphere above. The major atmospheric circulations and nearly all weather phenomena are found between the

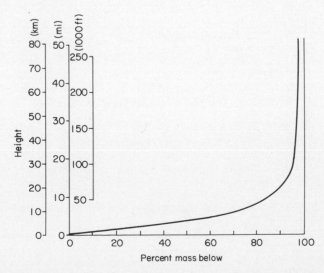

FIG. 16-4. Percent of total atmospheric mass below altitudes up to 80 km (50 miles).

surface and the tropopause, a layer containing nearly 90% of the atmosphere's mass. Today's jet aircraft cruise with about 80% of the atmosphere beneath them, and nearly a third of the atmosphere lies below the crestlines of mountain ranges in the western United States.

Taking the ratio of the thickness of the 90% atmosphere (12 km) to the radius of the earth (6370 km), the result is about 0.0019. The figure for the 50% atmosphere is about 0.0008. Thus the atmosphere is an exceedingly thin blanket surrounding earth, like the outer skin of an onion or the tissue paper wrapping of a new baseball. Seen on this scale, the life-supporting blanket of air is pitifully limited.

The average sea level atmospheric pressure is 1.04 kg cm^{-2} (14.7 lb in.$^{-2}$). Multiplication by the surface area of earth $[4\pi \ (6.37 \times 10^8 \text{ cm})^2]$ yields the result that the mass of the atmosphere is about 5.1×10^{18} kg, or about 6 quadrillion tons. On this scale, the atmosphere is tremendously large.

As the mass of the atmosphere is large by man's everyday measures, so is the energy content and exchange rate in the atmosphere. Again as with mass, energy in the atmosphere is small judged by the standard of the energy available on earth from the sun. Table 16-1 compares the energy involved in various processes and events, with available solar energy as the standard.

In our lives, the energy of an H-bomb is awesome, equivalent to a thousand thunderstorms and a million tornados. On the other hand, an H-bomb is but one-millionth of the solar energy arriving on earth in a ten-day period. It contains but one-tenth the energy of a hurricane and one-hundredth the energy of a cyclonic storm, several of which may move across the United States in a week. The idea has been proposed that atomic weapons be used to disrupt threatening hurricanes. The energy in a bomb is one ten thousandth the energy of a hurricane.

TABLE 16-1

Relative Total Energy of Various Processes and Events on Earth [a]

Solar energy received per day[b]	1
Melting of snow cover during hemispheric spring[b]	10^{-1}
Annual world use of energy by man[b]	10^{-2}
Cyclone (low-pressure storm system)[b]	10^{-3}
Hurricane or typhoon[b]	10^{-4}
"Thermonuclear weapon" (H-bomb) (1954)[b]	10^{-5}
To remove Los Angeles' inversion by heating layer of air beneath it[c]	10^{-5}
Summer thunderstorm[b]	10^{-8}
Daily power generation of Hoover Dam, Arizona[b,c]	10^{-8}
Atomic bomb of 1945[b]	10^{-8}
To set the air layer beneath Los Angeles' inversion in motion at 2.5 m sec^{-1} (5 mph)[c]	10^{-8}
Tornado[b]	10^{-11}
Street lighting, one night, New York City[b]	10^{-11}
Single strong wind gust[b]	10^{-17}

[a] Solar energy received is 3.7×10^{21} cal day^{-1}.
[b] Based on Lettau and adapted from Sellers (1).
[c] Adapted from Neiberger (2).

TABLE 16-2

Summary of Atmospheric Scales of Size and Time

Scale name	Time scale	Horizontal scale	Vertical scale
Microscale	1 second–1 hour	1 mm–1 km	1 mm–10 m
Mesoscale	1 hour–$\frac{1}{2}$ day	1 km–100 km	10 m–1 km
Macroscale	$\frac{1}{2}$ day–1 week	100 km–hemisphere	1 km–20 km

The point of discussing trade-off rates of energy and mass is that, on the one hand, earth and atmosphere are small, tenuous, and finite. On the other hand, man in his daily manipulation of energy deals with amounts that are miniscule compared with quite ordinary weather phenomena. The declared objective of some to "control the environment" must be considered in this perspective.

IV. Size and Time Scales of Meteorology and Climatology

There are generally three intergrading size and time scales referred to in meteorology and climatology. Table 16-2 summarizes them. Processes and their results are identifiable at each scale, and in general they do not express themselves at a smaller scale if there is vigorous activity at a larger scale. Thus, for example, the rather large temperature differences that might result on the microscale from the major differences in sunniness of Fig. 16-2 are suppressed by cloud or wind generated on the meso- or macroscale.

In the remainder of this chapter, and in the later chapter on atmospheric motion, an attempt is made to discuss processes first on the local, or microscale, and then on the mesoscale and the macroscale. The behavior of the atmosphere in relation to problems of air pollution, or to any activity of man, is the resultant of a mixture of components operating at different size and time scales. When rather large differences in atmospheric properties are observed within short distances, it is likely that local processes are dominant and the mixture is rich at the microscale and lean at the mesoscale and macroscale. When, as in vigorous storminess, differences locally are minimal, the mixture is richest at the macroscale.

V. Determinants of Local Climate

Composition of the earth's atmosphere and the geometric relationships between sun and earth have already been cited as primary determinants of the annual and diurnal (daily) cycles of climate. The nature of the earth's surface was also cited. On the microscale, the analysis of the heat budget of the earth–atmosphere interface provides a means for examining the role of the earth's surface in determining local climate. The statement of the heat budget concept (also called the energy budget) is a statement of continuity for heat in a specified system during a specified time: (Input) + (Outflow) + (Storage) = 0.

(a)

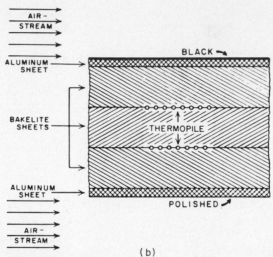

(b)

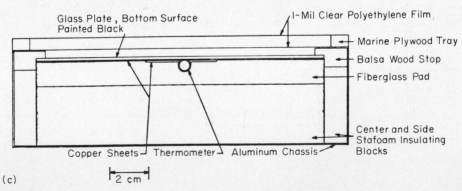

(c)

FIG. 16-5. Types of radiometers. (a) Commercially available unit. (b) Cross section of sensor plate. When the top is blackened and the bottom polished (as shown), the instrument is used as an allwave hemispherical radiometer $(S_i + I_i)$; when the top and the bottom are blackened, the instrument is a net radiometer $(S_i - S_o + I_i - I_o)$. (c) Cross section of an "economical radiometer." The unit as shown is a hemispherical sensor; a pair of units facing up and down become a net radiometer. After "Physical Climatology" by W. D. Sellers, The University of Chicago Press, 1965. For additional details, see Section 22-V.

At midday, with clear skies and typical summer conditions, the energy input at the interface is a balance of the several streams of radiation coming to and leaving the surface. This net radiant transfer will be discussed in more detail in the following chapter. The input of net radiation is dissipated in several ways. Some flows into the soil or water beneath and is stored there. Some of the input is utilized in the evaporation of water at the interface. The remainder is dissipated in the convective heating of air flowing across the surface. If there is substantial vegetation present, a few percent of the input may be involved in the photosynthetic processes going on. Neglecting this usually small biological component, the balance for the interface may be written

$$S_i - S_o + I_i - I_o \pm G \pm L_v E \pm H = 0 \qquad (16\text{-}6)$$

where S is the flux of solar (shortwave) radiation, I the flux of infrared (longwave) radiation, G the flux of heat into the soil or water beneath, L_v the latent heat of vaporization for water, E the evaporation rate, and H the flux of sensible heat by convection into the airstream. The subscripts refer to the incoming (i) and outgoing (o) radiation. A flux to the interface is taken as positive and away negative. The units for all terms are energy-area^{-1}-time^{-1}. The incoming shortwave flux S_i may be measured with solarimeters (as in Fig. 16-3). If the shortwave absorptivity of the surface, a_s, is known then $S_o = (1 - a_s)S_i$, and $S_i - S_o = a_s S_i$. By day, this component is positive.

The net flux of longwave radiation is usually negative since the incoming component from the sky is smaller than the back radiation from the surface. Water vapor content and temperature of the lower atmosphere, together with height, amount, and type of cloud are the primary factors which determine the longwave flux (I_i) from the sky. Since the earth's absorptivity for longwave is nearly 1.0, the sum ($a_s S_i + I_i$) constitutes the energy that must be dissipated by the interface through the other processes represented by ($-I_o \pm G \pm L_v E \pm H$).

Estimates of the flux I_i may be obtained from empirical relationships involving only data from near the interface (see Chapter 17). More elaborate estimates involve use of measurements of temperature and moisture obtained from soundings above the surface. Finally, radiometric instrumentation is available with which to measure components of the radiation balance directly, either singly or in groups. Several of these instruments are shown in Fig. 16-5.

Partitioning of the energy ($a_s S_i + I_i$) among the four dissipative processes, determined in large part by the nature of the interface itself and the materials just beneath it, is the way in which the heat budget acts as a determinant of local climate. For example, in a dry microclimate most of the dissipation will take place through the course of the day by means of the flux of back radiation I_o and sensible heat to the airstream H. Flux to the soil G is less important, and that to evaporative dissipation less important still. These relationships are shown for a dry, midsummer microclimate in Fig. 16-6a, where the dissipation term I_o is included within the net radiation term. As would be expected, accompanying air temperature is high and relative humidity low at midday. In contrast, Fig. 16-6b shows analogous results from a pine forest during a period of plentiful moisture. Primarily because the surface absorptivity of a tree canopy a_s is larger than that of the pasture in Fig. 16-6a, the

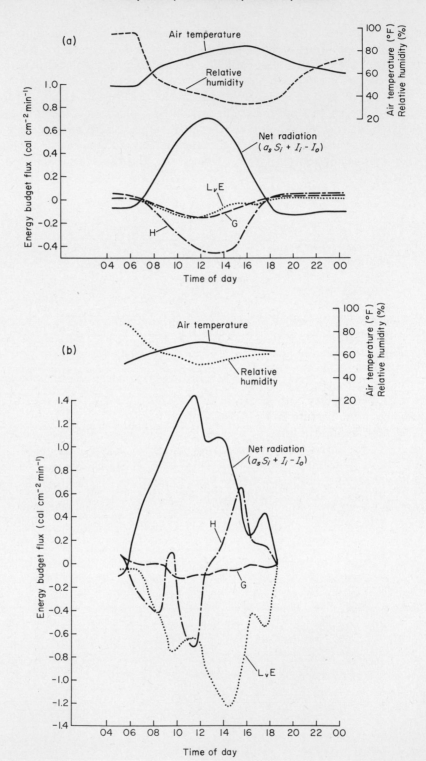

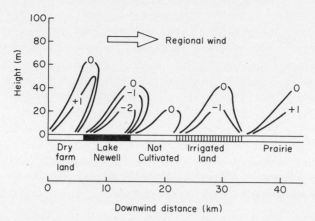

FIG. 16-7. Local superheating (°C) and cooling in the lower atmosphere due to a variety of surfaces, in southern Alberta (5).

net radiation at midday is larger. Cloudiness produced midday variability in the net radiation. Beneath the canopy, the soil flux G is small. Before noon, both H and L_vE represent dissipation of heat from the canopy. After noon, however, warm air brings heat from outside the tree stand, and the H term appears as a heat source for the canopy. The large heat dissipation by evapotranspiration L_vE keeps air temperature and humidity both moderate and relatively constant as compared with the values above the pasture in Fig. 16-6a.

All of the four dissipative terms $(-I_o \pm G \pm L_vE \pm H)$ are dependent upon the temperature of the interface. Thus that temperature changes in response to the radiant heat load on it, until a balance has been achieved. In the process of matching dissipation to heat load, air temperature and humidity are in large measure determined. In the process, energy from the sun and sky are accepted by the interface and reintroduced into the atmosphere. The interface acts as a heat exchanger, and the different transfer rates at different places across the earth's surface ultimately produce motion and other atmospheric behavior. When motion becomes vigorous enough the contrasts between places are reduced and the macroscale processes dominate over the microscale. Figure 16-7 shows something of the variety of local conditions produced by the nature of the underlying surfaces and relatively quiet macroscale conditions. As will be shown in later chapters, the variety of local surfaces and the atmospheric behavior they produce may have a great deal to do with the effectiveness of the atmosphere in dispersing contaminants. In and near urban and industrialized areas, the variety of local surfaces may become very complex and the resulting atmospheric behavior difficult to analyze and predict.

FIG. 16-6. Diurnal march of energy budget components, air temperature, and relative humidity (a) above a pasture in Nebraska (3), (b) in a pine forest in Australia (4).

VI. Other Determinants of Climate

On the mesoscale and macroscale, the spatial relationships of land and water, and the location of major topographic barriers constitute the major determinants of climate in addition to those already mentioned.

Open water has a relatively constant surface temperature compared with that of land surfaces nearby. Because water absorbs shortwave radiation beneath its surface, because its heat capacity is large (large amount of heat to raise the temperature of a given volume), because convective motion carries excess heat quickly to or from the surface layer, and because high evaporation rates tend to cool the surface, open water is usually warmer than nearby land at night and during winter, and colder by day and during the summer especially in midlatitudes.

In addition to the relative constancy of surface temperature of open water and its effect on atmospheric motion and behavior, ocean currents often modify further the temperature contrasts found across shorelines on the eastern and western margins of continents. Figure 16-8 shows a generalized conception of surface circulation in the oceans and the resulting modification of temperature on land. The circulation

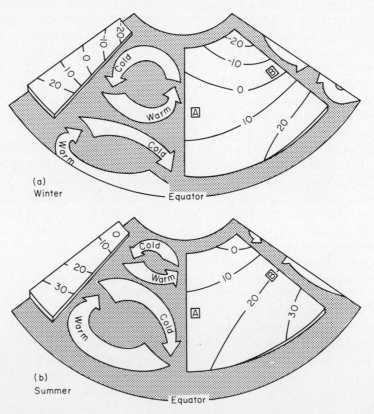

FIG. 16-8. Generalized surface circulation in the oceans, and seasonal mean temperature (arbitrary units) on land, northern hemisphere in (a) winter and (b) summer.

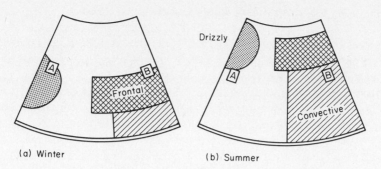

FIG. 16-9. Generalized precipitation types over a standard continent in the northern hemisphere. Compare with Fig. 16-8.

arrows are labeled "cold" and "warm" to indicate temperatures relative to the oceanic temperatures found elsewhere at the same latitude. A result is the divergence of mean temperatures (isotherms) on the western continent and their convergence on the east. The effects at individual localities are shown, for example at "A" on the west coast (Annual range of mean temperature 13°–5° = 8°) and at "B" on the east coast (range 20°–(−4°) = 24°). Major mountain ranges oriented north–south, such as the North American Rockies and the South American Andes, reduce the oceanic influence on climates downwind to the east. The result is generally lower temperatures in winter and higher temperatures in summer than would be observed if the mountains were not there. The modification results not only from the interference with eastward movement of maritime air, but also from the ease of equatorward movement of cold air (winter) and of poleward movement of warm air (summer).

East–west mountain ranges, such as the Alps in Europe and the Himalayas in southeast Asia, tend to impound cold air on their poleward slopes in winter and to promote deep inland penetration of storms from oceans to the west.

Regionally, different types of precipitation tend to predominate in a given season. Figure 16-9 adds the idea of regional-seasonal precipitation types to the temperatures shown in Fig. 16-8 for a standard continent. The "drizzly" type found on the west coast is associated with the "10" isotherm found near the divergence of offshore currents—north in summer and south in winter. The precipitation results largely from the gentle uplift of airstreams coming onshore from the west, and rising over coastal mountains and/or colder, denser air at the surface.

The "frontal" type is also associated with the "10" isotherm and the convergence of contrasting air masses from north and south along that line. The behavior of the large-scale storms, which result from converging air masses and yield frontal precipitation, is discussed in Chapter 20. Suffice it to say here that the precipitation is quite intermittent, alternating regularly with sunny skies every day or so.

Finally, the "convective" precipitation type is found in the areas where deep, warm, moist currents of air flow onshore from tropical oceans, there to experience vigorous convective motion (see later discussion of Figure 19-11). Thus, a west coast station such as "A" will experience few downpours, and will receive most of its precipitation during winter in many hours of relatively low intensity. Whereas, a station such as "B" on the east coast will receive its precipitation more equally between seasons, at moderate intensities during winter, and during a relatively few periods of high intensity during summer. Combining Figs. 16-9a and 16-9b, one sees the basis for deserts and other dry areas in the southwestern and northeastern corners of the (northern hemisphere) continent, where indeed they are found in North America and Asia, as well as in the southern hemisphere counterparts of South America, Africa, and Australia.

VII. Weather versus Climate

Weather is the state or condition of the atmosphere at a moment or during a few hours, and the variations in the state during a few days or a month. Climate is the typical mixture of weather to be expected in a region during a certain period of the year. In both, variability is important. Weather changes quickly. Climate changes slowly but contains a great variety of weather.

Although average values of weather elements do contain important information, they often mask major differences when time relationships are ignored. Table 16-3, for example, shows that Salem, Oregon, and Harrisburg, Pennsylvania, have very similar annual averages of temperature and precipitation. Comparisons of mid-

TABLE 16-3

Various Measures of Annual and Seasonal Temperature and Precipitation for Salem, Oregon, and Harrisburg, Pennsylvania

	Salem		Harrisburg	
	Metric	English	Metric	English
Temperature				
Annual mean	11.3°C	52.4°F	11.8°C	53.3°F
July mean	19.0	66.1	24.6	76.2
January mean	3.6	38.5	−0.4	31.3
July–January difference	15.4	27.6	25.0	44.9
Precipitation				
Annual mean	1040 mm	41.8 in.	920 mm	37.7 in.
July mean	10	0.4	89	3.5
January mean	170	6.7	71	2.8
July–January difference	−160	−6.3	18	0.7
June–August mean total (Summer)	58	2.3	269	10.6
Dec–February mean total (Winter)	492	19.4	178	7.0
Summer–Winter difference	−434	−17.1	91	3.6

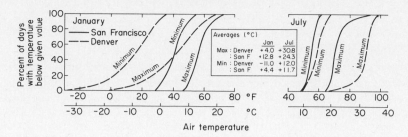

FIG. 16-10. Cumulative frequency distributions of maximum temperature and minimum temperature for San Francisco, California and Denver, Colorado in January and July, 1960–1964 (6).

winter and midsummer periods, however, show large and important differences in both elements. Salem is near the west coast and experiences much smaller annual variations of temperature. Harrisburg, near the east coast, has much smaller annual variation in precipitation, as well as major differences in its timing.

Just as annual averages can conceal important seasonal differences, monthly averages can conceal the extent to which weather varies and the nature of the climatic mixture at a station. Figure 16-10 shows information on temperature for stations typifying midlatitude climates which exhibit relatively low variability (San Francisco, California) and relatively high variability (Denver, Colorado). This summary of daily maximum and minimum temperatures shows the type and magnitude of variation that monthly mean values cannot.

References

1. Sellers, W., "Physical Climatology," 272 pp. Univ. of Chicago Press, Chicago, Illinois, 1965.
2. Neiberger, M., *Science* **126,** 637–645 (1957).
3. Lettau, H., and Davidson, B., eds., "Exploring the Atmosphere's First Mile," Vol. 1, 376 pp. Pergamon, Oxford, 1957.
4. Denmead, O., *Agr. Meteor.* **6,** 357–371 (1969).
5. Holmes, R., *Agron. J.* **62,** 546–549 (1970).
6. "Climatological Data of the United States by Sections." U.S. Department of Commerce, Weather Bureau.

Suggested Reading

Barry, R. G., and Chorley, R., "Atmosphere, Weather, and Climate," 320 pp. Holt, New York, 1970.
Day, J., and Sternes, G., "Climate and Weather," 407 pp. Addison-Wesley, Reading, Massachusetts, 1970.
Flohn, H., "Climate and Weather," 253 pp. McGraw-Hill, New York, 1969.
Geiger, R., "The Climate near the Ground," 4th edition, 611 pp. Harvard Univ. Press, Cambridge, Massachusetts, 1965.
Griffiths, J. F., "Applied Climatology," 118 pp. Oxford Univ. Press, London and New York, 1966.

Questions

1. (a) Develop a general procedure for calculating the time of sunrise as a function of latitude, date, slope azimuth, and slope inclination.
   (b) Use the procedure to prepare a graph of the time of sunrise, as a function of date, for a slope of 45° facing southeast, at 45° north latitude.
   (c) What is the time of *sunset* on the slope in part (b) on July 22?

2. On a slope of 70° facing an azimuth of 60° east of south at 50° north latitude, estimate the hour and amount of the greatest solar radiation intensity for June 22 and December 22 in the absence of an atmosphere. *Hint*: Will it be at noon? (*Note*: Save these calculations for Question 17-7.)

3. (a) What is the mean atmospheric pressure at 20 km?
   (b) What is the sea level atmospheric pressure due to oxygen?
   (c) What is the 20 km atmospheric pressure due to nitrogen?

4. (a) With a 50% efficiency of transmission, how many Hoover dams would be needed to supply the energy used by man?
   (b) How many to light New York City for a year?
   (c) If one could dissipate a typhoon by applying 1% of its energy at the right time and place, how many A-bombs would it take?

5. (a) Consider the heat budget at noon in Fig. 16-6a. If the pasture is horizontal, the latitude is 45° N, and the date is July 22, what is the ratio of the observed net radiation to the theoretical incident solar radiation in the absence of an atmosphere?
   (b) What is the ratio for 1600 hours?

6. On the standard continent of Fig. 16-8, map the mean annual temperature range (summer minus winter). How does your map compare with your knowledge of the climates of the continent you live on?

7. Referring to Fig. 16-10, estimate the following probabilities:
   (a) The January minimum temperature at Denver is 10°F or less.
   (b) The January maximum temperature at Denver is above freezing.
   (c) The July maximum temperature at San Francisco is above 22°C.
   (d) The July maximum temperature at Denver is below 90°F.
   (e) The July maximum temperature at San Francisco is between 20° and 25°C.

Chapter 17

RADIATION IN THE ATMOSPHERE

In the previous chapter the point was made that the earth–atmosphere interface is the lower atmosphere's primary heat exchanger. Through the various surfaces underlying the atmosphere, heat and moisture are injected into the air. The fact that major spatial and temporal differences exist in injection rates across the earth's surface results in pressure differences within the atmosphere. Pressure differences produce motion—vertical and horizontal—and the motion produces variations in the multitude of variables leading to "weather."

Consideration of problems of resource management in the atmospheric environment often centers on the three-linked chain—source–transfer–receptor. The atmosphere not only carries pollutants from source to receptor; it also affects the source processes and the behavior of the receptors. As obvious examples, the atmosphere has a major influence on the rate of consumption of fuel for domestic heating; likewise the atmosphere exerts partial control over the wetness, and thus the suceptibility to damage by soluble pollutants, of both living tissue and nonliving materials.

The concept of "Spaceship Earth" (1) is that of our planet and its atmosphere, moving in space as a closed system, except for the input of sunlight and the output of heat—both input and output being forms of radiant heat transfer. With the formation of our atmosphere, generation of its weather, and the maintenance of life within it dependent on the processes of energy transfer, it is clear that understanding of these processes is essential for one intent upon the task of avoiding misuse of the atmospheric resource of earth.

I. The Black Body

Examination of the processes of radiant heat transfer in the air environment begins with an understanding of the base of reference for most radiation processes: the black body. A material whose properties are describable in theory, but which does not (as far as we know) exist in nature, the black body has as its basic characteristic the fact that it absorbs all electromagnetic radiation that is incident upon it. Thus the ratio of absorbed to incident energy—the *absorptivity*—for a black body is 1.0.

Unless it is at the zero of Kelvin (absolute) temperature, the materials of a black body will emit radiation. The energy will be distributed on the spectrum of electromagnetic radiation according to *Planck's distribution law for emission:*

$$E_\nu d\nu = c_1 \nu^3 \left[\exp\left(c_2\nu/T\right) - 1\right]^{-1} d\nu \tag{17-1}$$

which is a function of the frequency of the radiation ν, and the Kelvin temperature of the black body T. The dependent variable E_ν is the amount of radiant flux (energy·area^{-1}·time^{-1}) from the black body in the frequency band from ν to $(\nu + d\nu)$.* Figure 17-1 shows the distribution function on a linear scale of frequency for two temperatures. Frequency and wavelength, λ, are related by the expression $\lambda\nu = c$, or $\lambda = c/\nu$. Because many discussions of radiation are carried on in terms of wavelength rather than frequency, the nonlinear scale of wavelength in Fig. 17-1

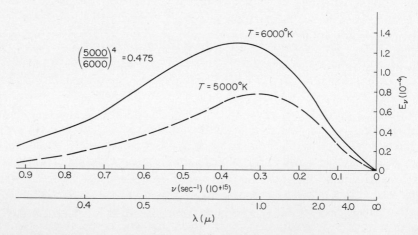

FIG. 17-1. Planck's distribution on a linear scale of frequency and nonlinear scale of wavelength, for two temperatures, 6000° and 5000°K.

* The values of the constants in Eq. (17-1) are—$c_1 = 2\pi h/c^2$ and $c_2 = h/k$, where $h = 6.55 \times 10^{-27}$ erg·sec, Planck's constant; $c = 3 \times 10^{10}$ cm/sec, the speed of light; and $k = 1.37 \times 10^{-16}$ erg/deg, Boltzmann's constant.

has larger values to the right, where the zero of frequency is found. The reason for showing Planck's distribution as a function of frequency is that according to another law by Planck, the energy of a photon, E^* ,and thus the energy of a monochromatic stream, is proportional to the frequency; $E^* = h\nu$, where h is Planck's constant. Thus the display in Fig. 17-1 is an energy diagram in which a unit of area, regardless of shape or location, represents the same amount of energy in any part of the diagram (2–4). This property will be useful presently.

The energy emitted by a black body and contained in the waveband from λ to $(\lambda + d\lambda)$ may be evaluated by $E_\lambda = cE_\nu/\lambda^2$. Also, as suggested in Fig. 17-1, the frequency of maximum emission decreases as the temperature of the black body decreases (5) according to *Wien's Displacement Law:*

$$\nu_{max}/T = 5.9 \times 10^{10} \text{ sec}^{-1} \text{ deg}^{-1} \qquad (17\text{-}2)$$

The analogous expression in terms of wavelength is $T\lambda_{max} = 2900$ μdeg. Thus, for a black body with temperature 6000°K, near that of the sun, $\nu_{max} = 35.4 \times 10^{13}$ sec^{-1} = 0.85 μ and $\lambda_{max} = 0.49$ μ.

The integral of Eq. (17-1) over frequency and over an entire hemisphere yields the expression for allwave radiant flux, from a unit area of black body, known as the *Stefan-Boltzmann Law*

$$F = \sigma T^4 \qquad (17\text{-}3)$$

where the Stefan-Boltzmann constant σ has the value 0.817×10^{-10} cal cm^{-2} min^{-1} deg^{-4}. Thus, for example, the area under a curve such as in Fig. 17-1, is proportional to the fourth power of the temperature associated with the curve. A doubling of the absolute temperature results in a multiplication of the flux by 16; an increase of 1% in the absolute temperature produces an increase of 4% in the flux. As shown in Fig. 17-1, F approximately doubles as T increases from 5000° to 6000°K.

II. Solar Radiation and Terrestrial Radiation

The sun is approximately a black body with temperature near 6000°K. The solar flux is $(4\pi R_\omega^2)(\sigma T_\omega^4) = 56 \times 10^{26}$ cal min^{-1}, where R_ω is the sun's radius, and T_ω is the effective temperature of the sun's surface. By the inverse square relationship of diminishing intensity, this flux has been reduced to the solar constant, 2.0 cal cm^{-2} min^{-1} = $(\sigma T_\omega^4)(R_\omega/R_d)^2$ after traveling the approximately 93 million miles $(1.5 \times 10^8$ km) to the outer limits of earth's atmosphere. During the travel, however, the relative distribution of the sun's radiant energy by frequency has scarcely changed. Only the flux intensity has changed. Accordingly solar energy arriving at the earth's atmosphere appears as in Fig. 17-2.

The solar flux is intercepted by the earth–atmosphere cross section of area πR_e^2, where R_e is the radius of the spherical system. At any moment, therefore, the average solar flux over the entire surface area of $4\pi R_e^2$ is one fourth of the solar constant, 0.5 cal cm^{-2} min^{-1}. According to recent interpretation of satellite radiometric data (6) the mean flux intensity to space from the earth–atmosphere system is 0.33 cal cm^{-2} min^{-1}, so that for the "Spaceship Earth" to be in radiative equilib-

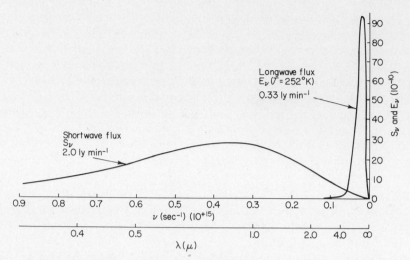

FIG. 17-2. Distribution of solar flux and global flux on a linear scale of frequency.

rium, the global absorptivity must be near $(0.33/0.5) = 0.66$. If 66% of the solar flux is absorbed, 34% must be reflected and the global *albedo* (fraction of incident sunlight reflected) is 0.34.

Using the Stefan-Boltzmann relationship, the effective radiant temperature of the earth–atmosphere system is $(0.33/\sigma)^{1/4}$, or 252°K. The second curve in Fig. 17-2 describes the spectral distribution from a black body at 252°K, which is essentially the distribution of the global flux to space. According to the above discussion, the area under the curve for the solar flux in Fig. 17-2 must be equal to $(2/0.33) = 6.06$ times that under the curve for global flux.

In Fig. 17-2 may be seen the small overlap between the two distributions and the distinct separation into two qualities of radiation. This gives rise to the terms *shortwave (solar) radiation* and *longwave (terrestrial or global) radiation*, as if they were quite separate streams. The division between the two is taken to be 4 μ (4 × 10^{-4} cm). The lower limit of solar radiation is taken to be 0.15 μ and the upper limit of terrestrial radiation 80 μ. These limits include about 99% of each of the two streams of radiant energy.

III. Absorptivity, Emissivity, and Kirchhoff's Law

The absorptivity a has been defined above as the fraction of incident radiation which is absorbed by the materials on which it is incident. Similarly, the fractions which are reflected and transmitted by the material are, respectively, the *reflectivity* r and the *transmissivity*, τ. Since these are the only alternatives for the disposition of incident radiation, $a + r + \tau = 1.0$. As noted in the previous section, the albedo is a special case of the reflectivity in which the incident radiation is the solar beam.

Any of the three terms describing the passive reactions of a material to incident radiation may be expressed for individual frequencies or wavelengths, for frequency or wavebands, or for entire spectra. For example, the transmissivity of a material at frequency ν_1 is τ_{ν_1}; while the absorptivity of a material for the solar spectrum, or bandwidth from 0.15 to 4.0μ is $a_s = a_{0.15-4.0\ \mu}$.

Examine now the physical meaning of the effective absorptivity of a material over a frequency band from ν_1 to ν_2. The energy arriving from all surrounding sources in the small band from ν to $(\nu + d\nu)$ is of the form $E_\nu'd\nu$ [see discussion of Eq. (17-1)]. Of the incident energy in the band ν to $(\nu + d\nu)$, the amount absorbed is $a_\nu E_\nu'd\nu$, and over the larger band of interest, the total amount of energy absorbed is the integral

$$\int_{\nu_1}^{\nu_2} a_\nu E_\nu'd\nu$$

Since over the band ν_1 to ν_2 the total amount of incident ratiation is the same integral without the a_ν, the effective absorptivity for the band is, by the definition of absorptivity, the ratio

$$a_{\nu_1-\nu_2} = \int_{\nu_1}^{\nu_2} a_\nu E_\nu'd\nu \bigg/ \int_{\nu_1}^{\nu_2} E_\nu'd\nu \tag{17-4}$$

In the top of Fig. 17-3 is shown the monochromatic absorptivity, a_ν, for each of the major absorbing gases of the atmosphere in the long-wave spectrum between 4.0 and 80 μ. In the lower portion of Fig. 17-3, the light curve represents the distribution of flux from the earth's surface with mean temperature 286°K (7).* The heavy curve results from multiplication of E_ν' by a_ν at each frequency, and the shaded area below the heavy curve represents the energy absorbed by the atmosphere after it leaves the underlying surface. Since the figure is a true energy diagram (see Section 17, I), the shaded area represents the value of the integral in the numerator of Eq. (17-4) and the area under the light curve the value of the integral in the denominator. The long-wave absorptivity of the atmosphere with clear sky is the ratio of these two areas—approximately 0.75. Note in particular that the ratio of the shaded area to the total area in the upper portion of Fig. 17-3 marked "Atmosphere" is *not* the long-wave absorptivity of the atmosphere, as seen by Eq. (17-4).

The *emissivity* ϵ of a material is the ratio of the energy emitted by the material to the energy emitted by a black body at the same frequency (or wavelength) and temperature. By definition, then, the emissivity of a black body is 1.0 at any frequency (wavelength), over any band, and in any spectrum. A corollary is that a black body emits the maximum amount of energy physically possible for any combination of frequency and temperature.

An additional important law governing radiant heat transfer is *Kirchhoff's Law* which states that the absorptivity of a material is equal to the emissivity of the material for the same frequency (wavelength) and temperature:

$$[a_\lambda = \epsilon_\lambda]_T \quad \text{or} \quad [a_{\nu_1-\nu_2} = \epsilon_{\nu_1-\nu_2}]_T \tag{17-5}$$

---

* This temperature is for the earth's surface—one part of the earth–atmosphere system. The value 252°K used earlier is for the entire system.

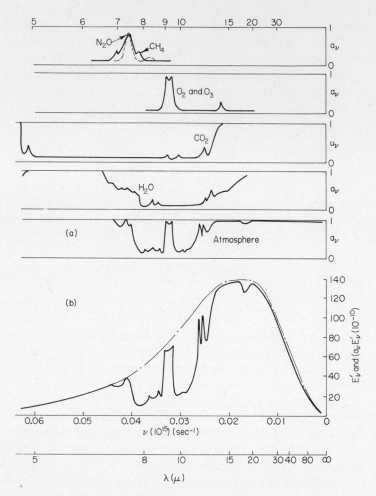

FIG. 17-3. (a) Absorption spectra of CH_4, N_2O, O_2 and O_3, CO_2, and H_2O over the band 4–80 μ, and the resulting spectrum for the atmosphere. (b) Emission and absorption of terrestrial radiation on a linear scale of frequency. Adapted from Fleagle and Businger (8).

According to this law, then, the spectra in the top portion of Fig. 17-3 are also emission spectra for the atmospheric gases, and the shaded area in the bottom of the figure represents the distribution of energy emitted by the lower atmosphere.

IV. Transmissivity, Depletion, and Beer's Law

Of considerable concern in atmospheric science, and especially as it relates to air pollution, is the manner in which radiation is depleted as it passes through a partially absorbing medium, such as the gases of the atmosphere. As might be expected on the basis of the nature of many similar physical phenomena, the amount of depletion that takes place is proportional to the amount incident and to the path length through (and/or amount of depleting material present in) the medium. More exactly, it has been found empirically that for wavelength λ, the change in intensity, $dR_{\lambda o}$, of the incident beam, $R_{\lambda o}$, as it passes through a layer of thickness dz and at an angle θ from the normal to the layer, is

$$dR_{\lambda o} = (-k_\lambda \cdot dz \cdot \sec \theta) R_{\lambda o} \qquad (17\text{-}6)$$

where the *absorption coefficient* k_λ (with units length^{-1}) describes the physical nature of the medium at the wavelength in question. The pathlength is $dz \cdot \sec \theta$ through the layer, and integration through an entire atmospheric layer of depth z gives the fractional depletion

$$R_\lambda / R_{\lambda o} = \exp (-k_\lambda \cdot z \cdot \sec \theta) \qquad (17\text{-}7)$$

an equation known as *Beer's Law*. The relationship may be expressed for the entire shortwave spectrum by means of a calculated effective coefficient for incident short-wave radiation k_s. Thus fractional depletion of the solar beam by scattering, reflection, and absorption is

$$S_n / S = \exp (-k_s \cdot z \cdot \sec \theta) \qquad (17\text{-}8)$$

If the effective transmissivity of the zenith path $(\theta = 0)$ is taken to be $\tau = \exp (-k_s z)$, then

$$S_n / S = \tau^{\sec \theta} \qquad (17\text{-}9)$$

where S_n is the intensity of the solar beam on a surface normal to the beam beneath the layer, and S is the normal solar flux above the layer. The processes of depletion of radiation by a partially absorbing medium are certainly not as simple as suggested by this discussion. As noted, depletion consists of absorption, scattering, and reflection. Pollutants in the atmosphere deplete in complex ways, related to the nature of the pollutants and their size distribution in the case of a solid aerosol (9) (see also Chapter 21). Nevertheless the basic framework for viewing the processes of depletion given here is a useful one for many analyses and for understanding this part of the atmosphere's energy balance.

As an example of calculations using Beer's Law in combination with other information, one may deduce from the data in the famous graph of Hewson and Longley (10) that a cloud deck has an effective absorption coefficient (assuming $Z = 45°$ as a "mean") of the order of 0.02–0.005 m^{-1}. The exact value depends on the kind of cloud and its structure, but the physical meaning is that each meter-thick layer of the cloud deck depletes the solar radiation entering the layer by between 2% and 0.5%.

Figure 17-4 shows the relationship of fractional transmission, S_n / S, to cloud layer thickness as just described. Likewise the upper right portion of Fig. 17-4 shows calculations based on the definition just given for effective transmissivity. The results show that the fraction of solar radiation which would penetrate the cloud from a sun directly overhead would be between 0.94 and 0.67 for a layer 20 m thick and between 0.61 and 0.14 for a layer 100 m thick. The implications for analysis of the energy budget of the earth–atmosphere interface, and thus of atmospheric behavior, should be clear. The implications of Beer's Law for analysis of visual range in the atmosphere are discussed in Chapter 21.

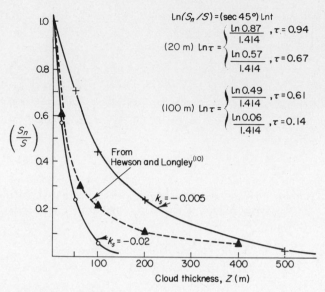

FIG. 17-4. Fractional transmission of the solar beam through clouds having absorption coefficients k_s of 0.02 and 0.005 m^{-1}, with angle of incidence $Z = 45°$.

V. The Estimation of Net Longwave Radiant Flux in the Lower Atmosphere

As mentioned in the preceding chapter, the flux of longwave radiation from the lower atmosphere, I_i, may be obtained from appropriate instrumentation (Fig. 16-5) or from calculations based on empirical methods. The longwave flux from a cloudless sky has been related in several ways (7) to the heat and moisture contents of the lower atmosphere. A representative relationship attributed to Brunt is

$$I_{i, \text{ clear}} = \sigma T_a{}^4(b + ce^{1/2}) \tag{17-10}$$

where the air temperature near the surface is T_a (°K) and the vapor pressure of the air near the surface is e (in millibars). Constants in the equation are the Stefan-Boltzmann constant σ [Eq. (17-3)] and the empirical constants b and c which typically assume values of 0.6 and 0.05, respectively. One millibar of pressure is 0.75 mm Hg. The units of the flux I are cal cm^{-2} min^{-1} when these values for the constants are used.

The presence of clouds may be accounted for by another empirical relationship

$$I_{i, \text{ cloud}} = I_{i, \text{ clear}}(1 + Kn_c{}^2) \tag{17-11}$$

where K is related to the height and type of cloud as in Table 17-1, and n_c is the amount of cloud cover in tenths (0 for clear sky and 10 for overcast). Combining a version of Eq. (17-3)—$I_o = \sigma T_a{}^4$—with Eqs. (17-10) and (17-11) yields an approximate basis for estimating net longwave flux at the earth–atmospheric interface

$$I_i - I_o = \sigma T_a{}^4[(b + ce^{1/2})(1 + Kn_c{}^2) - 1] \tag{17-12}$$

TABLE 17-1

*Typical Values of the Empirical Constant K in Eqs. (17–11)
and (17–12) Related to Altitude of Clouds* [a]

Cloud type	Altitude (typical) (m)	K
Cirrus	12,200	0.04
Cirrostratus	8,400	0.08
Altocumulus	3,700	0.16
Altostratus	2,100	0.20
Stratocumulus	1,200	0.22
Stratus	500	0.24
Nimbostratus or fog	0 to 100	0.25

[a] After Sellers (7).

For an air temperature of 22°C (295°K), I_o is 0.62 cal cm^{-2} min^{-1}. If the vapor pressure is 16 mbar, and the cloud-related factors have values $K = 0.2$ and $n_c = 0.5$, then the net longwave flux is

$$I_i - I_o = (0.62)\{[(0.6) + (0.05)(16)^{1/2}][1 + (0.2)(0.5)^2] - 1\}$$

or -0.09 cal cm^{-2} min^{-1}, about 14.5% of the flux I_o. Sellers notes (reference 7, page 53) that even in dry air the net flux rarely exceeds 30% of I_o, and that the climatological averages for the United States are about 23% in January and 18% in July.

Figure 17-5 shows the nature of the diurnal patterns of the radiation balance at

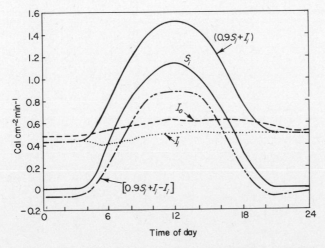

FIG. 17-5. Diurnal patterns of components of the radiation balance at the earth's surface. Incoming shortwave flux S_i, incoming longwave flux I_i, and back radiation from the surface I_o, are adapted from Gates' Fig. 11 (11) based on data from Hamburg, Germany, during June by Fleischer.

a typical midlatitude location. The three basic components of incoming shortwave flux S_i incoming longwave flux I_i and the back radiation from the interface I_o are taken from Gates (11). On the reasonable assumption that the absorptivity of the surface is 0.90, the heat load is shown as $(0.9S_i + I_i)$. [See discussion of Equation (16-6).] Finally the net radiant heat flux $(S_i - S_o + I_i - I_o)$ is obtained as $(0.9S_i + I_i - I_o)$.

Several things about this display of radiation balance are typical of most midlatitude situations. First, of course, S_i is zero at night. Second, in summer particularly, I_o is slightly larger than I_i in magnitude, the net longwave flux, $(I_i - I_o)$, being similar to the results of the calculation above based on Eq. (17-12). Third, the midday magnitude of I_o is similar to that in the same calculation above. Fourth, the daily totals of S_i and I_i are similar. Finally, the midday net radiation is about ten times the magnitude of the nocturnal net radiation, and opposite in sign. The symmetry of the solar radiation curves from Gates (11) indicates that the sky was essentially cloudless on the day of observation.

VI. The Greenhouse Effect

The atmosphere, like most gases, has a different band transmissivity for shortwave radiation than for longwave. More specifically, its shortwave transmissivity is greater than its longwave transmissivity. The result, as will now be shown, is that the longwave flux density leaving the surface exceeds the shortwave flux density entering the atmosphere, and the surface temperature is therefore higher than it would be if the atmosphere were not so opaque in the longwave spectrum. Because the atmosphere's differential transmissivity plays the same role as the differential transmissivity of the glass in a horticultural greenhouse, the effect of raising the mean temperature of the earth's surface is called the "greenhouse effect."* To understand the nature of this result, consider the simplified model of radiation balance shown in Fig. 17-6. Here the three-layered system of space, atmosphere, and earth

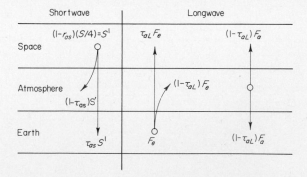

FIG. 17-6. A simplified model of radiative flux in an earth–atmosphere–space system.

* Fleagle and Businger insist that comparison of the atmosphere and a horticultural greenhouse in this regard is misleading (see reference 8, page 153).

is shown with the solar radiation fluxes on the left side and the longwave fluxes on the right. The model assumes that the earth is a black body ($a_e = 1.0$) and that the atmosphere is semitransparent but not reflective ($\tau_a = 1 - a_a$). The real atmosphere is reflective, so for this model the shortwave flux S' is considered as that entering the top of the atmosphere ($S' = (1 - r_{as})(S/4)$, where S is the solar constant). Finally, the model does not take account of the nonradiative transfer which takes place by convection between earth and atmosphere (see later).

With a format of the balance (income) = (outflow) for each of the three layers of the system, the balances are

$$\text{Space:} \quad \tau_{aL}F_\epsilon + (1 - \tau_{aL})F_a = S' \tag{17-13}$$

$$\text{Atmosphere:} \quad (1 - \tau_{aL})F_\epsilon + (1 - \tau_{as})S' = 2(1 - \tau_{aL})F_a \tag{17-14}$$

$$\text{Earth:} \quad (1 - \tau_{aL})F_a + \tau_{as}S' = F_\epsilon \tag{17-15}$$

where symbols are as follows.

Subscripts: a = atmosphere, e = earth, s = shortwave, L = longwave
τ = transmissivity = $(1 - \text{absorptivity})$, or
absorptivity = $(1 - \text{transmissivity})$ for the atmosphere,
Fluxes: S' = shortwave not reflected by the atmosphere, $F_a = \sigma T_a{}^4$, longwave
from the atmosphere; $F_\epsilon = \sigma T_\epsilon{}^4$, longwave from the earth.
Temperatures: T_a = mean temperature of the atmosphere, T_ϵ = mean
temperature of earth surface

Finally, recall Kirchhoff's Law prescribes that the atmosphere's longwave absorptivity, $(1 - \tau_{aL})$, equals its longwave emissivity.

One of the three balance equations is redundant, containing only information contained in the other two. Therefore, consider only the balances for atmosphere and earth. In addition, define the ratio $K = F_\epsilon/S'$, and eliminate F_a between the two balance equations for atmosphere and earth. The result is

$$K = \frac{F_\epsilon}{S'} = \frac{\sigma T_\epsilon{}^4}{S'} = \frac{(1 + \tau_{as})}{(1 + \tau_{aL})}, \, 0 < \tau_{as}, \tau_{aL} < 1 \tag{17-16}$$

From this relationship it is quite clear that when τ_{as} exceeds τ_{aL} as it does in the real atmosphere, then K exceeds unity and the flux density from the earth's surface exceeds that entering the atmosphere. The more τ_{as} exceeds τ_{aL}, the greater will be $F_\epsilon = \sigma T_\epsilon{}^4$ relative to S', and thus the larger will be the mean temperature T_ϵ. This is the greenhouse effect.

VII. The Radiation Inversion

One of the more important results of radiant heat transfer in the atmosphere, so far as air pollution is concerned, is the radiation inversion. An *inversion* is a vertical thermal structure in which the temperature increases in the atmosphere as altitude increases (see Section 18-III). This superposition of warmer air above colder air, as

TABLE 17-2

Numerical Simulation of the Formation of a Nocturnal Temperature Inversion by Radiant Heat Loss at the Earth's Surface[a]

Time step (about 1 hour)	Layer No.	Starting heat (arbitrary units)	Heat lost		Heat gain		Lost to space	Net loss	Ending heat
			Up	Down	Up	Down			
1	4	1000	90	90	90	90	—	0	1000
	3	1000	90	90	90	90	—	0	1000
	2	1000	90	90	90	90	—	0	1000
	1	1000	90	90	90	90	—	0	1000
	Sfc	1000	100	—	—	90	10	10	990
2	4	1000	90	90	90	90	—	0	1000
	3	1000	90	90	90	90	—	0	1000
	2	1000	90	90	90	90	—	0	1000
	1	1000	90	90	89.1	90	—	0.9	999.1
	Sfc	990	99	—	—	90	9.9	9	981
3	4	1000	90	90	90	90	—	0	1000
	3	1000	90	90	90	90	—	0	1000
	2	1000	90	90	90	90	—	0	1000
	1	999.1	89.9	89.9	88.3	90	—	1.5	997.6
	Sfc	981	98.1	—	—	89.9	9.8	8.2	972.8

4	4	1000	90	90	90	90	—	0	1000
	3	1000	90	90	90	90	—	0	1000
	2	1000	90	90	89.7	89.7	—	0.3	999.7
	1	997.6	89.7	89.7	87.6	90	—	2.0	995.6
	Sfc	972.8	97.3	—	—	89.7	9.7	7.6	965.2
5	4	1000	90	90	90	90	—	0	1000
	3	1000	90	90	90	90	—	0	1000
	2	999.7	90	90	89.7	90	—	0.3	999.4
	1	995.6	89.7	89.7	86.8	90	—	2.6	993.0
	Sfc	965.2	96.5	—	—	89.7	9.7	6.8	958.4
6	4	1000	90	90	90	90	—	0	1000
	3	1000	90	90	89.9	90	—	0.1	999.9
	2	999.4	89.9	89.4	89.4	90	—	0.4	999.0
	1	993.0	89.4	89.4	86.2	89.9	—	2.7	990.3
	Sfc	958.4	95.8	—	—	89.4	9.6	6.4	952.0

[a] Simulation rules

"Surface" layer (Sfc)

1. Loss upward = 10% of starting heat
   i. Loss to space = 1% of starting heat
   ii. Loss to layer above = 9% of starting heat

Layers 1–4

1. Loss to layer above = 9% of starting heat
2. Loss to layer below = 9% of starting heat

Notes: (a) Only the surface layer loses heat to space. (b) The surface layer neither loses heat to nor gains heat from the underlying soil.

will be discussed in some detail in the next chapter, is a major deterrent to vertical motion and thus to dilution of contaminants in the lower atmosphere.

A radiation inversion usually occurs in relatively stagnant air at night, when the chief mode of heat transfer is by longwave radiation at and near the earth–atmosphere interface. Through the interworkings of the Stefan-Boltzmann and Kirchhoff laws and the differential absorptivities of air and solid materials at the interface, the air nearest the interface cools first and most rapidly, followed by the layer of air just above, and then by the layer next above that, and so on as long as skies are clear, winds are calm, and until the sun rises.

Table 17-2 is the result of a numerical simulation intended to show the processes involved in production of a radiation inversion at the interface. The model for the simulation consists of four layers of air above the "surface" layer. The surface layer radiates and absorbs as a black body to the air layer above it and to "space" lying outside the five-layered system. Each air layer radiates and absorbs with (arbitrarily chosen) absorptivity and emissivity of 0.9, or 90%. The table is arranged in time steps (each the order of 1 hour), the first at the top. Within each time step, the layers are arranged as they are in the natural system—surface beneath lowest layer number 1, which is in turn below layer number 2, and so on. All layers begin the simulation with equal heat contents (heat units are arbitrary), and with the assumption that their thicknesses (masses) are such that all layers begin with the same temperature. For this reason, then, the column of heat contents at the right of each time group shows also the vertical structure of temperature at the end of that time segment.

All columns involving heat content and transfer are taken to represent a unit of horizontal area, and the temperatures are such that a unit area would radiate, according to Eq. (17-3), an amount of heat equal to 10% of the heat content of the volume adjacent to the unit area and within the layer. For the surface layer, therefore, 10% of the starting heat content is radiated upward in a unit time step, 0.9 of that being absorbed by the air layer above the surface, and the remaining 0.1 being lost to space according to the Kirchhoff Law. Also according to Kirchoff's Law, each air layer emits, during each time step, 0.9 of the amount a black body would emit at the temperature of the layer. The simulation rules are given below in the footnote to Table 17-2.

Aside from tracing each step in the simulation, appreciation of the following comments will aid understanding of the model.

1. The column marked Starting heat for a Time step duplicates the column marked Ending heat for the step before.

2. The sum of all entries under Net loss for a Time step equals the entry Lost to space for that step, within the margin of rounding error. For example under step 5, $(0.3 + 2.6 + 6.8)$ equals 9.7, the number of heat units lost to space from the surface layer during step 5.

3. The entry Lost up for an air layer equals three other entries: (a) the entry Lost down for the same layer, (b) Gain down for the layer below, and (c) Gain up for the layer above. For example in step 5 again, the 89.7 under Lost up for

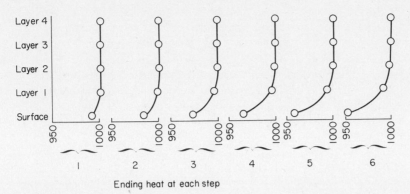

Fig. 17-7. Sequence of results from the simulation of Table 17-2, column headed Ending Heat.

layer 1 equals the same value for Lost down of layer 1, Gain down for the surface layer, and Gain up for layer 2. Furthermore the 89.7 is equal to $(0.9)(995.6/10)$ rounded.

4. Because of note (b) in the footnote, the surface layer has no entries under Lost down and Gain up. The entry Lost up is 10% of Starting heat, and equals the sum of Lost to space for the surface layer and Gain up for layer 1. In step 5 again, $(965.2/10) = 96.5 = (86.8) + (9.7)$. With this arrangement, then, the entry Net loss is the negative of the algebraic sum of the four entries headed Heat lost and Heat gain for that layer.

Although the numbers for heat units in Table 17-2 are arbitrarily chosen, they stand in realistic proportion to the losses at each step; and the value of the emissivity for the air layers is realistic in this context. Thus the simulation involves realistic numbers and gives realistic results for time steps the order of one hour each and air layers the order of 10–20 m in thickness. Figure 17-7 shows the results from Table 17-2 plotted from the entries under Ending heat. These curves are quite similar in appearance to a sequence of temperature profiles through the course of a night. Clearly, the surface layer cools first and most rapidly as the depth of the inversion layer increases slowly through the night. As the sun rises the following morning, shortwave radiant exchange begins to dominate, and the air column is heated from below as the inversion layer disappears.

VIII. Global Radiation Balance

In summary of the relationships and processes regarding radiant heat transfer, it will be useful to consider the global radiation balance. This tracing of the relative amount of energy flowing along each of the several pathways in the earth–atmosphere system is seen in Fig. 17-8. Here values are all expressed as percentages and thus are related to an arbitrary 100 units of shortwave radiation entering at the top of the atmosphere. All are averaged for the whole earth over an entire year. The

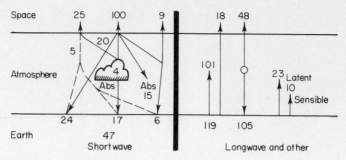

FIG. 17-8. Global radiation and heat balance. After Lowry (4).

100 units as seen above amount to 0.5 cal cm^{-2} min^{-1}, or 263 kcal cm^{-2} year^{-1}. Conditions at any one place and time might be quite different as regards the values associated with each pathway, but all the processes would be active and the energy balance at least as complex as is shown.

The diagram shows the same three components as Fig. 17-6: space, atmosphere, and earth. To the left is shortwave energy, and to the right longwave and the non-radiative fluxes of sensible and latent heat.

Of the 100 entering units, 34 are returned to space as shortwave radiation. The albedo, therefore, is 0.34. Of the 19 units absorbed within the atmosphere, four are absorbed by clouds and 15 by molecules and aerosols of the atmosphere. These molecules and aerosols scatter 6 units toward earth in addition to the 9 units scattered back to space.

TABLE 17-3

Heat Budget for the Earth's Surface, Averaged for the Entire Surface for an Entire Year[a]

Arriving	
Direct sunlight	24 units
Diffuse sunlight (through clouds)	17
Scattered sunlight (blue sky)	6
Longwave flux from atmosphere	105
	152
Leaving	
Absorbed by the atmosphere in "opaque wavelengths" (see Fig. 17-3b)	106 units
Directly to space in "transparent wavelengths" (see Fig. 17-3b)	18
Latent heat flux (evaporation)	23
Convection	10
	152

[a] After Lowry (4).

Of the shortwave energy reaching the earth's surface, 47 units are absorbed and 5 units reflected, giving a mean albedo of about 0.10 for the interface. The global albedo of 0.34 is a weighted mean of the 0.10 for the interface and the much larger value (about 0.85) for cloud tops. This may be seen by any modern air traveler in the blinding reflection from clouds and the comparatively dull appearance of most parts of the surface underneath.

The longwave diagram to the right shows 153 units being radiated from the atmosphere—105 to the earth's surface and 48 to space. In addition, the earth's surface itself is radiating 119 units—18 directly to space (see Table 17-2) and 101 to the atmosphere. The 153 units emitted by the atmosphere and the 119 units by the earth's surface involve all sides of the globe, whereas the shortwave incident energy of 100 units involves only the sunlit side of the globe at any one time.

To satisfy oneself that the diagrams do indeed represent a balance, one may check to see that the sums associated with arrows whose heads terminate in any one of the three portions of the system are equal to the sums associated with the arrows originating in that same portion. For example, the budget for the earth's surface is shown in Table 17-3.

References

1. Fuller, R., "Operating Manual for Spaceship Earth," 143 pp. Southern Illinois Univ. Press, Carbondale, Illinois, 1968.
2. Wald, G., *Science* **150**, 1239 (1965).
3. Various writers replying to Wald, *Science*, **151**, 400–403 (1966).
4. Lowry, W., "Weather and Life: An Introduction to Biometeorology," Chapter 3. Academic Press, New York, 1970.
5. Widger, W., *Bull. Amer. Meteorol. Soc.* **49**(7), 724–725 (1968).
6. VonderHaar, T., and Suomi, V., *Science* **163**, 667–669 (1969).
7. Sellers, W., "Physical Climatology," 272 pp. Univ. of Chicago Press, Chicago, Illinois, 1965.
8. Fleagle, R., and Businger, J., "Introduction to Physical Meteorology," 346 pp. Academic Press, New York, 1963.
9. McCormick, R., and Baulch, D., *J. Air Pollut. Contr. Ass.* **12**, 492 (1962).
10. Reproduced, among numerous other places, as Fig. 1.10 in Barry, R., and Chorley, R., "Atmosphere, Weather and Climate," 320 pp. Holt, New York, 1970.
11. Gates, D., "Agricultural Meteorology," Chapter 1, *in* Meteorological Monographs, Vol. 6. American Meteorological Society, Boston, 1965.

Suggested Reading

Charney, J., Radiation, *in* "Handbook of Meteorology." (F. Berry, E. Bollay, and N. Beers, eds.). McGraw-Hill, New York, 1945.
Gates, D. M., "Energy Exchange in the Biosphere," 151 pp. Harper, New York, 1962.
Goody, R. M., "Atmospheric Radiation," 436 pp. Oxford Univ. Press (Clarendon), London and New York, 1964.
Kondratyev, K., "Radiative Heat Exchange in the Atmosphere," 411 pp. Pergamon, Oxford, 1965.
Kondratyev, K., "Radiation in the Atmosphere," 912 pp. Academic Press, New York, 1969.

Questions

1. (a) By actual calculation, confirm the values of $E_{\nu\ max}$ for $T = 5000°K$ and $6000°K$ given in Fig. 17-1.
   (b) For a black body of temperature $300°K$, calculate ν_{max} and λ_{max}.
2. Calculate the ratio of the energy radiated by a material of emissivity 0.9 and temperature $27.3°C$, and the energy radiated by another material of emissivity 0.5 and temperature $0°C$.
3. Calculate the percentage increase in the energy emitted by a material of emissivity 0.8 when its temperature increases from $40°$ to $100°F$ if its emissivity does not change with this heating.
4. Estimate the new effective radiant temperature of the earth–atmosphere system if satellite radiometric data show a mean flux intensity of 0.4 cal cm^{-2} min^{-1}.
5. Using Eq. (17-4) and Fig. 17-3b, estimate the absorptivity of the atmosphere for the band $8-13\ \mu$.
6. If the earth emits so little energy at the frequency of the maximum solar emission, why does it absorb so much solar energy?
7. (a) For the surface and hours determined in Problem 16-2, estimate the rate of absorption of solar energy for each date if the surface has an albedo of 20% and the atmosphere has an effective zenith transmissivity [Eq. (17-9)] of 80%.
   (b) Are these the maximum rates of absorption on these days? If not, calculate the times and rates of maximum absorption.
8. (a) Using Eq. (17-12) and assuming a cloudless atmosphere with effective zenith transmissivity 0.8 and surface vapor pressure 14 mbar, estimate the absorptivity of the pasture in Fig. 16-6a. Hint: if latitude is $45°N$, date is July 22. Figure 16-6a gives values of T_a.
   (b) Estimate the pasture's *albedo* if there had been a cloud cover of $\frac{2}{10}$ cirrostratus.
9. Using Eq. (17-16) for the greenhouse effect, and assuming a constant value of the atmosphere's longwave transmissivity at 25% (Fig. 17-3b), plot the mean temperature of the earth's surface as a function of the atmosphere's shortwave transmissivity over its full range from 0.0 to 1.0. *Hint*: For S use the satellite radiometric data of Section II or the information in Fig. 17-8.
10. Using the same simulation rules of Table 17-2, obtain the ending heat of the surface layer after Time step 3 if the absorptivity of the air layers is reduced from 90 to 70%.

Chapter 18

ATMOSPHERIC THERMODYNAMICS

The energy that drives atmospheric motion enters the atmosphere partly through radiant heat transfer and partly through sensible and latent heat transfer by convective processes near the earth–atmosphere interface. As the atmosphere receives heat by various processes, its motion then redistributes the heat obeying the laws of thermodynamics.

Students of air pollution have a basic need to understand the redistribution of heat in the atmosphere. The need arises because the redistribution involves wind and vertical motion, so important in the transport of pollutants. In addition, the redistribution is related to widespread and persistent suppression of vertical motion beneath subsidence inversions and to the natural cleansing processes associated with precipitation.

Atmospheric scientists analyze these thermodynamically based processes of heat distribution, change of state, and motion by means of a group of nomographs. There are several such diagrams (1), each designed with a particular set of problems in mind. Probably the best known among meteorologists, and therefore the one most likely to be encountered by air pollution control personnel in conversations with meteorologists, is the *pseudoadiabatic chart* (see Appendix B). With it meteorologists make conversions involving thermodynamic variables, or make rudimentary forecasts of such elements as height of cloud base. In this chapter some of the concepts and discussions may seem to the reader not to be connected directly with problems of air pollution, for example those about the several processes of cloud formation. Two objectives are intended for inclusion of such materials. First, it is considered essential that the reader become familiar with the pseudoadiabatic diagram, and these discussions broaden that familiarity. Second, the point of view in these chapters with respect to the atmosphere is that full understanding of the atmosphere as a resource is preferred to partial understanding of a few narrow concepts related only to today's acute air pollution problems.

I. Thermodynamics of Dry Air

The basic relationship depicted by the pseudoadiabatic chart is the one connecting the temperature and the pressure of dry air undergoing a constant energy (adiabatic) process. Because of this the chart has temperature and pressure coordinates. The relationships of the adiabatic process and its representation on the chart may be derived as follows.

Begin with the equation of state for a gas

$$p\beta = R_d T \tag{18-1}$$

where p is pressure, β volume per unit mass, R_d the gas constant for dry air, and T the absolute temperature (°K). Add to this foundation the First Law of Thermodynamics

$$dH = dU + dW \tag{18-2}$$

where dH is a change in the heat energy of a unit of mass of air, dU is a change in the internal energy of the unit mass, and dW is the work done by or on the unit mass. If the sample of dry air is enclosed in a chamber with one wall a frictionless piston, adding heat dH will result in an expansion with no change of pressure. That is, the heat will be converted directly and completely to work $(p \cdot d\beta)$. Hence

$$dH = dU + p \cdot d\beta \tag{18-3}$$

But if one maintains a constant volume of the sample while adding heat dH, there will be a change of temperature, the amount of which depends on the specific heat of the dry air at constant volume

$$dH = dU = c_v \cdot dT \tag{18-4}$$

Combining (18-3) and (18-4) and substituting from the differential form of Eq. (18-1) gives

$$dH = (c_v + R_d)dT - (R_d T/p)dp \tag{18-5}$$

Note that when the sample undergoes isobaric temperature changes $(dp = 0)$, the specific heat of dry air at constant pressure c_p equals $(c_v + R_d)$ by definition. Dividing by T

$$(dH/T) = c_p(dT/T) - R_d(dp/p) \tag{18-6}$$

In the adiabatic process being analyzed, $dH = 0$. Also the value of $(R_d/c_p) = 0.288$. With these and integration of (18-6), the result is *Poisson's Equation for Dry*

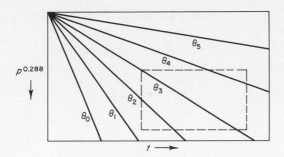

FIG. 18-1. Lines of equal potential temperature on the coordinate system of the pseudoadiabatic chart.

Air:

$$(T/T_o) = (p/p_o)^{0.288} \tag{18-7}$$

Defining *potential temperature* θ as the temperature any parcel of dry air would have if brought adiabatically to 1000 mbar of pressure (about 1 atm)*, then

$$\theta = T(1000/p)^{0.288} \tag{18-8}$$

On coordinates of T and $p^{0.288}$, Eq. (18-8) is a family of straight lines, each member associated with a value of θ. The pseudoadiabatic chart has such coordinates, the pressure ordinate inverted so that "up" on the diagram will be the same as "up" in the atmosphere. The family of lines for potential temperature are sketched in Fig. 18-1.

A portion of the diagram in Fig. 18-1, such as the one enclosed by the dashed lines, is the basis of the nomograph in Appendix B. It includes the values of the temperature and pressure encountered in the lower atmosphere—temperatures between -50 and $+50°C$ and pressure between 1050 and 400 mbar. Meteorologists call the lines of constant θ "dry adiabats." They are process lines for adiabatic motions of dry air. In addition, within an error of at most about 3%, area on the chart is proportional to the energy in a partial or cyclic thermodynamic change or process. This feature will be used presently.

II. Thermodynamics of Moist Air

The *saturation vapor pressure of water* e_s, is the partial pressure of water vapor in equilibrium with a liquid water surface at the same temperature. The functional relationship between temperature and saturation vapor pressure is given by the curve marked Relative humidity = 100% in Fig. 18-2 (also see Appendix C). For purposes of this discussion e_s is a function of temperature only. As a rule of thumb, e_s expressed in millibars doubles for each 11°C increase in temperature. Removed from a liquid water surface, air most commonly contains less than it would

* Standard sea level pressure is 1013.26 millibars (mbar).

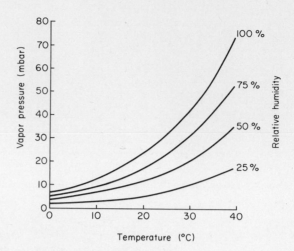

FIG. 18-2. Saturation vapor pressure of water as a function of temperature, and the relationship of relative humidity to vapor pressure and temperature.

at saturation. The *ambient vapor pressure* e is then less than e_s, and the condition is defined by the *relative humidity* r

$$r = (e/e_s) \qquad (18\text{-}9)$$

Because of this, a scale of relative humidity may be included in a second nomograph, the *psychrometric chart*, in Fig. 18-2. This second nomograph will be considered again presently.

The ratio of the mass of water vapor to the mass of dry air in a sample is the *mixing ratio, m*. It follows that the mixing ratio is also the ratio of densities of the moist and dry components of moist air

$$m = (M_v/M_d) = (\rho_v/\rho_d) \qquad (18\text{-}10)$$

where M is mass in a sample, ρ is the density in a sample, v refers to the vapor component, and d to the dry air in the sample. Since by definition $\beta = 1/\rho$, the equations of state [Eq. (18-1) above] for the two components of a sample are

$$e = \rho_v R_v T \qquad \text{and} \qquad (p - e) = \rho_d R_d T \qquad (18\text{-}11)$$

where R_v is the gas constant for water vapor and p is the total pressure in the sample. The value of the mixing ratio for saturated air ($r = 100\%$) having temperature T and pressure p is the *saturation mixing ratio, m_s*:

$$m_s = \frac{(e_s/TR_v)}{(p - e_s)/(TR_d)} = \frac{(0.622e_s)}{(p - e_s)} \qquad (18\text{-}12)$$

with the ratio of the gas constants equal to 0.622. Because of this dependence of m_s on temperature and pressure, a set of lines for values of m_s may be superimposed on the dry adiabats of the pseudoadiabatic chart, as suggested in Fig. 18-3. Note that because e_s is usually only about 2–4% of p, the approximation follows that relative humidity $r \doteq (m/m_s)$.

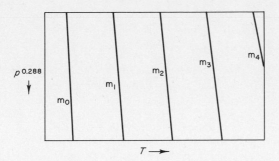

FIG. 18-3. Lines of equal saturation mixing ratio on the coordinate system of the pseudo-adiabatic chart.

Adiabatic expansion of moist air is a reversible process for only as long as any condensate that forms is not removed. As a practical matter in meteorology, however, it is convenient to approximate the process of the expansion of saturated air (i.e., lifting to lower pressure) by specifying that the condensate is lost to the rising parcel as precipitation—a pseudoadiabatic process. Since the heat of condensation supplements the internal energy as a source of heat for the work of expansion, the drop in temperature for a given drop in pressure will not be as great as in the case of dry air. Thus lines in the third set on the pseudoadiabatic chart describe this process and are called "wet adiabats." Because of the retarded drop in temperature with falling pressure, these lines are more vertical than the dry adiabats, as shown in Fig. 18-4.

As may be seen in Fig. 18-4 and in Appendix B the wet adiabats have a slope approaching that of the dry adiabats near the top of the chart. This is because the progressive removal of condensate with ever-greater lifting results in increasingly dry air. On the complete chart of Appendix B, the dry adiabats are each labeled with the value of the potential temperature θ in degrees Celsius. The wet adiabats are labeled with the value of the potential temperature of the dry adiabat which they approach asymptotically at the origin shown in Fig. 18-1. The lines of constant m_s are labeled with values of the saturation mixing ratio expressed as grams per

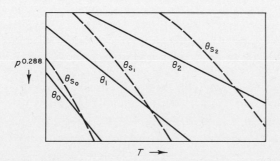

FIG. 18-4. Lines describing the wet or pseudoadiabatic process of rising saturated air on the coordinate system of the pseudoadiabatic chart.

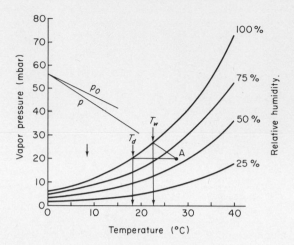

FIG. 18-5. The relationships of dewpoint temperature and wet bulb temperature to dry bulb temperature, vapor pressure, and relative humidity. Refer to Appendix C for further information.

kilogram, or parts per thousand ($\%_0$). Applications of the pseudoadiabatic chart will be presented beginning in the next section.

There are two other measures of atmospheric moisture often used in meteorological analysis. When air is cooled isobarically (without pressure change), and without addition or subtraction of moisture, condensation occurs at the dewpoint temperature T_d. Thus for example, the dewpoint associated in Fig. 18-5 with the temperature-vapor pressure combination at point A is found at the same vapor pressure and temperature T_d.

When water is added to unsaturated air and the sample is then cooled by evaporation of the added water the result of continual addition of water will be the saturation of the sample and cessation of further cooling at the *wet bulb temperature, T_w.* At this juncture the vapor pressure of the sample e_w has a value higher than the ambient vapor pressure of the sample e prior to the additions of water. Of course the value of T_w is less than the original dry bulb temperature T. The relationship among these variables is expressed by

$$T_w = T - (0.622L_v/c_{pd})[(e_w - e)/p] \qquad (18\text{-}13)$$

where L_v is the latent heat of vaporization of water, and c_{pd} is the specific heat of dry air at constant pressure. Thus, in addition to vapor pressure and temperature, the wet bulb temperature is a function also of the total pressure (barometric pressure) of the sample. This relationship is sketched in Fig. 18-5, where the effect of pressure in varying the *wet bulb depression,* $(T - T_w)$ is shown by the change of slope of the diagonal lines with change in pressure. Appendix C shows the relationship in greater detail.*

---

* If $K = (0.622\ L_v/pc_{pd})$, Eq. (18-13) becomes $T = (T_w + Ke_w) - Ke$. On coordinates of T and e (the psychrometric chart) this is a family of straight lines with slope $-K$ and intercepts on the line $e = 0$ of $(T_w + Ke_w)$. Thus for a given value of $(e_w - e)$, a decrease in p increases K and increases the value of the intercept as shown on the chart.

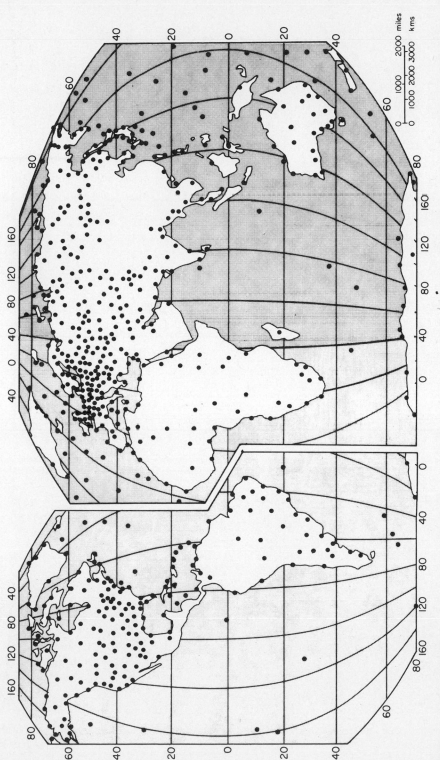

Fig. 18-6. World network of radiosonde stations in 1972. Ten stations are not shown: One at 14° W–8°S; nine in the mid-Pacific area 160°E–140°W, and 30°N–30°S. Each dot represents a station at which an upper air sounding is made each day at 0000 hr GMT, at 1200 hr GMT, or both. The source of this chart is The Secretary-General, World Meteorological Organization, Geneva, Switzerland.

FIG. 18-7. Radiosonde sensor-transmitter used by the U.S. National Weather Service. Photo courtesy of NOAA.

III. Lapse Rate and Vertical Stability

In meteorology, the rate at which the air temperature decreases with height is known at the *lapse rate* of temperature:

$$\gamma = -dT/dz \qquad\qquad (18\text{-}14)$$

where vertical is denoted by the coordinate z. Knowing the lapse rate in the atmosphere enables the meteorologist to understand much of the recent history of the air and to make certain predictions about its behavior in the hours just ahead. To learn about the thermal structure of the atmosphere meteorologists maintain a network of stations from which instrument packages, called *radiosondes*, are attached to balloons and released on regular schedules, usually once every 12 hours. The network of these stations, shown in Fig. 18-6, is a worldwide program which routinely makes vertical soundings of temperature, moisture, pressure, and wind as a basis for constructing various maps, or "snapshots," of the state of the atmosphere at an instant. The radiosonde used in the United States is shown in Fig. 18-7. Each sonde radios information to the release station from aloft. The coded information is transformed and plotted automatically to produce pictures of the vertical structure of temperature, moisture, pressure and wind above the station. A sketch of a temperature sounding is shown in Fig. 18-8. Soundings of moisture (Section 18 - V) and wind (Chapter 20) will be discussed subsequently.

In Fig. 18-8, the atmosphere exhibits a layering, each layer with a single lapse rate. In the lowest layer AB the lapse rate is greater (greater decrease of temperature with increasing height) than in any of the layers. Next above, in layer BC, the lapse rate is slightly less, and in the layer above D, still less. These three layers exhibit positive lapse rates—$\gamma > 0$. In the layer CD, however, the lapse rate is negative, marking a *temperature inversion*. As will be seen presently, a great deal may be inferred about the capacity of a layer of atmosphere to dilute or concentrate pollutants when a sounding such as this is available.

To relate the temperature sounding, and thus the lapse rate, to the *vertical*

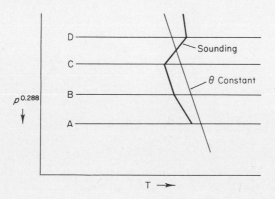

Fig. 18-8. Temperature sounding compared with a dry adiabat on coordinates of the pseudo-adiabatic chart.

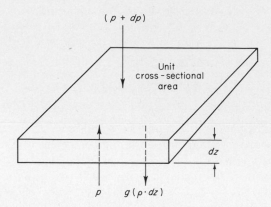

FIG. 18-9. Vertical forces acting on a unit volume of air with thickness dz and unit cross-sectional area.

stability, or tendency for vertical motion, begin with the balance of forces on a unit volume of the atmosphere as sketched in Fig. 18-9. The pressure on the bottom face is p, and the pressure on the top face is $(p + dp)$, while an additional downward force is due to the acceleration of gravity g acting on the mass $(\rho \cdot dz)$. If these three forces are in balance and the volume is at rest in the vertical

$$p - (p + dp) - \rho g \cdot dz = 0 \qquad \text{or} \qquad \partial p / \partial z = -\rho g \qquad (18\text{-}15)$$

Known as the *hydrostatic equation* for the atmosphere, this relationship says that, for the atmosphere to be vertically at rest or without any tendency for vertical motion, the density of the air at any level must bear a particular relationship to the rate of pressure change with altitude at the same level.

Rewritten, the hydrostatic equation says

$$- g - (1/\rho)(\partial p / \partial z) = 0 \qquad (18\text{-}16)$$

Similarly, if a parcel of air has the same pressure as its surroundings but a different density ρ_p Newton's Second Law implies that the resultant force per unit mass gives to the parcel a vertical acceleration

$$dV/dt = -g - (1/\rho_p)(\partial p / \partial z) \qquad (18\text{-}17)$$

where $V = dz/dt$, the vertical speed. Eliminating the pressure gradient between (18-16) and (18-17) gives

$$dV/dt = (g/\rho_p)(\rho_e - \rho_p) \qquad (18\text{-}18)$$

where subscript p refers to the parcel and e to the environment depicted by the sounding, which is presumed to be at rest. Using the equation of state in the form $p = \rho R_d T$, an alternate form of Eq. (18-18) becomes

$$dV/dt = (g/T_e)(T_p - T_e) \qquad (18\text{-}19)$$

These last two equations say what common experience has already said: if a parcel of air is brought into an environment where its density is less than that of its surroundings (or its temperature greater) at the same pressure, it will experience a positive (upward) acceleration and have a tendency to rise through the environmental air. Conversely with a higher relative density and lower relative temperature, it will tend to descend under influence of a negative acceleration.

By now the reader should have a feeling for the relationship between lapse rate and vertical stability. To make the matter explicit, consider expressions for the temperatures of two air parcels found at a distance $(z - z_o)$ above a reference level z_o. The environmental temperature is T_e, while that of a parcel lifted adiabatically from z_o is T_p

$$T_e = T_{o,e} - \gamma_e(z - z_o) \qquad \text{and} \qquad T_p = T_{o,p} - \gamma_d(z - z_o)$$

where γ_e is the lapse rate of the environment (sounding) and γ_d is the *dry adiabatic lapse rate*, $+1.0°C$ $(100 \text{ m})^{-1}$, associated with the lines of constant potential temperature on the pseudoadiabatic chart. If the lifted parcel was at environmental temperature at z_o, then $T_{o,e} = T_{o,p}$, and substitution for the two temperatures in Eq. (18-19) gives

$$dV/dt = (gz/T_e)(\gamma_e - \gamma_d) \qquad (18\text{-}20)$$

Since (g/T) is always positive, any displacement for which $\gamma_e > \gamma_d$ has an acceleration the sign of which is that of the displacement: $+z$ is upward and $-z$ is downward. This condition, in which any displacement is enhanced by the vertical temperature-density structure, is called *thermal instability*, and is seen in the lowest layer AB of Fig. 18-8. In an inversion vertical motion is suppressed because accelerations are in the direction opposite to that of displacement. Thus the relationships

$$
\begin{array}{lll}
\text{Unstable, or thermal instability when} & \gamma_e > \gamma_d & \\
\text{Neutral, or thermal neutrality when} & \gamma_e = \gamma_d & (18\text{-}21) \\
\text{Stable, or thermal stability when} & \gamma_e < \gamma_d &
\end{array}
$$

By direct extension of the criteria (18-21), comparable criteria relating stability to the lapse rate of potential temperature are

$$
\begin{array}{lll}
\text{Unstable when} & (d\theta/dz) < 0 & \\
\text{Neutral when} & (d\theta/dz) = 0 & (18\text{-}22) \\
\text{Stable when} & (d\theta/dz) > 0 &
\end{array}
$$

which agree with (18-21) by inspection of a sounding plotted on the pseudoadiabatic chart.

IV. Calculations with Surface Variables

The starting point for any use of the pseudoadiabatic chart is the set of observations of thermodynamic variables obtained at a weather station. At the standard weather station measurements are obtained of the dry bulb temperature T and the wet bulb temperature T_w. From a standard mercurial barometer in the weather station office an observation of barometric pressure p is obtained.

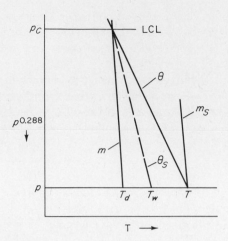

Fig. 18-10. Sketch relating values of thermodynamic variables at the earth's surface on the pseudoadiabatic chart.

The relationships among the three observations T, T_w, and p on the pseudoadiabatic chart are shown in Fig. 18-10. The value of the ambient mixing ratio, m, is found at the intersection of the dry adiabat ("θ"-Line) through T and the wet adiabat ("θ_s-line") through T_w, at a lower pressure p_c. The dewpoint temperature is then found at the intersection of the ambient m-line and the station pressure line p. The saturation mixing ratio at the station m_s is the value at T on the station pressure line. Thus beginning with T, T_w, and p, one determines m, m_s, T_d, and relative humidity $r = m/m_s$.

Here is a specific numerical example to aid understanding of these basic calculations. The results may best be checked on the pseudoadiabatic chart in Appendix B. If $T = 22°C$, $T_w = 15°C$, and $p = 1000$ mbar, then $m = 8$ ‰ (gm/kg) found at an intersection near $p_c = 850$ mbar. Following the line $m = 8$ down to the station pressure line, the dewpoint temperature is found—$T_d = 10.4°C$ approximately. Since the station saturation mixing ratio is 17 ‰, the relative humidity is about 8/17, or 47%.

The same conditions may be described by means of the psychrometric chart, Appendix C. The "point of state" on this chart is located at the intersection of the vertical line through $T = 22°C$ and the diagonal line ("sea level" $\doteq$ 1000 mbar) associated with $T_w = 15°C$ (see instructions on the chart). The relative humidity at this point may be read from the set of curving lines to be about 47%. The dewpoint temperature T_d is found at the intersection of $r = 100\%$ and the horizontal line through the point of state—10.4°C (see instructions). On the psychrometric chart the mixing ratio is not presented.

V. Convection, Condensation, and Cloud Formation

Also beginning with p, T, and T_w, a meteorologist can estimate the level at which clouds would form—the cloud base—if the air at the surface were to undergo forced lifting. Most readers of this text will, of course, not be concerned with making weather or cloud forecasts; but the following discussion should aid in a proper understanding of the pseudoadiabatic chart so fundamental in most discussions with practicing meteorologists.

When air is forced to rise, as on the windward side of a mountain range, it follows the process line marked on the pseudoadiabatic chart by the dry adiabat through the point of state (p, T). Upon rising, the air becomes saturated at the pressure p_c (Fig. 18-10) where the ambient mixing ratio line intersects the process line. In the numerical example above, this pressure is about 850 mbar. On the twin scales to the right of Appendix B, the U.S. Standard Atmosphere (on which aircraft altimeters are based) relates pressure and altitude. Thus, the cloud base with forced lifting—*the Lifting Condensation Level, LCL*—is about 1.45 km, or 4800 ft MSL. From this sketch one sees that the more moist the air before ascent, the lower the cloud base for clouds formed in this manner, and vice versa.

A more commonly encountered process of cloud formation occurs in the lower atmosphere when strata of air, with differing heat and moisture contents, are forced to mix by mechanical turbulence generated during movement over a rough surface such as low hills or a city. This process involves no addition or subtraction of heat or moisture—only a redistribution within the mixed layers. The conditions are presented in Fig. 18-11, where the initial temperature soundings are the same in both halves of the figure, but the dewpoint soundings are dry in one case and moist in the other.

In both parts of Fig. 18-11, a warmer air layer (larger values of θ) overlies a cooler layer. Upon mixing adiabatically, the layers produce a single layer with potential temperature θ_m, whose value is such as to equalize the areas ABEFA and BCDB in the figure. With a dry dewpoint sounding (Fig. 18-11a) the mixing yields a single layer with mixing ratio m_m the value of which equalizes the shaded areas bounded by the initial sounding and the line m_m—areas to the right of m_m

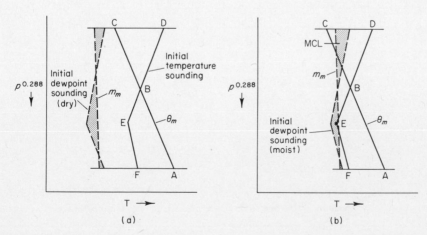

FIG. 18-11. Initial and final conditions for mixing of adjacent air layers (a) relatively dry and (b) relatively moist.

equal areas to the left of m_m. With a dry sounding, θ_m and m_m do not intersect within the single mixed layer; as a result no clouds may be expected to form during the mixing. With a moist initial dewpoint sounding, however, (Fig. 18-11b), cloud bases may be expected at the *Mixing Condensation Level MCL* within the mixed layer. Actually, the final mixed sounding would exhibit a wet adiabatic layer (constant θ_s) above the *MCL*, requiring a minor correction in the area BCBD for equalization with ABEFA. In any case, it is clear that a sounding is required for this analysis whereas none was required for the estimate of the *LCL* in Fig. 18-10.

Following the ideas in this discussion of the *MCL*, the reader may check his understanding by reference to the contents of Table 18-1. Here data on the temperature and dewpoint soundings are added to the data in the numerical example of the *LCL*, and an estimate of the *MCL* is obtained. As before, the reader should refer to the pseudoadiabatic chart in Appendix B when checking these calculations. The Table shows that the *MCL* at about 880 mbar (1.2 km or 3800 ft *MSL*) results from mixing a relatively cool dry stratum below with a relatively warm moist one above. The mixture might also have been between a cool moist stratum below and a warm dry one above and have the same outcome. This latter combination often occurs at west coastal stations and results in the "high fog"—stratocumulus clouds—so familiar to residents there. In the San Francisco Bay area, for example, the cloud base is much lower than that in the example of Table 18-1—more nearly 975 mbar or about 1000 ft—because the initial sounding is much more moist. But the process as described is essentially the same.

With respect to problems of air pollution, the process of adiabatic mechanical mixing described here transforms a stable air column, in which a temperature inversion otherwise suppresses vertical mixing, to a neutral air column in which contaminants are much more readily dispersed in the vertical. At this point, recall that the processes resulting in the *LCL* and the *MCL* are adiabatic, the former related to lifting of air without mixing, the latter to mixing without lifting.

A still more common process of cloud formation, more widely observed, is depicted in Fig. 18-12. Here initial soundings of temperature $(T_o - A)$ and dewpoint

TABLE 18-1

Sample Calculation (Refer to Appendix B) Showing Formation of Cloud Base at the Mixing Condensation Level MCL [a]

Pressure p(mbar)	Initial sounding				Mixed sounding			
	T(°C)	θ(°K)	T_d(°C)	m(gm/kg)	T(°C)	θ(°K)	T_d(°C)	m(gm/kg)
850	26	313.5	25	24	—	—	—	—
900	26	308	24	21	23	305	21.5	18
950	25	302.5	20	16	27.5	305	22	18
1000	22	295	15	8	32	305	23	18

[a] The MCL forms at the intersection of the dry adiabat $\theta = 305$°K and the mixing ratio line $m = 18$ gm/kg, which is found at $p_c = 880$ mbar.

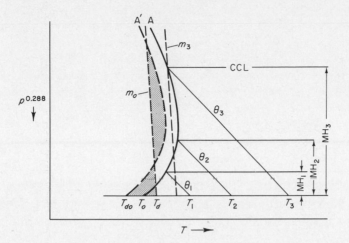

Fig. 18-12. Changes in the atmospheric sounding due to convective heating at the surface, and the resulting formation of cloud bases at the convective condensation level, CCL.

$(T_{do} - A')$ are shown as they might appear at sunrise, with a radiation inversion near the surface. Again, the sounding as shown represents the bottom two or three thousand feet of atmosphere above the station at which the sounding was obtained.

As the sun shines on the surface and the air above it is heated by convection (the H term in the heat budget of Chapter 16), the lower portion of the temperature sounding is transformed to progressively greater heights. When the surface temperature has increased to T_1 several hours after sunrise, the sounding theoretically is dry adiabatic (θ_1) up to the intersection with the original sounding, and unchanged up to A. When the surface temperature has risen to T_2 some hours later, the sounding in theory has a deeper adiabatic layer (θ_2); and so on to T_3 and θ_3. At each of these three times the lower atmosphere has been transformed by convective heating to the mixing height (MH); and the height associated with the maximum surface temperature T_M (either forecast or observed) is termed the *Maximum Mixing Height* (MMH), or sometimes the maximum mixing depth.

If the initial dewpoint sounding had an equivalent mean mixing ratio, m_o (estimated by area equalization as above), and no moisture were added to the sounding during the progress of the convective heating from below, clouds might be expected to form with bases somewhere above the top of the sounding in Fig. 18-12. If, as is more commonly the case, evaporation at the surface, accompanying the surface heating ($L_v E$ in the heat budget), had added moisture to the sounding, the dewpoint sounding would approach a greater mixing ratio, such as m_3, and clouds would form with bases at the *Convective Condensation Level CCL*. This process, of course, involves the addition of heat to the air column and is not adiabatic. As far as dispersion of pollutants is concerned, however, the result is the same as in the case of adiabatic mechanical mixing: the lower atmosphere no longer suppresses vertical motion as it did before the mixing began.

In summary, Fig. 18-13 shows sketches of the three cloud forming processes dis-

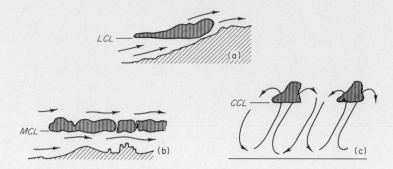

FIG. 18-13. Three types of cloud formed by processes discussed in Section 18-V.

cussed, and the clouds formed. Clouds above the *LCL* (stratus) have flat and continuous bases as the air is lifted upward by passage over major obstacles. Clouds above the *MCL* (stratocumulus) have relatively continuous but slightly roughened bases as the air is mixed mechanically by passage over hills and cities. Finally, clouds above the *CCL* (cumulus) appear over the portions of convective cells where heated air is rising, and do not appear over the portions where subsiding air completes the cells' circulation.

VI. The Subsidence Inversion

As just noted, mass continuity requires that in the atmosphere, as in any fluid, convective and cellular motion involving rising, cooling air with cloud formation implies that somewhere there must be subsiding, warming air with a suppression of cloud formation. Microscale cells smaller and macroscale cells larger than the kilometer-sized cells in Fig. 18-13c are constantly being formed and dissipated in the atmosphere.

The temperature-moisture conditions in the ascending and subsiding limbs of a convection cell are shown in Fig. 18-14a. An individual parcel of air rising from

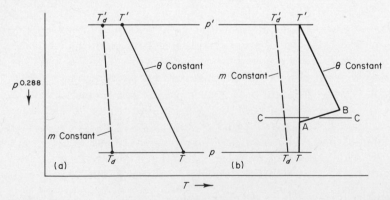

FIG. 18-14. (a) Parcel process lines in convection cells without condensation. (b) Formation of a subsidence inversion.

pressure p, where its temperature and dewpoint are T and T_d, will have temperature and dewpoint T' and T_d' at pressure p' if no condensation takes place. Since in this figure the two process lines do not intersect, no condensation takes place. As the parcel subsides toward higher pressure in the descending limb of the cell, its temperature and dewpoint return to their original values at pressure p. During the rise, the saturation mixing ratio decreases while the ambient mixing ratio remains constant; thus the relative humidity (m/m_s) increases. It then decreases again during subsidence.

In Fig. 18-14b, an initial temperature sounding TAT' is isothermal and stable. The dewpoint sounding is $T_d T_d'$ with a constant mixing ratio. If macroscale processes are such that air subsides only above the level C—C while the air below C—C remains in position and unheated, the result is a sounding TABT', exhibiting a strong inversion layer between A and B. Here the air subsiding from pressure p' to the level C—C must be replacing air previously below it and above C—C. Various circumstances in which this occurrence is observed will be discussed later. The resulting inversion, which exhibits a sharp increase in temperature and decrease in relative humidity, is based above the surface (unlike the radiation inversion) and is called a *subsidence inversion*.

VII. Potential Instability

The macroscale subsidence leading to a subsidence inversion implies that somewhere else there is macroscale lifting of a comparable magnitude. As we have seen, with the proper vertical distribution of heat and moisture, subsidence leads to dry, stable air which suppresses cloud formation and the dispersion of pollutants. Similarly, under proper conditions of heat and moisture, macroscale lifting of an entire

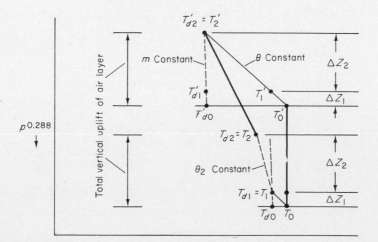

FIG. 18-15. Changes in a potentially unstable atmospheric sounding as the entire layer is lifted.

air layer can produce unstable conditions leading to rapid and effective dispersion of contaminants and formation of clouds and precipitation. This happens most readily under conditions of *potential instability* when a relatively stable sounding exhibits distinctly higher mixing ratios below than aloft, as in Fig. 18-15.

The initial conditions of the air layer are temperature and dewpoint T_0 and T_{d0} at the base of the layer and T_0' and T_{d0}' at the top. Thus, the temperature sounding (shown as isothermal) is $T_0 T_0'$. Note the "dewpoint spread" $T_0 - T_{d0}$ is much less at the base than $T_0' - T_{d0}'$ at the top, indicating dry air aloft.

The entire layer is lifted, as for example riding up a mountain range or over an underlying wedge of denser air, by a height increment of Δz_1. Note this involves height changes (not pressure changes) being equal at the base and at the top of the layer. During this first incremental lift, Δz_1, the base temperature decreases along the dry adiabatic process line to T_1 while the base dewpoint decreases along the m-line to T_{d1}, at which point saturation is reached at the base of the layer so that $T_1 = T_{d1}$. At the dry top of the layer, however, temperature and dewpoint decrease to T_1' and T_{d1}', but they are not equal and saturation is not yet obtained.

With the base of the layer now saturated, further lifting will result in wet adiabatic cooling along the process line θ_s at the base and along the line θ at the top. When saturation is reached at the top of the layer, and thus everywhere in the layer by now, the sounding is between $T_{d2} = T_2$ and $T_{d2}' = T_2'$. The sounding has been changed by the lifting from one with a zero lapse rate (isothermal) to one with a positive (less stable) lapse rate.

VIII. Conservatism of Atmospheric Properties

Implicit in the discussions just above has been the idea that under certain conditions and processes certain variables associated with an air parcel change while

TABLE 18-2

Conservatism (C) and Nonconservatism (NC) of Six Properties with Respect to Three Processes

Property	Process[a]		
	A	B	C
θ	NC	C	NC
m	C	C	NC
T_d	C	NC	NC
e	C	NC	NC
r	NC	NC	C
T_w	NC	NC	NC

[a] A = Isobaric warming-cooling without condensation.
B = Dry adiabatic without condensation.
C = Wet adiabatic lifting.

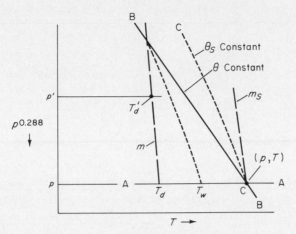

FIG. 18-16. Relationships among six properties and three processes given in Table 18-2.

others do not. A property is said to be conservative with respect to a specified process if it remains invariant under that process.

Table 18-2 gives a list of six properties and indicates for three processes if they are conservative (C) or nonconservative (NC). The first property, potential temperature θ, refers to the heat content of the air parcel, while the remaining five refer to the moisture content. Figure 18-16 will facilitate discussion.

Begin with process A, isobaric warming and cooling, which takes place along process line A-A. Any change of temperature will imply a change in potential temperature, so the combination is NC. In the absence of any condensation (line A-A to the right of T_d) both m and T_d will be invariant: C. So long as the total pressure p and the mixing ratio do not change, the vapor pressure will not change, and it is C. The relative humidity r, on the other hand, is the ratio of m and whatever value of m_s is associated with the point of state (p, T). Since m_s changes with θ, r is not conservative NC. Finally, since T_w must lie on the θ_s determined by m and θ, and since θ changes, T_w is NC.

Process B is the dry adiabatic expansion and compression discussed in, for example, Fig. 18-14. By definition this is a line of constant θ, and so long as the point of state does not move to the left of m—no condensation—m will remain C. During a major pressure change, as from p to p', the value of the dewpoint decreases only from T_d to T_d'. Thus, although technically not conservative under process B, the dewpoint is nearly invariant for minor changes of p, as in day-to-day pressure changes at a weather station (constant elevation). Thus, for example, dewpoint rather than wet bulb temperature is reported along with dry bulb in the hourly data transmission from a weather station located at an airport. Consideration of the other three properties with respect to process B will be left to the reader.

Process C takes place along line C-C and is restricted to expansion: that is, to decrease in pressure. The wet adiabat C-C crosses dry adiabats, so θ is NC. Since relatively humidity is always 100% under this process, and therefore r is C, the mixing ratio is m_s and changes with any drop in temperature. Vapor pressure must change with any change in total pressure, as in Process B, so e is NC. In Process C, T_w equals the air temperature, so T_w is also NC.

IX. Cross-Sectional Analysis—Air Masses

In discussions just above, various thermodynamic processes and changes in air layers have been considered. The idea has been implicit that the vertical structure of heat and moisture in the columns is typical of other columns nearby; and in fact that there is considerable horizontal homogeneity in the air across significantly large areas. This horizontal homogeneity implied in Figs. 18-11, 18-12, 18-14, and 18-15 is actually realized in the atmosphere in what are called *air masses*. The processes in these four figures are observed in the lowest kilometer or so of the atmosphere—the upper mesoscale or lower macroscale—but air masses often exhibit their characteristic structures through several times that depth—up to 3 or 4 km (700–600 mbar; 10,000–13,000 ft).

When a volume of air remains semi-stationary over a relatively homogeneous surface, or when it moves slowly across the surface, it exchanges heat and moisture with the underlying surface. After several days or a week, an equilibrium is approached in which the vertical exchanges, or fluxes, are reduced as initial contrasts between air and surface disappear.

If the surface is a heat source with respect to the air above, heat is mixed upward until the air mass approaches a constant potential temperature throughout a rather great depth—the greater the original contrast, the greater the depth (see Fig. 18-12). If the surface is a heat sink with respect to the air, the processes of radiant heat exchange lead to a deeper and deeper radiation inversion (Table 17-2 and Fig. 17-7).

If the underlying surface is a moisture source with respect to the air mass, the air will exhibit a deeper and deeper moist layer within the constraints which the temperature structure imposes through condensation and cloud formation. The air mass is rarely a moisture source with respect to the surface over periods of 4–7 days required to form the equilibrium and horizontal homogeneity.

Table 18-3 gives the basic scheme, adopted over much of the western world, for describing and designating air masses (2). Figure 18-17 shows the locations and ex-

TABLE 18-3

Terminology of Air Mass Designations

Nature of underlying surface	Latitudinal zone of source region		
	Tropical = T	Polar = P	Arctic = A
c = Continental	cT = Mainly summer in North America	cP = Mainly winter	cA = Winter only
m = Maritime	mT	mP	mA = Summer only

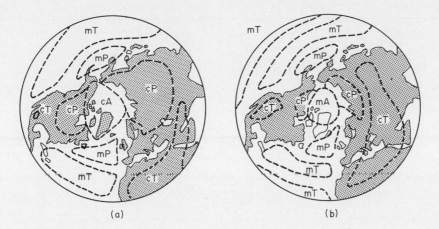

Fig. 18-17. Source regions for Northern Hemisphere air masses in (a) winter and (b) summer.

tents of the principal air mass *source regions* in the northern hemisphere. Of particular note is the virtual disappearance of the *cT* source during winter in North America, and the Arctic Ocean's change from a continental source in winter to a maritime source during summer, because of the appearance of open water in summer.

As air masses during formation take on a predictable vertical structure by interaction with underlying surfaces, so they are modified predictably when they move out of their source regions over different surfaces. A stable air mass passing over a

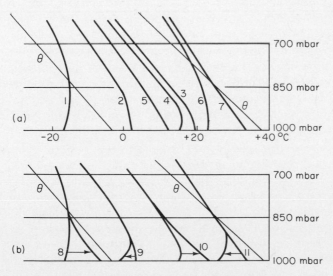

Fig. 18-18. (a) Characteristic soundings in major air masses. Winter: 1 = cP; 2 = mP; 3 = mT; summer: 4 = cP; 5 = mP; 6 = mT; 7 = cT. (b) Typical modifications of air masses during passage out of their source regions. 8 = cP over Great Lakes; 9 = mP over Northwest or northern Europe; 10 = cP over prairies or steppes; 11 = cT over Mediterranean.

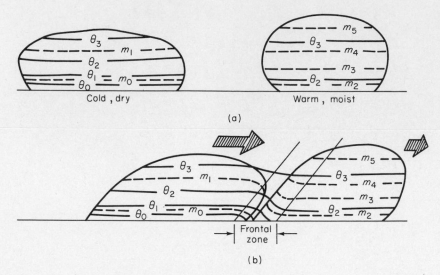

FIG. 18-19. (a) Schematic view of air mass structure. (b) The contrasts evident with air mass convergence along the frontal zone.

relatively warm surface outside its source region becomes relatively warm and unstable in its lower layers in much the same way as shown in Fig. 18-12. If the new surface is also a moisture source, as are the Great Lakes with respect to cP (continental, mainly winter) air from Canada in early winter, the result is gusty, turbulent weather with heavy precipitation falling from cumulus clouds. If a relatively warm, moist air mass moves over a cold surface, as when mP (maritime, mainly winter) air from the Gulf of Alaska moves inland during the winter in the Pacific Northwest, there is stabilization and condensation into stratus clouds in the lower layers as air is cooled to the dewpoint by the flux of heat to the cold surface. Because of the nature of buoyancy, there is modification through a deeper layer in the air being destabilized over the Great Lakes than in the air being stabilized over the Northwest. Figure 18-18 shows typical characteristics of the major air masses in the format of the pseudoadiabatic chart. Also shown are several typical results of air mass modification.

The principal reason for considering air masses and their modification in this chapter is to set the stage for an appreciation later of the macroscale storm systems of midlatitudes (see Chapter 20). Broadly speaking, the latitude (see Fig. 16-8) at which the oceanic gyres converge in western oceans and diverge in eastern oceans is the dividing line between polar and arctic source regions to the north and tropical source regions to the south. As shown, this division shifts northward in summer and southward in winter, as suggested also in Fig. 18-17. Again broadly speaking, air masses move out of their source regions toward the zone of convergence—called the *polar front*—and become eddies in a turbulent, eastward-moving macroscale stream in the atmosphere. The structure of a cold, dry, stable air mass as it approaches a warmer, less stable air mass is suggested in Fig. 18-19. The differences in structure are emphasized by showing heat and moisture content in terms of two properties that are conservative with respect to dry adiabatic motion—potential temperature and mixing ratio.

X. Cross-Sectional Analysis—Dispersion of Pollutants

A direct application of the conservatism of some properties arises from such vertical cross sections as shown in Fig. 18-19. Over short periods of time, say 1–2 days, there is a tendency for air, and any pollutants embedded in it, to move along surfaces of constant potential temperature—*isentropes.* Figure 18-20 shows another cross-sectional analysis of potential temperature for a mesoscale situation involving a shoreline and nearby hills or mountains.

The conservatism of potential temperature θ under dry adiabatic mixing insures that the analysis of isentropes in Fig. 18-20 depicts the status of heat content in the cross section. Thus the figure shows the atmosphere to be warmer inland from the shoreline and upward above the coastal strip. The introduction of heat into the atmosphere above sea level by way of heat exchange on the mountains results in the existence of a heat source relative to a horizontal transect through the cross section. Vertical isentropes mark areas of thorough vertical mixing where the lapse rate is dry adiabatic, as in the lightly shaded area of Fig. 18-20a. Areas in which θ decreases with altitude (heavily shaded) exhibit superadiabatic lapse rates due to strong surface heating. Areas in which θ increases with altitude are the rule, and mark regions in which lapse rates are less than dry adiabatic. The steep vertical gradient of θ over the coastal strip at night represents a radiation inversion (3).

If pollutants are injected into the atmosphere at point A (above shoreline) during the day, and move along an isentrope to point B before nighttime, they will

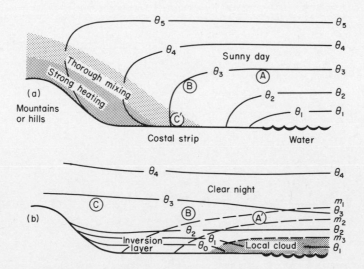

Fig. 18-20. Mesoscale cross section of potential temperature across a shoreline and mountain range (a) during a sunny day; (b) the following night.

probably move along the same isentrope to point C during that night. However, if the pollutants are injected at point A' during the night, and move to point B before daybreak, their expected movement during the following day will be to the surface near C'. As will be discussed in a later chapter, the pollutants will undergo considerable diffusive dilution en route during the times suggested in Fig. 18-20, but their movement should nonetheless be generally in the manner shown.

Cross-sectional analysis of mixing ratio—also conservative for vertical mixing—suggests areas in which stratus cloud formation might be expected offshore (Fig. 18-20b). The reader will recognize that the data requirements for routine analyses of this type are beyond the limits of most programs, but the intent in mentioning the concept at this point, in the context of thermodynamics, is to begin preparing the reader for an understanding of the complexity of the atmospheric transport of pollutants when the process is viewed on all the time and space scales described in Chapter 16.

References

1. For example, see Chapter 2 of Haltiner, G. J., and Martin, F. L., "Dynamical and Physical Meteorology," 470 pp. McGraw-Hill, New York, 1957.
2. For example, see Barry, R., and Chorley, R., "Atmosphere, Weather, and Climate," pp. 119–129. Holt, New York, 1970.
3. For examples of this form of mesoscale cross-sectional analysis, see Cramer, O. P., and Lynott, R. E., *Bull. Amer. Meteorol. Soc.* **42**, 693–702 (1961), and *J. Appl. Meteorol.* **9**(5), 740–759 (1970).

Suggested Reading

Byers, H. R., "General Meteorology," 540 pp. McGraw-Hill, New York, 1959.
Hess, S. L., "Introduction to Theoretical Meteorology," 362 pp. Holt, New York, 1959.

Questions

1. In order to compare the ordinate scale of the pseudoadiabatic diagram with a logarithmic scale, plot $p^{0.288}$ versus $\ln p$ over the range $400 \leq p \leq 1000$.
2. Complete the blanks in each of the sets of values:

	T	T_w	T_d	e_s	e	r	p	m_s	m
(a)	—	—	15°C	—	—	50%	1000 mbar	—	—
(b)	25°C	—	—	—	20 mbar	—	1000 mbar	—	—
(c)	20°C	—	—	—	—	40%	—	20‰	—
(d)	—	15°C	10°C	—	—	—	800 mbar	—	—

3. (a) Estimate, with Eq. (18-15) and Fig. 16-4, the distance (m) between pressure surfaces 10 mbar apart at Denver, Colorado, "The Mile High City." Compare this estimate with the standard atmosphere on the right margin of the pseudoadiabatic diagram.
   (b) Do the same for Amsterdam, Holland.

4. (a) Calculate the direction and magnitude of the acceleration per unit mass on an air parcel with temperature 10°C that has been compressed isothermally in a dry adiabatic atmosphere from 800 to 1000 mbar.
   (b) If the compressed parcel becomes saturated with moisture while assuming the environmental temperature at 1000 mbar, what would its temperature be if it were then lifted back to 800 mbar?
5. What is the *Lifting Condensation Level* (*LCL*) for each of the four conditions of Question 2?
6. The morning sounding shows the following:

$$p = 1000 \text{ mbar} \quad 950 \quad 900 \quad 850 \quad 800 \quad 750 \quad 700 \text{ mbar}$$
$$T = \quad 7°C \quad 8 \quad 8 \quad 7 \quad 5 \quad 3 \quad 0°C$$

   (a) If the afternoon maximum temperature at the station (1000 mbar) is 20°C, what will be the *Maximum Mixing Height* (*MMH*)?
   (b) If the mixing ratio below the *MMH* is 8‰ at the time of the maximum temperature, will there be a cloud base in the layer? If so, at what *CCL*?
   (c) What is the maximum possible mixing ratio in the layer below the *MMH* without cloud formation in the layer?
7. Develop scales (for example, a series of marks on a straight edge) with which to locate and mark all layers of air on a cross section such as Fig. 18-20a with a vertical scale of 1 km = 10 cm, having lapse rates between dry adiabatic and wet adiabatic.

Chapter 19

ATMOSPHERIC PRECIPITATION

Precipitation of water from the atmosphere, in various forms, is the product of a complex set of processes that represent much of the atmosphere's natural self-cleansing mechanism. Rain, snow, sleet, hail, and other forms of precipitation may seem to the casual observer as simply "the thing that happens when there are clouds and wind." On the contrary, precipitation is not only the result of an intricate web of processes in the hydrologic cycle; it is part of the cycles of many substances of which particulate and gaseous aerosols are of interest in this text.

In this chapter the reader should obtain a feel for the complexity of the web, though by no means a definitive summary of it. The reader should obtain a feeling for the nature of the processes at issue as a basis for understanding and evaluating theories and proposals that continue to be put forth concerning the role of air pollutants in changing the behavior of the atmospheric resource, on all scales from micro to macro. The effects of precipitation on sulfur dioxide released from a tall industrial stack is an example (see Section 19-V). The suspected effects of pollutants on rainfall patterns over and downwind of major urban areas is another. Then too, many of the concepts in this chapter permit a better understanding of the "wet" technology of air pollution control.

The ingredients, processes, and products of the web of precipitation phenomena are depicted in Fig. 19-1. Materials are denoted in boxes, and processes are denoted by text accompanying connecting arrows. The diagram was suggested by the much

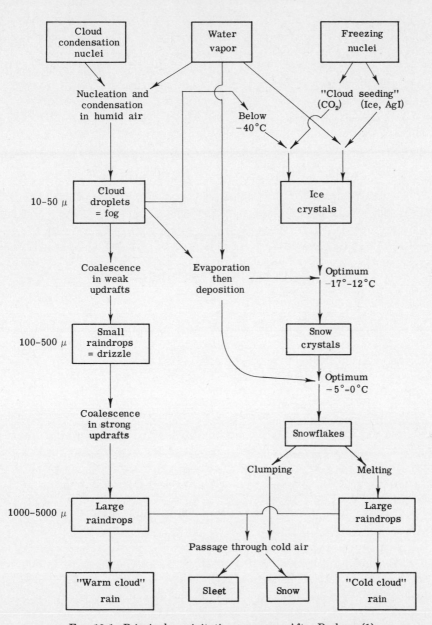

FIG. 19-1. Principal precipitation processes. After Braham (1).

more definitive diagram of Braham (1), and will serve as the basis for most discussions in this chapter.

I. Cloud Droplets and Fog

By a process not yet completely understood, small hygroscopic particles, called *cloud condensation nuclei*, are "activated" in the presence of humid air so that water molecules wet the exteriors of the nuclei. The 30 to 1000 activated nuclei per cubic centimeter are a minor fraction of the total number present and available for activation in supersaturated air. Subsequently more water vapor condenses on the wetted nuclei, which are usually between 0.01 and 1.0 μ in diameter. The ultimate size of such a growing cloud droplet is determined by the size and composition of the cloud condensation nucleus and by the degree of saturation of the air in which it grows.

Raoult's Law states that the relative humidity of the air at the surface of a salt solution where evaporation equals condensation is given by

$$r_{sa} = M_w/(M_w + M_s) \qquad (19\text{-}1)$$

where M_w is the mass of water and M_s the mass of solute per unit volume of solution. At the surface film on a wetted cloud nucleus, therefore, condensation will tend to occur as long as r, the ambient relative humidity of the air exceeds r_{sa}. The rate of condensation tends to be greater the more r departs from r_{sa}.

A countervailing tendency to any condensation of water vapor on small droplets is produced by the increased value of the equilibrium relative humidity at the surface of droplets with large curvature (small diameter). The equilibrium relative humidity as a function of this curvature is given by

$$r_c = \exp\ (Ks/d') \qquad \text{or} \qquad \ln r_c = Ks/d' \qquad (19\text{-}2)$$

where s is the surface tension of the droplet, d' is the diameter, and K is a constant involving the temperature and density of the liquid in the droplet.

Thus there is a tendency for droplet growth by condensation if r is larger than r_{sa}

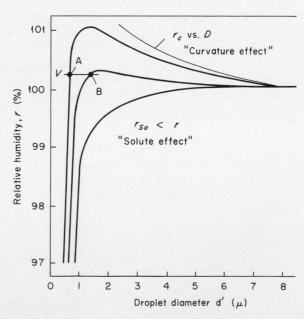

Fig. 19-2. Equilibrium relative humidity related to droplet diameter for droplets of three initial sizes.

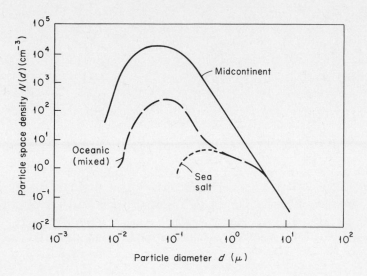

FIG. 19-3. Size distributions (space density) of natural atmospheric particulate aerosols. After Junge (2).

over a salt solution, but a tendency for evaporation if r is smaller than r_c over a curved water surface. Whether condensation proceeds or ceases is determined in a manner shown in Fig. 19-2. Here the droplet diameter at first activation is suggested for three droplets in the lower left corner. If all three have essentially the same film thickness of water, and therefore about the same salinity in the film, the largest of the droplets has the largest condensation nucleus. All three grow rapidly because r_{sa} is less than r (100.25% in this example) by an amount greater than r_c exceeds r.

As the droplets grow, and the salinity of the film decreases, r_{sa} increases (Raoult's Law), while r_c decreases. The effective equilibrium relative humidity, which is a balance between r_{sa} and r_c, is related to droplet diameter along each one of the three curves beginning at small values of d'. For all three, a large droplet with a small curvature and weak film solution has an equilibrium relative humidity that approaches 100%.

Finally, it may be seen in Fig. 19-2 that the ultimate size of a growing droplet is determined in large measure by the degree of supersaturation of the air in which it grows. Thus for the smallest droplet pictured the diameter of its greatest possible growth is associated with point A, where the equilibrium relative humidity equals the ambient relative humidity r. For the middle-sized droplet the greatest droplet diameter is associated with point B. The curve for the largest of the three droplets illustrated does not rise above $r = 100.25\%$, so the droplet will continue to grow, though less and less rapidly, to diameters exceeding several microns.

Figure 19-3 shows the spectrum of sizes and relative numbers of atmospheric aerosol particles, the most numerous being in the size range required of cloud condensation nuclei. In this diagram due to Junge (2), the ordinate is the *particle space density* $N(d) = dN/d[\ln d]$, where N is the total concentration of particles of diameter less than d. Thus analogously to the radiant energy in the spectrum of Fig.

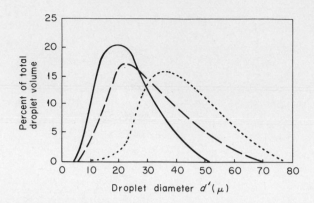

FIG. 19-4. Typical size spectra of cloud droplets in three regions. For interpretation of the ordinate, see question 5. (———), Southwestern U.S.; (– –), Central U.S.; (· · ·), Caribbean. After Day and Sternes (3).

17-1, the number of particles per unit volume in the band between d and $(d + \Delta d)$ is $\Delta N = N(d) \cdot \Delta[\ln d]$ with units (particles per unit volume per unit bandwidth), or cm^{-3} as shown on the ordinate in the figure. For example, at $d = 1$ micron (10^0 on the abscissa), $N(d) \doteq 60$ cm^{-3} (midcontinent) and at $d = 10^{-1}$ micron, $N(d) \doteq 1.6 \times 10^4$ cm^{-3}. For a bandwidth of $\ln d = 0.05$, in the first case $\Delta d \doteq 0.051$ and in the second $\Delta d \doteq 0.005$. Thus, in the band between 1.000 and 1.051 μ, one may expect to find $\Delta N = (60)(0.05) = 3$ particles cm^{-3}. In the band between 0.10 and 0.105 μ, one may expect to find $\Delta N = (1.6 \times 10^4)(0.05) = 800$ particles cm^{-3}. The particles range in origin through airborne dust and tiny salt crystals entering the air from the foam of breaking ocean waves. Through the process described by Fig. 19-2, a mixture of different sized nuclei will grow into cloud droplets having different diameters, but limited generally to the range 10–50 μ (Figs. 19-1 and 19-4).

Various meteorological events lead to the supersaturation of air and the formation of cloud droplets. If warm air passes over a distinctly colder surface, or if cold dry air passes over a warmer water surface, the resulting ambient relative humidity may slightly exceed 100%. Similarly, if air is forced rapidly aloft, the adiabatic cooling from reduced pressures may produce supersaturation, especially if the lifting rate is large enough to keep cooling ahead of the warming by radiation from air surrounding the rising parcel. Finally, the radiative cooling described in connection with Fig. 17-7 may well produce supersaturation and cloud droplet formation. If the rapid production of cloud droplets occurs in air lying on the earth's surface, the result is called *fog*. The fog will thin or dissipate if the source of supersaturating vapor is cut off—by removal from a water surface or by entrainment of dry air into the fog bank—or if the ambient relative humidity is lowered by the addition of heat, as in the morning sun.

In a fog bank, or in a cloud, the spectrum of droplet sizes tends to remain relatively constant with time, the spectrum depending on the sources (and therefore the size spectrum) of the original nuclei. Figure 19-4 suggests the nature of these droplet spectra.

II. Coalescence, Drizzle, and Warm Cloud Rain

All of the airborne materials mentioned—dust particles, salt crystals, and cloud droplets—"fall toward earth." However, if the air surrounding them is an updraft the speed of which equals their falling speed, they remain at the same height relative to the earth. More generally any falling object has a terminal velocity V_t which is the speed whose associated drag on the object equals the difference between the gravitational and buoyant forces on the object. In still air, the buoyant force is essentially zero; but for the object held aloft in an updraft, it is not.

Figure 19-5 shows the terminal velocity of a water drop as a function of its diameter.* According to Byers (reference 4, p. 148) drops with diameters less than about 80 μ attain velocities that follow Stokes' Law

$$V_t = (gD^2/18\mu)(\rho - \rho_a) \tag{19-3}$$

while velocities for larger drops are obtained experimentally. Experimental results include the effects of the deformation which the drops undergo when acted on by the various forces in their environment. In Stokes' Law, g is the acceleration of gravity, μ is the viscosity of air, ρ is the density of the drop, and ρ_a the density of air.

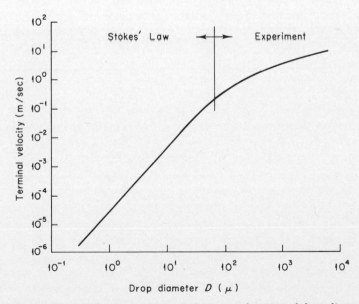

FIG. 19-5. Terminal velocity of water drops as a function of drop diameter.

* These values differ from those in Fig. 3-2 because of the smaller density and greater deformability of water drops as compared with the particles in that presentation.

As an example from Fig. 19-5, V_t for drops with $D = 10$ and 200 μ are 0.006 and 0.9 m/sec, respectively. In the absence of an updraft, the larger drop would pass the smaller droplet in their race toward earth at a relative speed of 0.894 m/sec. But in an updraft of 0.5 m/sec the droplet would be racing upward at 0.494 m/sec, while the drop fell closer to the earth at 0.4 m/sec.

With drops moving vertically relatively to each other in a cloud, with or without updrafts, collisions occur between drops. Updrafts increase a drop's residence time in a cloud and thus increase the likelihood it will collide with another drop. When a drop falls through the air, it sweeps out a volume of air. The volume must separate to allow the drop to pass through it, and smaller droplets in the volume tend to follow the air in which they are embedded. Because of their inertia, their electrical charges, or their responses to local diffusive processes, some of the droplets cross streamlines and collide with the passing drop. The fraction of droplets in the swept volume that eventually collides with the sweeping drop is called the *collision efficiency* (or target efficiency). If the sweeping drop has a turbulent wake, it may actually collide with droplets brought in on the rear from outside the swept volume. In this case the collision efficiency may exceed 1.0. Figure 19-6 shows typical efficiencies as a function of the diameters of the water drop and droplet. Very small droplets, with low inertia, seldom collide with a drop of any size.

Upon collision, most droplets coalesce with or become part of the drop, increasing its volume, surface area, and diameter. Thus by *coalescence* the larger droplets in a spectrum such as those in Fig. 19-4 grow at the expense of smaller droplets into the range of diameters between 100 and 500 μ (Fig. 19-1). These small raindrops contain between about 10,000 of the larger maritime droplets and a million of the smaller continental droplets. This estimate arises by noting that for the simple case in which droplets are all of diameter d' and drops all of diameter D, the number of droplets per drop is $(D/d')^3$.

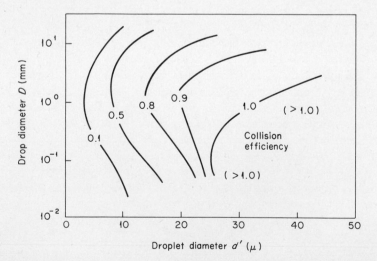

Fig. 19-6. Typical collision efficiencies of drops and droplets. After Engelmann (5).

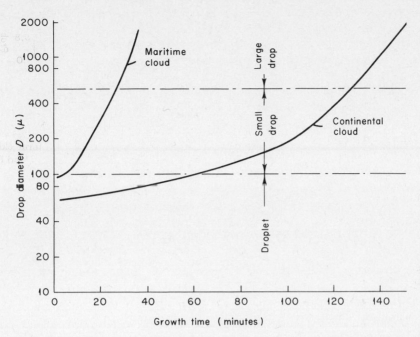

FIG. 19-7. Calculated size–time relationships for drop growth by coalescence, related to initial cloud droplet spectra. After Day and Sternes, from Braham (3).

With continued residence in a cloud and further opportunity for coalescence, a small rain drop (100–500 μ) grows into a large rain drop (1000–5000 μ) and reaches the surface as *warm cloud rain*. Calculations by Braham (reference 3, p. 247) show typical residence times (Fig. 19-7) involved in reaching the small cloud drop size range, then growing into the large cloud drop size range, as related to the initial size spectrum of nuclei and droplets.

The arrival of small rain drops at the earth's surface is termed *drizzle*. This occurrence is conditioned by two basic and opposing factors. First, the longer the drop stays in the cloud (residence time proportional to updraft strength or cloud thickness) the more subject it is to growth by coalescence into a large rain drop. Second, the longer the drop stays in the drier air beneath the cloud, the more subject it is to evaporative erosion (which is inversely proportional to surface relative humidity and directly proportional to the height of the cloud base). Figure 19-8a, for example, shows that Findeisen's calculations predict a 10 km fall is necessary for a large raindrop to be eroded to the size of drizzle under the below-cloud conditions specified in the figure.

A typical relationship among cloud base, cloud thickness and the probability of rain or drizzle at the surface is shown in Fig. 19-8b. As expected, the greater residence times associated with thicker clouds increase the probability that precipitation will reach the ground as rain rather than as drizzle. The cloud bases represented in Fig. 19-8b are too low for the effects of evaporative erosion to be of consequence, but the unexpectedly lower probability of rain from low clouds (base 0–700 m) suggests

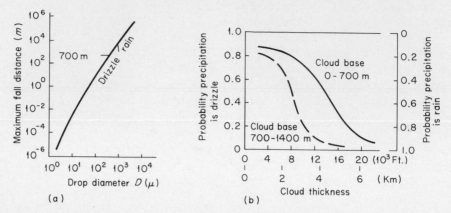

FIG. 19-8. (a) Maximum fall distance possible without disappearance by evaporation, related to drop size under specified below-cloud conditions (reference 3, page 245.) $r = 90\%$; $P = 900$ mbar; $T = 5°C$. (b) Probability that precipitation from a precipitating stratiform cloud will reach the ground as drizzle, related to cloud thickness and height of cloud base, in Ireland (6).

systematically lower updraft velocities in lower clouds. The point of this discussion is that certain precipitation phenomena such as drizzle are the result of very particular and relatively rare sets of circumstances.

III. The Wegener–Bergeron–Findeisen Process, Snow, and Cold Cloud Rain

To this point, the discussion of drizzle and rain has involved the sequence of events on the left side of Fig. 19-1. On the right side is displayed the more frequently occurring sequence involving subfreezing processes. In a sense the formation of ice crystals is the cold process analogue of the formation of cloud droplets—both involve the combination of "raw materials" from the top line of Fig. 19-1 into the basic building units of precipitation.

Some ice crystals are formed by the spontaneous freezing of cloud droplets which have been cooled to temperatures well below freezing by natural or artificial means. The closer a droplet's temperature comes to $-40°$ (C or F), the greater the probability it will freeze spontaneously. In one form of "cloud seeding" swarms of droplets are artificially cooled to near $-40°$ when chips of frozen carbon dioxide (dry ice) are introduced into their midst in a supercooled stratus cloud. The CO_2 is not essential, but under the proper circumstances greatly increases or hastens the production of ice crystals.

Another sequence leading to production of ice crystals is based on the fact that, at temperatures between $-40°$ and $0°C$, the equilibrium pressure of water vapor is less over an ice surface than over a water surface. The difference increases to 0.25 mbar in the range between $-15°$ and $-10°C$, being smaller above and below that range. The result, as pointed out by Wegener in 1911 (3), is that water molecules will evaporate from droplets and then undergo deposition (passage directly from vapor to ice) on any ice crystals in the immediate environment. These ice crystals form on *sublimation nuclei*, (or freezing nuclei) which are in turn introduced to supercooled clouds either naturally or artificially. A particular crystalline structure is required for a particle to act as a freezing nucleus. Certain platey silicate minerals borne aloft from the surface are suitable, as may be particles of extraterrestrial origin (3). It may be, too, that fragments of fractured ice crystals, from spontaneous freezing, act as freezing nuclei for other ice crystals. Finally, in a second form of cloud seeding, materials with the proper crystalline structure—notably silver iodide (AgI)—are introduced into supercooled clouds. Deposition of water vapor on any of these classes of nuclei results in formation of ice crystals.

As cloud droplets grow into drops by coalescence, so ice crystals grow into snow crystals by continued deposition of vapor at the expense of evaporation from the supercooled droplets nearby. The shape taken by ice and snow crystals depends upon the temperature of the environment in which they are first formed, while the moisture content of the environment determines the rate of growth. Figure 19-9 suggests the dependence of shape on temperature, and in particular shows the dendritic shape formed in the range between −12° and −17°C. In part because the ice–water vapor pressure difference is greatest here, and in part because the large surface area per unit mass on dendritic crystals provides more surface for deposition, the growth rate of crystals is several times larger in a cloud of these temperatures than in a cooler or a warmer cloud.

Through the process first described by Bergeron and Findeisen in the 1930's (3), snow crystals grow in a supercooled cloud to the size of snowflakes (Fig. 19-1). As snowflakes grow large enough to fall into warmer air, water films may form on their surfaces. The branched, barbed, dendritic flakes offer the maximum chances for aggregation, or "clumping." The presence between −5° and 0°C of a water film, which freezes when flakes contact each other, produces maximum chance of bonding flakes once they have clumped. Thus for snowflakes the optimal combination of temperatures for this form of coalescence is between −17° and −12°C in the region of formation and between −5° and 0°C in the region of coalescence. If flakes are melted before they reach the earth's surface, they arrive as large rain drops

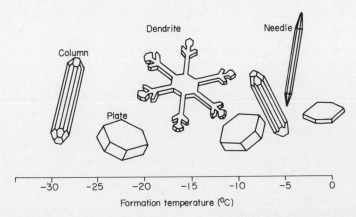

Fig. 19-9. Shapes of ice and snow crystals that form at different subfreezing temperatures.

(*cold cloud rain*). Otherwise they arrive as flakes of various shapes or, if they refreeze in cold air at the surface, as *sleet*.

IV. Types of Precipitation-Producing Storms

As suggested above, the size of a raindrop is determined by the residence time in the cloud, which is in turn related to the strength of the updraft, and by the below-cloud history of evaporative erosion or fragmenting in turbulent air. Of these determinants, the strength of the updraft seems to be most important. In addition, the size of the drops that reach the surface rather than the number of drops largely determines the rate of precipitation measured. Rainfall rate and spectra are related in Fig. 19-10.

What produces the updrafts necessary for drop and flake growth in clouds? Broadly speaking, gentle updrafts are produced by the uplift of an entire air layer as it moves slowly over an underlying surface or obstacle. Movement over terrain or over an air mass of greater density are the common forms of gentle uplift (Fig. 18-13a) and, under the proper conditions of moisture distribution, the uplift produces increased instability in the lifted layer (Fig. 18-15). A wind speed of 1 m/sec (about 2 mph) up a 1% slope for example, would support cloud droplets of several tens of microns on an updraft of 10^{-2} m/sec (see Fig. 19-5).

More vigorous updrafts are produced locally by the convergence of airstreams, for example in rough terrain; or by rapid heating of cold air that moves suddenly out over a relatively warm surface. These thermal updrafts produce cellular motion such as that in Fig. 18-13c. The movement of arctic cA air southward over the unfrozen American Great Lakes in early winter, or the movement of tropical mT air from warm waters over dry-hot land, as for example in Florida during the summer,

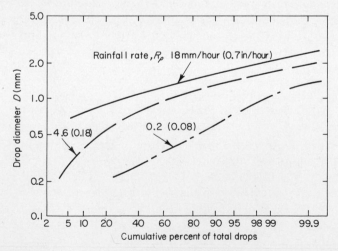

Fig. 19-10. Rain drop spectra related to precipitation rate, on logarithmic normal probability coordinates. After Engelmann, from Kelkar (reference 5, p. 213).

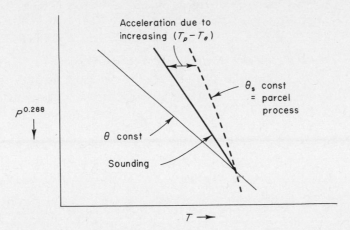

Fɪɢ. 19-11. Release of latent instability in a moist layer leading to strong updrafts locally.

are typical situations exhibiting moderate updrafts in cellular systems. When the air mass being heated is also being lifted, the updrafts are enhanced by being superimposed on the gentler upward motion. These updrafts are commonly of several meters per second, which would easily support small raindrops near 200-μ diameter.

The most vigorous updrafts of all are found in air columns where heat of condensation produces buoyant accelerations that enhance vertical motion already present. The sounding $T_2 - T_2'$ in Figure 18-15 has a positive lapse rate and is saturated at all levels as a result of the layer's being lifted. Under such conditions, a locally induced updraft near the base of the layer would release latent instability, as shown in Fig. 19-11, which would produce vertical accelerations and result in increasingly greater updraft speeds. Thus, a parcel of air moving up through the layer is increasingly warmer than its surroundings and thereby more and more accelerated [see Eq. (18-19)]. This motion tends to persist up to the levels where drier air is found, and deceleration is experienced. Though air pollutants may not have a major effect on updrafts, the next section suggests updrafts and rainfall rates have major effects on air pollutants.

V. Scavenging of Pollutants by Precipitation

Having examined the processes which produce precipitation, we may now better understand the atmosphere's natural self-cleansing by which precipitation removes pollutants from the air. Engelmann (5) suggests that this so-called *scavenging* involves (i) the delivery of material to the scavenging site, (ii) the in-cloud capture of materials called *rainout* (or snowout), and (iii) the below-cloud capture of materials called *washout*. Materials are delivered to clouds by the organized vertical motions, of all scales, described in this and preceding chapters.

Washout of particulate materials involves their capture below clouds by falling raindrops, a process the rate of which is described by the *washout coefficient* W_p.

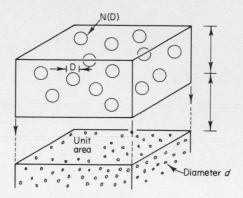

FIG. 19-12. Schematic representation of washout and the derivation of the washout coefficient.

Figure 19-12 shows the process in sketch form. The space density of raindrops in the size range from diameter D to diameter $(D + dD)$ is $N(D)$, while the flux density of these drops, $VN(D)$, is the number of such drops to pass through unit horizontal area in unit time, where V is the terminal velocity of the drops.

The cross-sectional area of a drop with diameter D is A, so the total area swept out by the flux $VN(D)$ is $AVN(D)$. Finally, the collision efficiency $E(d, D)$, which gives the fraction of the total swept area which was effective in capturing particles, is a function of the sizes of the particles and the drops (Fig. 19-6). Thus, the washout coefficient is defined as

$$W_p = \int_0^\infty E(d, D) AVN(D) \; dD \qquad (19\text{-}4)$$

The units of $N(D)$ are drops cm^{-3} (diameter increment, dD)$^{-1}$. Thus the units of W_p are (time)$^{-1}$. Engelmann gives typical values of the washout coefficient as a function of particle diameter and rainfall rate, R_p as shown in Fig. 19-13.

The meaning of the washout coefficient may be appreciated by a calculation based on the expression for the fractional depletion of the particle concentration X_o, to a value X during a washout period of length t

$$\text{X/X}_o = \exp{(-W_p t)} \qquad \text{or} \qquad \ln{(\text{X/X}_o)} = -W_p t \qquad (19\text{-}5)$$

For a rainfall rate of 2 mm/hour (about 0.08 in./hour, or 1.9 in./day) and particulate material with typical diameters near 8 μ (e.g., wetted sea salt nuclei, as in Fig. 19-3), $W_p = 3 \times 10^{-4}$ sec^{-1}. Thus, the time required for half the material to be rained out would be $t = \ln{(0.5)}/ - (3 \times 10^{-4}) \doteq 3210$ second $\doteq 39$ minutes. For 90% of the material to be rained out would require about 2 hours. Clearly, more intense rainfall cleanses the atmosphere better, but smaller particles make the cleansing more difficult. As will be seen in the next chapter, however, clean air following precipitation is often more attributable to a change of air mass than to washout alone.

According to Engelmann (5), evidence is conflicting about the scavenging of particles by snow. Although the large cross-sectional area of large flakes seems to make capture more efficient, electrical charges on flakes and particles seem to make flakes give up captured particles as they approach the earth's surface.

The scavenging of gaseous pollutants by washout is expressed reasonably well by the relationship

$$W_p = (5 \times 10^{-4})YR_p^{0.65} \quad \text{or} \quad \ln(W_p/Y) = \ln(5 \times 10^{-4}) + 0.65 \ln R_p \quad (19\text{-}6)$$

for highly soluble and reactive gases such as bromine, where W_p is again in $(\text{sec})^{-1}$, Y is the gas's diffusivity (cm^2/sec), and R_p is the rainfall rate in (mm/hour). For less reactive gases such as iodine, the washout coefficient is several orders of magnitude smaller than the above expression would indicate. The very few estimates available for the washout coefficient of gases by snow indicate that for any given solubility and reactivity (i.e., for any given gas), W_p is about 10^{-2} of the corresponding value for rain. This is attributed (5) to the low adsorption rates for gases on ice surfaces.

Recall that a certain rainfall rate roughly prescribes a certain spectrum of raindrop sizes (Fig. 19-10). If that mixture of drops is retained within a cloud by updrafts rather than falling from the cloud, one would expect the same magnitude of scavenging efficiency for pollutants within the cloud as for the same drops falling past the pollutants below the cloud. The rainout efficiencies for this process are very nearly the same as those indicated in Fig. 19-13 for the appropriate combinations of drop spectra (i.e., rainfall rate) and particle diameter. The results analogous to those from the above example with Eq. (19-5) are that, in roughly 18 minutes, half the 8-μ particles in a cloud having the drop spectrum described by the middle line in Fig. 19-10 would have been captured. (W_p for $R_p = 4.6$ mm/hour and $d = 8\,\mu$ is about 6.5×10^{-4}). The estimation of rainout effects is complicated, however, by the fact that what appear to be clouds containing the same air and drops through

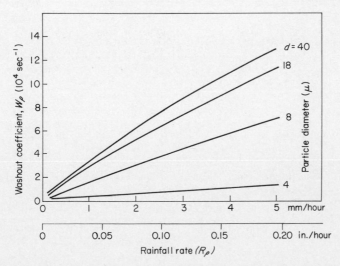

Fig. 19-13. Typical values of the washout coefficient W_p as a function of rainfall rate and diameter of the scavenged particle. After Engelmann (5) from Chamberlain.

a period of time are actually ever-changing systems into which air is being constantly entrained along with its fresh, "unwashed" pollutants. Whatever the actual magnitudes of rainout processes, it is well recognized that very small particulates (e. g., 0.1–0.5 μ) have very large atmospheric residence times and are only under extraordinary circumstances removed from the atmosphere by the processes described here. This may have great implications for possible climatic modification by air pollutants, as will be discussed in Section 21-V.

With an understanding of the materials in this chapter the reader may appreciate a recently suspected phenomenon. Tall stacks perhaps 10^2 m high emit sulfur compounds from power plants using sulfur-bearing coal as fuel. When the plumes from these stacks are wet (or are exposed to drizzle) and contain drops in the size range of a 200-μ diameter, in which the sulfur gases are dissolved, the drops exhibit terminal velocities near 1 m/sec (Fig. 19-5). Thus allowing for evaporative erosion of the drops as they fall from the plume downsteam they may descend near the surface in a time the order of 5 minutes. According to Findeisen's estimates (Fig. 19-8a) these drops could fall only about 10^2 m before evaporating. Thus, despite construction of a tall stack at the power plant (in order to keep high concentrations of sulfur above the surface) the sulfur may actually be washed down to very nearly ground level several hundred meters downwind under rather ordinary circumstances. It is this sort of "order-of-magnitude" feeling for pollutant–atmosphere interaction that this chapter has intended to give the reader.

References

1. Braham, R. R., *Bull. Amer. Meteorol. Soc.* **49**, 343–353 (1968).
2. Junge, C. E., "Air Chemistry and Radioactivity," 382 pp. Academic Press, New York, 1963.
3. Day, J., and Sternes, G., "Climate and Weather," 407 pp. Addison-Wesley, Reading, Massachusetts, 1970.
4. Byers, H., "Elements of Cloud Physics," 191 pp. Univ. Chicago Press, Chicago, Illinois, 1965.
5. Engelmann, R., *in* "Meteorology and Atomic Energy" (D. Slade, ed.), pp. 208–221. U. S. Atomic Energy Commission, Washington, D.C., 1968.
6. After Mook, C., *Bull. Amer. Meteorol. Soc.* **36**(9), 490 (1955).

Suggested Reading

Battan, L., "Cloud Physics and Cloud Seeding," 144 pp. Doubleday, Garden City, New York, 1962.
Battan, L., "Harvesting the Clouds," 148 pp. Doubleday, Garden City, New York, 1969.
Blanchard, D. C., "From Raindrops to Volcanoes," 180 pp. Doubleday, Garden City, New York, 1967.
Davis, K., and Day, J., "Water, the Mirror of Science," 195 pp. Doubleday, Garden City, New York, 1961.
Mason, B. J., "Clouds, Rain, and Rainmaking," 145 pp. Cambridge Univ. Press, London and New York, 1962.
U.S. Atomic Energy Commission. "Precipitation scavenging. Proceedings of the Symposium on precipitation scavenging, June 2–4, 1970." Available as CONF-700601, National Technical Information Services, Department of Commerce, Springfield, Virginia 22151.

Questions

1. (a) In which bandwidth $\Delta d = 0.01\ \mu$ is the concentration of midcontinent particulate aerosol 500 cm^{-3}?

   (b) In which bandwidth, Δd, beginning at $d = 0.2\ \mu$, are there on the average 2 sea salt particles cm^{-3}? Contrast this with the results of part (a).

2. (a) In $\frac{1}{2}\%$ supersaturated air, can the particles in question 1(a) grow as droplets to the 10-μ diameter size?

   (b) Can the particles in 1(b)?

3. Discuss the effects of air temperature and atmospheric pressure (i. e., altitude) on the terminal velocity of water droplets.

4. (a) In an updraft of 0.3 m sec^{-1}, what diameter droplet will be carried aloft at 0.1 m sec^{-1}?

   (b) For what diameter droplet will the relative velocity of the droplet in part (a) be $+0.25$ m sec^{-1}?

5. In Fig. 19-4, does the diameter that represents droplets of the largest "percent of total droplet volume" also represent the diameter with the largest number of droplets? *Note*: The values on the ordinate in Fig. 19-4 hold true when the curve connects central values of bands 5 μ wide beginning at 0, 5, 10, ...

6. (a) A small drop grows by the coalescence of 100,000 of the single-sized maritime droplets (Fig. 19-4) which are most numerous. What is the diameter of the drop? *Note*: These size determinations should have been made in question 5.

   (b) What size would the drop be if it had grown from the most numerous "central U.S." droplets?

7. (a) What is the collision efficiency of the drop and droplet in 6(a)?

   (b) In 6(b)?

8. With the data presented in Fig. 19-10, check the validity of the statement in Section 19-IV that "the size of drops . . . , rather than the number of drops largely determine the rate of precipitation."

9. Check the units of the washout coefficient, W: (time)$^{-1}$ to your satisfaction.

10. Calculate an estimate of the time required to wash out 90% of the particles of diameter 5 μ in a drizzle of 0.5 mm (hour)$^{-1}$.

11. Estimate from Eq. (19-6) the diffusivity of the gas which reacts quantitatively to washout in the same way as the particles in question 10. *Note*: As an order-of-magnitude reference, Y for still air is about 0.2 cm^2 sec^{-1}.

Chapter 20

MOTION IN THE ATMOSPHERE

Recall the stated intention in this book (Chapter 16) to discuss the mixture of atmospheric processes bearing on a problem, beginning with the microscale, then moving through the mesoscale to the macroscale. Recall also the idea that vigorous activity at a larger scale tends to suppress the full expression of patterns at any smaller scale. The plan in this chapter is to follow the progression from small, short-term patterns to the large persistent patterns of atmospheric motion.

As should be apparent by now to the reader, the processes at work in the atmosphere exhibit themselves in mixes. The formation of a particular cloud, the rainfall rate at a particular time at a certain station, or the generation of a particular wind gust are all the results of several, separate, definable, describable processes acting simultaneously. Thus, as an example, though we may be able to describe the several factors which appear to act on wind moving through rough terrain, it would be virtually impossible to predict beyond generality the statistical properties of the wind at a particular place within the terrain, let alone its exact behavior from one moment to the next. The student should be prepared at this point for the inescapable fact that study of atmospheric motion, and particularly the transport of pollutants by that motion, cannot reasonably be undertaken expecting "engineering precision" in either general or specific results.

I. The Microscale—Wind Profiles, Fluctuation, and Turbulence

Common experience attests to the fact that wind near the earth's surface does not blow steadily and smoothly from the same direction at the same speed. Wind gusts on face and body, the rippling of flags, and the erratic flight of paper scraps or dust particles make this fluctuation visible. This uneven flow is referred to as turbulence, and we often find it convenient to think of the motion as being made up of a basic smooth flow with constant speed and direction, upon which are superimposed the random contributions of swirls, or "eddies," in a wide range of sizes.

The eddies of turbulent motion are of two basic types, caused by two different processes. Air forced to move past an object protruding into the windstream will tumble and turn on itself, producing eddies having sizes and speeds related to the average wind speed and to the shape and size of the object. The result is *mechanical turbulence*, and a detailed speed or direction record of such a motion looks like that in Fig. 20-1a. Day and night, hot or cold, the protruding object will produce a record very much like this so long as the basic wind speed is the same.

The other type of turbulent motion results from parcels of superheated air rising from the surface, and the descending motion of other parcels taking the place of the rising air. The result is *thermal turbulence*, which produces quite a different wind record, as shown in Fig. 20-1b. In most cases turbulence is a mixture of the two types, but under very special circumstances the air may move with essentially no turbulence at all in *laminar flow*. Such a motion would produce nearly a straight line as its wind record trace.

Broadly speaking, mechanical turbulence as associated with neutral thermal stability [where $\gamma_e = \gamma_d$ as in Eq. (18-21)] because the fluctuations produced mechanically by roughness elements in the windstream are neither suppressed nor enhanced by the thermal structure of the air. By the same kind of reasoning, thermal turbulence is associated with thermal instability ($\gamma_e > \gamma_d$) because any fluctuation tends to be enhanced, and laminar flow is associated with great thermal stability ($\gamma_e < \gamma_d$) because any fluctuations are immediately damped out. Clearly then, wind behavior and temperature structure are closely related. Again broadly speaking,

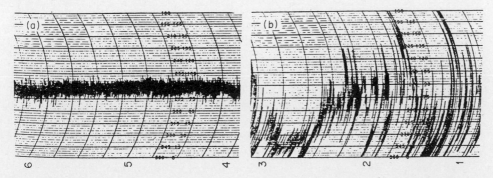

FIG. 20-1. (a) Typical wind direction record of mechanical turbulence. (b) Thermal turbulence. From Smith (1).

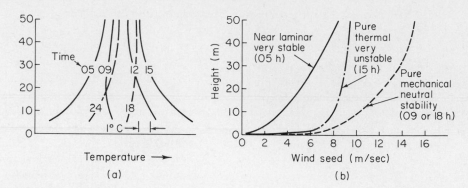

Fɪɢ. 20-2. (a) Generalized diurnal changes in temperature profiles near the ground. (b) Typical wind profiles associated with temperature structure and type of turbulence, as indicated.

Fig. 20-2 shows the typical diurnal changes in temperature profiles and accompanying wind profiles near the surface of the earth.

Because pure thermal turbulence and laminar flow occur only with light winds, following the rule set forth in the first paragraph of this chapter, wind behavior most of the time has a generous contribution of mechanical turbulence in the mixture. In fact the greater the wind speed, the more the temperature profile approaches the condition of neutral stability—the dry adiabatic lapse rate. Furthermore, since pure thermal turbulence is produced most of the time by strong surface heating under an intense solar heat load, the presence of clouds also produces a tendency toward neutral stability. Thus sun, temperature lapse rate, and wind behavior are all closely intertwined in a broadly predictable manner.

Although temperature lapse rates tend to reverse between day ($\gamma_e > 0$) and night ($\gamma_e < 0$), reflecting the fact that the surface is a source of heat by day and a sink by night, wind profiles near the surface always exhibit increasing winds with increasing height. This reflects the fact that, with respect to momentum, the surface always acts as a sink and the windstream aloft, a source. As a result, the mean wind speed U at a given height z may be related to the speed at a reference height, z_0, by an empirical result called the "log–log wind profile" or the "power law":

$$\ln U_z = \ln U_0 + a \ln (z/z_0) \qquad \text{or} \qquad (U_z/U_0) = (z/z_0)^a \qquad (20\text{-}1)$$

where the exponent a has a value that approaches 1 under very stable conditions (laminar flow), and approaches zero under very unstable conditions. For neutral conditions, a is near 0.15 in a profile spanning the range between about 10 and 100 m, and is about 0.5 in a profile between about 2 and 10 m under neutral lapse. Under any commonly encountered conditions a wind profile for the layer between about 2 and 100 m will plot very nearly a straight line on log–log coordinates (2).

Detailed records of wind speed, such as those in Fig. 20-1, are usually obtained from such instruments as the cup anemometer or the propeller-vane shown in Fig. 20-3a and 20-3b. Crosswind variations are obtained from the directional capability of a propeller vane (Fig. 20-3b) but in careful research on atmospheric turbulence, both crosswind and vertical components of wind fluctuation are obtained from such

FIG. 20-3. (a) Low-torque cup anemometer. (b) Propeller-vane speed-direction sensor. (c) Bivane sensor for horizontal and vertical angular departures from the average wind. (d) Tri-axial wind sensor, sensing wind speed, direction, and angular departures.

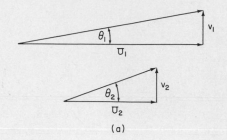

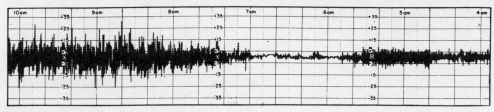

(b)

FIG. 20-4. (a) Relationships among mean wind speed, the turbulent vector component, and the angular departure. (b) Sample chart trace of vertical angular departure. The instrument was at 10m height. From Slade (reference 3, p. 51).

TABLE 20-1

Typical Values of Angular Departure (Degrees) in Turbulent Wind Fluctuations, Expressed as Standard Deviations, and Related to Height and to Thermal Stability of the Lower Atmosphere[a]

Height (m)	Turbulence parameter	Stability class				
		Very stable[b]	Stable	Neutral[c]	Unstable	Very unstable[d]
200	σ_θ	1°	3	7	12.5	25
	σ_ψ	0.5°	1.5	5	12	20.5
	$\sigma_\theta/\sigma_\psi$	2.00	2.00	1.40	1.04	1.27
100	σ_θ	1	3.5	7.5	12.5	25
	σ_ψ	1.5	2	5	9	17
	$\sigma_\theta/\sigma_\psi$	0.67	1.75	1.50	1.39	1.47
50	σ_θ	1.5	4.5	8	13	26
	σ_ψ	2.5	2.5	5.5	8.5	15.5
	$\sigma_\theta/\sigma_\psi$	0.60	1.80	1.45	1.53	1.68
10	σ_θ	7	7	8.5	14.5	26.5
	σ_ψ	4.5	3	6	8.5	14.5
	$\sigma_\theta/\sigma_\psi$	1.55	2.33	1.42	1.71	1.83

[a] After Slade (3).
[b] Associated with laminar flow.
[c] Associated with mechanical turbulence.
[d] Associated with thermal turbulence.

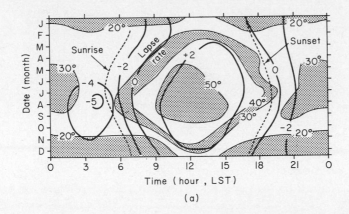

(a)

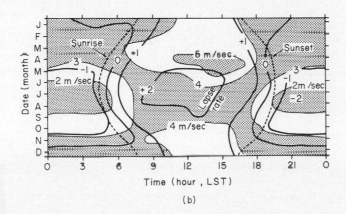

(b)

Fig. 20-5. (a) Mean lapse rate (°C/100 m) between 1.5 and 120 m, and mean wind direction range (about $5\sigma_\theta$, in degrees) related to date and hour, NRTS, Idaho. After Slade (reference 3, p. 37 and 52). (b) Mean lapse rate (°C/100 m) between 6 and 60 m, and mean wind speed (m/sec) related to date and hour, Ottawa, Ontario. After Munn and Stewart (4).

bidirectional (bivane) sensors as are shown in Fig. 20-3c. Finally, all three eddy components of fluctuation (downwind, crosswind, and vertical) may be obtained from an instrument such as the tri-axial anemometer, shown in Fig. 20-3d.

Crosswind and vertical angular departures in turbulent flow tend to be normally distributed.* The actual relationships connecting the angular departure, the mean wind speed U, and the eddy component which produces the angular departure are sketched in Fig. 20-4. The standard deviations of the crosswind angular departures, σ_θ, and the vertical angular departures σ_ψ are often used to describe their respective eddy components in a turbulent field. Table 20-1 suggests the way in which these

---

* The normal distribution is the bell-shaped curve described in any standard statistics text. Together with the mean value, the standard deviation of such a curve completely describes it, since the distribution has only these two parameters.

two measures of turbulent behavior are related to height and stability in the air layer near the earth's surface. In the table, values of σ_θ (crosswind fluctuation) decrease with height for all stability classes, while σ_ψ decreases with height only in the stable categories. Crosswind eddy components v tend to be constant with height, so the decrease in σ_θ is due to the increase in U with height, (since $\sigma_\theta U \approx \sigma_v \approx$ constant). The same reasoning holds for σ_ψ in stable and neutral conditions, but the large vertical eddy components produced by the cellular circulations of thermal turbulence produce an increase with height of σ_ψ in the unstable categories. Values of the ratio $\sigma_\theta/\sigma_\psi$ give a rough indication of the distribution of energy between the two components under the various circumstances described.

In summary of the interactions among solar heating, temperature structure, and wind behavior, consider Fig. 20-5, showing mean lapse rates, wind speeds, and crosswind variation as functions of date and hour. This figure combines two well-known results derived from observational data of the U.S. National Reactor Testing Station at Idaho Falls, Idaho (reference 3, pp. 37 and 52) and of the Meteorological Service of Canada (4). Although the two localities are quite different (rural Idaho and urban eastern Canada), the basic interactions among the variables are qualitatively very similar. These observations may be taken as a guide to microscale wind behavior related to diurnal and annual cycles, and to rural–urban differences.

II. The Microscale—Plume Behavior

Air pollutants are carried about in complex ways once they enter the moving, fluctuating atmosphere. Among the several ways to view and analyze this transport process, the one most often taken as a starting point is an examination of the pollutant plume issuing from a single chimney or stack. In the customary view of this *continuous point source*, the instantaneous and time-averaged behavior of a plume are shown in Fig. 20-6. Individual parcels of contaminants do not follow the path of the plume's "snapshot." Lower portions result from a temporary downdraft at the time of their emission; higher portions from updrafts. The "time exposure" of the plume widens somewhat with time, but keeps the same general appearance shown in the sketch. Mean concentration of pollutants has a roughly Gaussian (normal) distribution within any vertical cross section. Since the amount of pollutant material passing through any vertical crosssection at any moment equals the emission rate at the source, the area under the Gaussian curve must remain constant downstream. The result is as shown—mean concentrations are reduced downstream.

As suggested in the previous section, the behavior of a plume is sensitive to the interplay of sun, temperature structure, and wind. Air pollution meteorologists have categorized plumes according to the stability of the atmosphere above and below the stack height. Each of the five types shown in Fig. 20-7 has a descriptive name. In the figure, stability above the stack height decreases from left to right; stability below stack height decreases from top to bottom. As examples of conditions under which these types might occur, consider these. A coning plume is most probable under strong winds, or under overcast skies when surface heating is moderate. A fanning plume is most often observed under conditions of strong in-

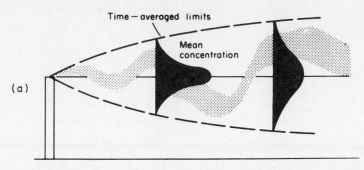

Fig. 20-6. (a) "Snapshot" and "time exposure" of a pollutant plume, showing cross sections of the Gaussian distribution of mean concentration. (b) Photograph of plume, instantaneous. (c) Same plume, time exposure. After Slade, from Culkowski (reference 3, p. 92).

version, usually just before sunrise after a calm, clear night. As the lower temperature sounding is modified by surface heating (see Fig. 18-12) the plume begins to fumigate. If the sun is bright and winds light, the plume will then tend to become a looping type. Models for predicting average concentrations within such a plume will be discussed in the next section.

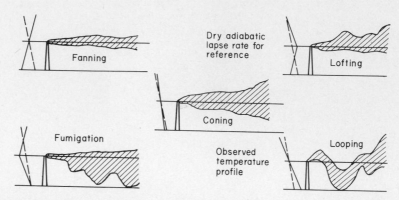

FIG. 20-7. Plume nomenclature related to thermal stability of the atmosphere below and above the source height. Below-source stability decreases from top to bottom. Above-source stability decreases from left to right.

Not all pollutants come from continuous point sources. In the language of air pollution meterologists, an explosion is an example of an *instantaneous point source*, while the plume from the pass of a crop-duster aircraft over a field is an example of an *instaneous line source*. A freeway choked with traffic is an example of a *continuous line source*. All of these source configurations are modeled by similar mathematical methods. *Area sources*, such as that in an extensive refinery complex with multiple sources of the same pollutants, or the center of a major city as a source of dust and smoke, require different approaches to analysis and prediction. When space and time dimensions of an analysis increase well beyond the microscale (Table 16-2), still other methods are required. Some of the approaches described in recent years will be presented in the following two sections.

III. Models of Diffusion and Dispersion from Point Sources

What was perhaps the first model for analysis of diffusion was built by Adolph Fick in 1855. The variations of his basic idea have been called Fickian diffusion, transfer theory, and K theory. The fundamental notion is that the time change in contaminant concentration at a point results from the existence of a gradient of concentration at that point, and that the diffusive behavior of the medium may be characterized by its diffusivity K_d. Thus

$$d\chi/dt = K_d \nabla^2 \chi \qquad (20\text{-}2)$$

where χ is the concentration of the pollutant and t is the time. For the steady state $(\partial\chi/\partial t = 0)$, with mean wind along the x-axis and no variation of mean wind with height, and with limited anisotropy, $(K_x \neq K_y \neq K_z)$, but $K_d \neq f(x, y, z, t)$, Eq. (20-2) takes the form

$$U(\partial\chi/\partial x) = K_x(\partial^2\chi/\partial x^2) + K_y(\partial^2\chi/\partial y^2) + K_z(\partial^2\chi/\partial z^2)$$

If the following boundary conditions are imposed

(1) $\chi \to \infty$ as $x \to 0$ Infinite χ at source

(2) $\chi \to 0$ as $x, y, z \to \infty$ Zero χ at great distance

(3) $K_z(\partial\chi/\partial z) \to 0$ as $z \to 0$ No downward transport into the earth

(4) $\int_0^\infty \int_{-\infty}^\infty U\chi(x, y, z)\delta y\delta z = Q, x > 0$ The rate of transport of contaminant

through any vertical plane downwind is constant and equal to the emission rate at the source Q

the approximate solution to (20-2) for a continuous point source is

$$\chi(x, y, z) = [Q/4\pi r(K_y K_z)^{1/2}] \exp \{[-U/4x][(y^2/K_y) + (z^2/K_z)]\} \qquad (20\text{-}3)$$

where $r = (x^2 + y^2 + z^2)^{1/2}$. The centerline of the plume lies along the x-axis, which is by definition the mean wind direction; the mean wind speed is U; the crosswind direction is y; and the vertical is z. Boundary condition (4) says the areas under the Gaussian curves in Fig. 20-6a are equal to Q and to each other.

Though this approach was originally formulated for, and had reasonable success with predicting diffusion by molecular processes, diffusion in the atmosphere is produced primarily by organized motion of eddies that have complex dependence on the variables depicted in Fig. 20-5 and also on other variables. Among other things, this model says that along the centerline of the plume ($y = z = 0$) the downwind decay of the concentration is independent of the wind speed U and is linearly inverse to x (since here $r = x$). This is not in good agreement with observations, which time and again have shown χ is inversely proportional to $(Ux^{1.76})$—the wind speed and a power of x.

The Fickian approach is described here mainly to put it in context with other methods for the benefit of the reader encountering these matters for the first time. It has historic interest; it is often encountered in the literature of some years ago; but it is seldom recommended for present-day operational use.

By far the most-used approach to modeling the continuous point source is assumption of the "Gaussian plume." The notion underlying this approach is that each particle of contaminant moves in random fashion through continuous time and space, independently of the presence of any other particles. The result of these independent, random movements superimposed on the underlying dispersive flow (signified by the mean wind U) is that when a particle from a source of strength Q has been carried downstream for travel time $t = x/U$, its probable departure from the x-axis must be accounted for, as must the departures of all other particles. Because the effluent of a time unit will be spread uniformly through U units of plume, the general form for the relationship will be

$$\chi(x, y, z) = (Q/U)F(y)G(z) \qquad (20\text{-}4)$$

where the horizontal and vertical diffusive functions F and G may be chosen by the modeler.* If the plume is constrained vertically within a layer L units thick, $G(z)$

* The authors are indebted to Dr. H. A. Panofsky for the form of this introductory equation and the vision to see its utility.

soon becomes simply $(1/L)$. In similar fashion a plume constrained horizontally by e. g., a valley W units wide will soon obey $F(y) \doteq (1/W)$. Thus the downstream concentration in a valley surmounted by an inversion will not fall below the value of $\chi \doteq (Q/ULW)$. Of course this is why pollution sources in valleys with frequent inversion lids are inescapably bad risks, especially because inversions in these circumstances are usually accompanied by low values of U.

When plumes are not constrained, the diffusive functions are usually adaptations of the double normal probability (Gaussian) surface. This surface is suggested by the normal curves in the plume of Fig. 20-6a.* The consequence of the assumption of these Gaussian departures is a plume in which the concentration is described at downstream point (x, y, z) by

$$\chi(x, \, y, \, z) \; = \; (Q/2\pi\sigma_y\sigma_z U) \; \exp \; \{[-y^2/2\sigma_y^2] + [-(z-H)^2/2\sigma_z^2]\} \quad (20\text{-}5)$$

where H is the stack height, or the effective source height above the ground surface. In making an effort to account for reflection of pollutants from the surface, rather than allowing them to be absorbed as is the case in Eq. (20-5), the model assumes final form by imagining a second "mirror image" source and plume at distance H below the surface, and adding the two equations—the one above and another identical except for replacement of H by $-H$. The result after addition is the *Gaussian plume model*:

$$\chi(x, \, y, \, z) \; = \; \left(\frac{Q}{2\pi\sigma_y\sigma_z U} \right) \left[\exp \frac{-y^2}{2\sigma_y^2} \right] \left[\exp \frac{-(z-H)^2}{2\sigma_z^2} + \exp \frac{-(z+H)^2}{2\sigma_z^2} \right] \quad (20\text{-}6)$$

Here x enters functionally since σ_y and σ_z are both increasing functions of x, as will be discussed presently. The correspondences to Eq. (20-4) should be clear.

Figure 20-8 shows the elements of the Gaussian plume related to the components of Eqs. (20-4) and (20-6). Because the actual plume meanders within the envelope

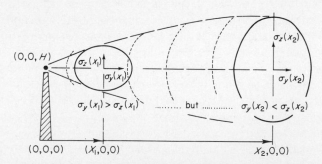

FIG. 20-8. Schematic diagram of the Gaussian plume from a continuous elevated point source. The origin of coordinates is at the base of the stack. Crosswind diffusion exceeds vertical at downwind distance x_1. At distance x_2, vertical exceeds crosswind in the case depicted.

* The double normal probability surface is described in any standard statistics text as something like

$$Pr(x, \, y) \; = \; (2 \, \pi\sigma_x\sigma_y)^{-1} \exp(-\tfrac{1}{2}) \; \{[(x-\bar{x})^2/\sigma_x^2] + [(y-\bar{y})^2/\sigma_y^2]\}$$

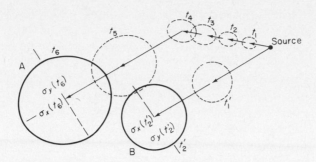

FIG. 20-9. The Gaussian puff model. Puff A is dispersed through time steps t_1, t_2, ..., t_6, and puff B through steps t_1' and t_2'. Step t_5 and t_1' are contemporaneous, as are steps t_6 and t_2'. In both puffs $\sigma_x(t)$ slightly exceeds $\sigma_y(t)$.

described by the figure and the model equation, the model describes the time-averaged concentrations in the plume, the averaging time being the order of 5 minutes under typical circumstances. Clearly, although the Gaussian plume model has the advantages of modest data requirements and the simplicity to be used in manual calculations, it fails to describe reality when terrain is complex or when meteorological conditions are changing over short periods of time the order of an hour.

Several of the special cases of the Gaussian plume model are described in Appendix D, along with various nomographs useful in desk-top solutions of the equation. Among the cases are those of the ground level plan of mean concentration $\chi(x, y, 0)$; the distribution of mean concentration along the ground level centerline of the plume $\chi(x, 0, 0)$; and the value of the maximum concentration expected at ground level $\chi_{max}(x, 0, 0)$. Numerous other adaptations are reviewed in *Meteorology and Atomic Energy* (3).

Although a complete set of models, based on statistical theory, was produced by Sutton and his co-workers for use in making estimates of diffusion from point and line sources, their details will not be presented here. They are conceptually very similar to the Gaussian plume model, and as such would add little to the reader's initial appreciation for the approach of diffusion modelers. Details may be obtained in several places already referred to in this book (5a, b).

The *Gaussian "puff" model* for dispersion from a continuous point source is less well known and more complex (requiring a computer of at least moderate capacity), but it probably describes physical reality better than either the Fickian or the Gaussian plume models. The essence of the puff model (6) is shown in Fig. 20-9. In it puffs of effluent are released at regular intervals (time steps), move downstream under constant conditions of wind speed and thermal stability (diffusivity), maintaining during the interval a three-dimensional Gaussian distribution of concentration within the puff volume. During each time step, each puff assumes and keeps the Gaussian puff configuration with its center determined by the trajectory since release and its in-puff distribution determined by the three diffusion parameters σ_x, σ_y and σ_z, which are increasing functions of travel time. Thus each puff has a center and a volume which are determined by mean dispersing wind speed U; by atmospheric stability; and by its own travel time from release t. Within the puff, the pollutant concentration is given by

$$\chi = \left(\frac{Q}{(2\pi)^{3/2}\sigma_x(t)\,\sigma_y(t)\,\sigma_z(t)} \right) \exp - \left[\frac{(x - Ut)^2}{2\sigma_x{}^2} + \frac{y^2}{2\sigma_y{}^2} + \frac{z^2}{2\sigma_z{}^2} \right] \quad (20\text{-}7)$$

which is clearly Gaussian, and in which the distances y and z are now measured from the puff center while x is measured along the trajectory from the source. In the Gaussian plume, the model accounts only for crosswind (σ_y) and vertical (σ_z) diffusion, whereas here downwind (σ_x) diffusion is also described within a puff. Because each puff (and therefore the entire plume of puffs) is free to wander in response to changing U, and not constrained to a single centerline, the diffusion parameters are given as functions of travel time t, rather than of downwind distance. These functions of $\sigma(t)$ are similar to those of $\sigma(z)$ shown in Appendix D.

In Fig. 20-9, the puff with the longer trajectory was first exposed to light winds and stable thermal conditions (probably fanning), followed by stronger winds from a different direction and greater instability (probably coning). The puff with the shorter trajectory was released at the time of wind shift and decreased stability. The three diffusion parameters need not be equal for a given travel time, and in Fig. 20-9, σ_x is shown slightly larger than σ_y. The two heavy lined puff envelopes are for the same time step, though one has a greater travel time (t_6) than the other (t_2').

When a plume leaves a source with significant momentum and/or buoyancy, due respectively to a rapid exit velocity and excess temperature relative to the ambient air, a correction must be made for the additional rise of the plume before calculations with analytical models produce reasonable results. Although there have been many "plume rise equations" proposed (reference 3, pp. 190–198) two in particular are given here as typical formulations. They were selected from the many as having the merits of good general accuracy, ease of calculation, and ability to depict clearly the two effects involved. The Davidson-Bryant formula is an example of the several empirical formulas:

$$\Delta H = 2r_{st}(V_{st}/U)^{1.4}[1 + (\Delta T/T_s)] \quad (20\text{-}8)$$

where

ΔH is the adjustment to the stack height (m)
r_{st} is the inside radius of the stack (m)
V_{st} is the exit velocity of the effluent, (m/sec)
U is the mean wind speed at stack height, (m/sec)
ΔT is the stack gas temperature minus the ambient air temperature (°K), and
T_s is the stack gas temperature (°K)

The term involving V_s represents the allowance for momentum of the effluent stream, while the term $(\Delta T/T_s)$ allows for buoyancy.

The same two effects appear in a formula of Briggs (reference 3, p. 192) which he derived by dimensional analysis, and which typifies the more theoretically based plume-rise equations

$$\Delta H = 2.6(F_p/Us)^{1/3} \quad \text{where} \quad F_p = gV_sr^2(\Delta T/T_s) \quad (20\text{-}9)$$

and the stability parameter making allowance for the thermal structure of the atmosphere is $s = (g/T)(\partial\theta/\partial z)$. Here g is the acceleration of gravity (m/sec^{-2}) (and is included to agree with (3) even though it disappears in (F_p/Us)), and θ is the potential temperature, as in Chapter 18. In a neutral atmosphere, where $s = 0$, a plume with any vertical momentum will not cease rising at all. For neutral conditions, Briggs says the plume rise may be related to downstream distance x by $\Delta H_x = 2.0(F_p x^2/U^3)^{1/3}$ and that the effective maximum rise will be near $\Delta H_{\max} = 10^3(F_p/U^3)$. The formulas of Briggs appear to be the most reliable over a wide range of conditions, although they require a temperature sounding whenever the lower atmosphere is nonneutral.

Rather than using analytical models, such as the Gaussian plume, for study of dispersion from point sources, Thompson (7) has examined the possibilities of numerical simulation for the purpose. In this model, Q "particles" are released from a source in a time unit. Each particle is acted upon for a time step (i.e., $1/Q$ of a time unit) by the turbulent field at the particle's location. The turbulent field may be described, at the choice of the modeler, to include various components of horizontal and vertical shear (i.e., $\partial U/\partial y$ and $\partial U/\partial z$], of buoyancy, and of anisotropy $(\sigma_\theta \neq \sigma_\psi)$. Specification of the field may then be changed for another series of units, and so on. Terrain may easily be simulated in the field specification, as may such obstacles as urban complexes. Thompson's study resulted in realistic pollutant configurations made explicit by projection of a motion picture consisting of the sequence of computer scope displays produced by the model. This approach, of course, requires very sophisticated computer equipment and programming.

IV. Models of Diffusion and Dispersion from Area Sources

Though solutions estimating downwind effects of individual sources are sometimes useful in effective control of air pollution, there is a more pressing need for models that treat multiple and area sources, their downwind effects, and perhaps more important, their control through zoning and regional planning. Several types of instructive and potentially useful models are available. Broadly they fall into two groups—those that make integral use of the Gaussian plume or puff, and those that do not. Panofsky calls these models which treat multiple and area sources "city models" (8), though their insights are not necessarily restricted to cities. They have several underlying notions or tactics in common.

The first underlying notion is what may be called the "forward–backward" principle the recognition of which most writers attribute to Gifford. Turner (9) states it approximately this way: "The relative concentration (χ/Q) at a receptor at $(0, 0, 0)$ due to a source at (x, y, H) where the x-axis points upwind is equal to the relative concentration at a receptor at $(x, y, 0)$ from a source at $(0, 0, H)$ where the x-axis points downwind." The idea is shown in Fig. 20-10a, where the contribution to the total pollutant load at point B due to the source at point A may be viewed (in plan) on a coordinate system with origin at A [as in Eqs. (20-3), (20-4), and (20-6)] or with origin at B. More specifically, the "backward" plume from a given receptor encompasses, and therefore defines, all the sources which make

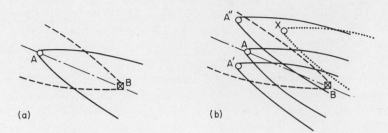

FIG. 20-10. (a) Relationship between source and receptor in the "forward–backward" principle of Gifford. (b) Relationship between receptor and several sources upwind. All plumes are shown in plan view, and all are the same shape.

a contribution to the concentration at the receptor. In Fig. 20-10b all plumes are the same shape. Sources at A, A′, and A″ are within the backward plume from receptor B. The source at X is not. Sources at A, A′, and A″, but not at X, contribute to the concentration at B.

The second underlying notion is what may be called "relative addition". Whether or not the source strengths Q are the same at all the indictable sources in Fig. 20-10b, the total concentration at B may be obtained by using a diffusion model to calculate the relative concentration (χ/Q) at B from each source, multiplying each relative concentration by its appropriate source strength, and adding the results. If the diffusion model in Eq. (20-6) is being used, for example, the relative concentration (χ/Q) is a function of source height H; dispersing wind speed U; level of atmospheric turbulence, as contained in the functional connections between x and σ_y, σ_z; and downwind position.

Models for which source inventory data and proper data on atmospheric conditions are available may be used to pursue several objectives—(a) to predict concentrations at any point downstream, (b) estimate relative contributions from any point upstream (i.e., find the most probable source of a certain contamination downstream), or (c) undertake a form of zoning, or source siting, to prevent worsening of a pollution problem. In the latter application, climatological statements of atmospheric conditions, rather than data for a particular period of time, are the appropriate input.

Four area-source models employing Gaussian diffusion explicitly are those by Turner (9), Clarke (10), Miller and Holzworth (11), and Bowne (12). Turner divides his landscape into mile squares, and assumes each square behaves as a continuous line source of length 1 mile oriented crosswind for the time unit of concern. The effect of this assumption is sketched in Fig. 20-11a, where the point source is effectively upsteam of the 1-mile line by a distance inversely proportional to σ_y, which describes plume spread as a function of turbulence. Downstream relative concentrations (χ/Q) are obtained from relationships such as Eq. (20-6) where σ_y and σ_z are functions of x according to nomographs similar to those in Appendix D. The forward–backward and the relative addition principles are employed to evaluate concentrations in the area modeled.

Clarke's model assumes the source landscape to be divided into sector areas, as in Fig. 20-11b. Downstream relative concentrations at the receptor of interest, located at the apex of the upwind sector are obtained by assuming vertical diffusion is Gaussian within the sector while concentrations are horizontally uniform within the sector. Account is taken of turbulence through a graphical relationship connecting σ_z and x, similar to that used by Turner.

Miller and Holzworth assume sources which are uniform within infinite crosswind strips, as in Fig. 20-11c. This results in the assumption that relative concentrations are uniform in the crosswind direction. Vertical concentrations are Gaussian out to the distance (or travel time) where reflection from the ground and from the top of the mixing layer (L units above the surface) begin to produce vertical uniformity. Beyond that distance, concentrations are proportional to $(1/L)$ as discussed above in connection with Eq. (20-4). Both Clarke and Miller and Holzworth employ the principle of relative addition, while of course the forward–backward principle is not relevant to these two models.

Bowne is very explicit in dividing his pufflike model into the dispersion and the diffusional phases. He employs all available wind velocity data from the area of concern to calculate a stream function defining the horizontal wind field at all points. Within a specified time period, this permits the prediction of a trajectory and endpoint representing the dispersion from each uniform, square source area. The dimension of the grid is chosen to suit the application. Once the effluent from a subarea has been dispersed downwind, nomographs relating σ_y and σ_z to travel time, vertical stability, and the variability of wind direction σ_θ permit the use of a Gaussian puff equation to diffuse the effluent about the trajectory's endpoint. Bowne recommends use of the forward–backward principle to assess the contributions of upwind sources at selected receptors rather than using the model to predict concentrations at all points in the grided area under study.

Whereas all the models just described permit changes of wind velocity with time, only Bowne's explicitly allows differences across the landscape at any one

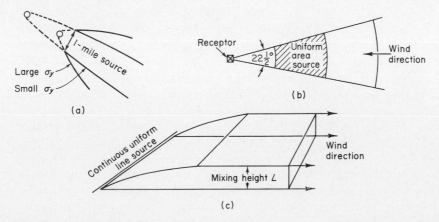

FIG. 20-11. Source configurations assumed by (a) Turner (9), by (b) Clarke (10), and by (c) Miller and Holzworth (11).

time. Thus it is implicit that his is receptive to mesoscale terrain effects. Similarly, Bowne's model accepts spatial variation in the factors leading to σ_y and σ_z at one time. These steps toward reality are also features of a set of non-Gaussian models by Reiquam (13, 14). His models conceive of the atmosphere over a region as divided into a three-dimensional grid of boxes, through which the mass of pollutant is moving. Mechanisms are assumed to operate which distribute pollutants uniformly—both horizontally and vertically—within each box during each time step. Box sizes and time steps are chosen at the user's discretion. Mass continuity of pollutants through the grid of boxes is assumed, and mass budgets for each box are kept.

In this *"box model"* approach for a single layer of boxes, the concentration of the pollutant in box n at the end of time step t is given by

$$\chi_{n,t} = [1/AL_{n,t}][r_{n,t}q_{n,t} + R_{n,t}Q_{n,t}] + [\rho_{n,t}\chi_{n,t-1}] \qquad (20\text{-}10)$$

where L is the mixing height (variable to reflect meteorological conditions across the airshed), A is the basal area of the box, q and Q are advected (imported) and local source strengths during the time step, and r, R, ρ are residual fractions defined by the geometry of the wind field. Figure 20-12 shows Reiquam's conception of the manner in which wind vectors enter Eq. (20-10) via formulation of the residual fractions. In his first paper (13), he presents verification of the "conventional configuration" of the model as a means of estimating the results of a known or postulated source configuration. In the second (14), he speculates on using the "inverse configuration" for assigning or siting sources in order to accomplish prescribed air quality goals in the airshed.

Moses (15) takes another non-Gaussian approach to estimating pollutant concentration at a specified receptor under particular meteorological conditions. This straightforward *Tabulation Prediction Scheme* simply sorts a lengthy record (2 or 3 years) of actual experience for a given pollutant in a given region, the sorting being according to sets of basic meteorological variables. For example, the scheme used at the Argonne Laboratory near Chicago divides wind direction, wind speed, ceiling height, air temperature, and time of day into classes. For each set of class values (e. g., direction 0°–140°, speed 4–7 knots, ceiling height above 7000 ft, air temperature 30°–39°F, and time of day 04–15 hour) all cases are assembled from

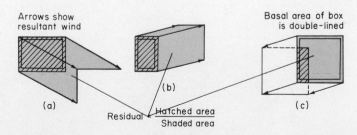

FIG. 20-12. Formulation of residuals in Eq. (20-10) as functions of resultant wind. (a) Residual r; (b) residual R; and (c) residual ρ. Wind vectors and box dimensions are shown as different for each to emphasize spatial variability possible in the model. After Reiquam (13, 14).

TABLE 20-2

Observed 24-hour Sequence of Wind at a Major Airport

Hour	Dir/Spd[a]	Hour	Dir/Spd	Hour	Dir/Spd	Hour	Dir/Spd
0000	27/3	0600	09/5	1200	16/10	1800	12/12
0100	26/2	0700	07/7	1300	18/8	1900	12/9
0200	26/3	0800	09/6	1400	17/9	2000	11/8
0300	20/2	0900	08/5	1500	12/12	2100	10/3
0400	17/2	1000	10/3	1600	10/9	2200	08/4
0500	16/4	1100	12/4	1700	11/9	2300	09/3

[a] Direction in tens of degrees (27 = 270°)/Speed in knots.

the records of air quality data, and the relative probabilities of various values of concentration of the pollutant of interest are estimated and tabulated. The tabulation may be produced on printed pages or in retrievable groups on magnetic tape. Clearly, the scheme is limited both to a particular region and to a program which has already amassed a record of experience. Moses asserts that a regularly (e. g., yearly) updated tabulation has served well in Chicago as (a) an adjunct calibration scheme for more elegant mathematical models, (b) a means for monitoring changes in source strength and location, and (c) a means for estimating by interpolation the expected experience at points between the sites at which air-quality measures were obtained for the basic record file.

Implicit in the discussions of these multiple and area source, or city models, is the importance of sequences in wind behavior. Table 20-2 gives wind speed and direction as actually observed at a major airport. Figure 20-13 shows these observations as a wind rose (frequency distribution) and as two 24-hour trajectories. The wind rose is a standard graphical tool of air pollution climatology, and is dis-

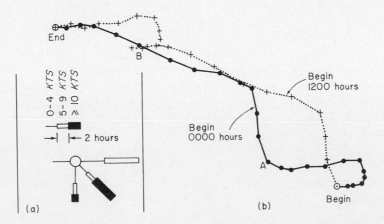

FIG. 20-13. (a) Wind rose for the data of Table 20-2. (b) Trajectories derived from data of Table 20-2. See text for explanation.

FIG. 20-14. Constant-level tetroon. The pressure altitude at which the tetrahedral-shaped, gas-filled package floats is approximately predetermined by the mass of gas placed in the non-expandable envelope before release. (For details, see reference 3, p. 298).

cussed in Appendix E. Though it contains some of the information about wind which is useful in applications to air pollution control, the trajectories in the figure make clear the importance of sequence to a realistic consideration of the dispersion of pollutants. (Construction of trajectories is also discussed in Appendix E.) Although one of the two trajectories was "made up" by reversing the order of the two 12-hour halves of the day, it is quite realistic, as many actual trajectory studies show (e.g., reference 3, pp. 176–179). Though the 24-hour resultant wind is the same for both trajectories, there are several important differences. For example, 12-hour travel for the trajectory beginning at midnight leads to point A, while 12-hour travel for a noon emission leads to point B. Stagnation of the effluent emitted at midnight is near the source; stagnation of the effluent emitted at noon is far to the northwest. The first trajectory experiences only minor "doubling back," while the second exhibits an actual retracing of a path near point B. Depending on what the land use is at each of the points on the map of Fig. 20-13, the sequence of wind events can be extremely important.

In their various assumptions, all of these models and graphical techniques necessarily blur reality slightly in order to enable estimates which are reasonable first-order approximations. The wind rose blurs reality by grouping data into speed and direction categories. Trajectory constructions blur reality by assuming that wind behavior is the same everywhere in and above a region during a time period, and by assuming that winds are constant during the time period. Even the use of the standard field method for studying trajectories by the tracking of constant-level tetroons (Fig. 20-14) blurs reality by assuming that air and pollutants remain in packets or parcels, without diffusion, during the course of dispersion. Whatever the shortcomings of these methods, the important role of sequence in atmospheric motion should be appreciated.

As with any group of models intended to address the same problems, the optimal choice of model depends upon the requirements and the capabilities of the user. Whatever precision the Gaussian plume equation may appear to give a model over that available from a box model may be only an illusion in situations involving terrain and/or major spatial variability of atmospheric conditions. In addition, the data requirements to characterize atmospheric conditions may be distinctly simpler for a box model. On the other hand, for short time intervals the uniform distribution within boxes required by the box model may be unrealistic for a chosen box size. Some of the models require information about mixing height, while others do not. Each model described here has an encouraging record of accuracy in actual prediction. All can probably be improved with care. The fundamental point here is that users now have a choice of models and approaches that did not even exist as recently as a decade ago. With them, many important questions may be answered meaningfully to a first approximation sufficient for intelligent use of an air resource.

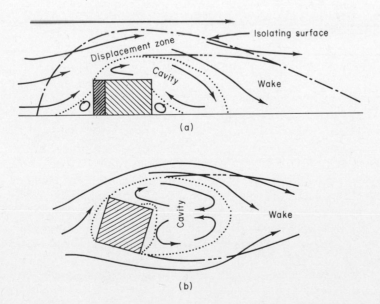

FIG. 20-15. Wind flow around a building under typical conditions of atmospheric stability and wind speed. (a) Side view, and (b) plane view. After Halitsky (reference 3, p. 221).

V. Flow around Obstacles

When an airstream moving over a flat, homogeneously rough surface encounters an isolated obstacle, such as a building or a hill, the windstream is deformed near the obstacle. Then it recovers its former characteristics at some distance downstream. This section is concerned with flow around such isolated obstacles, while the next section examines flow among many such obstacles.

An obstacle may permit deformed but laminar flow around itself, or it may produce turbulence mechanically or thermally in the windstream. Laminar flow is observed in a stable, non-buoyant windstream flowing around a streamlined obstacle. Little will be said about this relatively rare occurrence except that the obstacle must be rounded and have—in the case of a hill-like form—a very broad base relative to its height.

As the obstacle becomes less rounded, taller for its base, or warmer relative to the windstream, turbulence becomes more likely in the deformed flow. Figure 20-15 shows a generalized conception of flow around a typical obstacle like a building. Halitsky (reference 3, p. 221), conceives of the deformed flow as consisting of a displacement zone, a wake, and a cavity enclosed within an isolating surface several times taller, and many times longer downwind than the obstacle. For any wind direction, an obstacle with sharp edges and corners will produce surfaces of *separation*, across which there is no component of wind flow. In Fig. 20-15 these surfaces are shown by the pecked lines. The sharp feature is called a salient edge, and the separation occurs there when a viscous fluid is unable to "turn the corner" fast enough to maintain flow parallel to the surface (16). Separation surfaces are con-

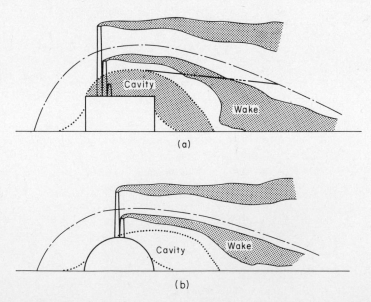

Fig. 20-16. (a) Distribution of pollutants injected within and outside the cavity. (b) Effect of streamlining the obstacle on design of an effluent stack.

FIG. 20-17. Smoke plume reaching the ground in the wake of a building. From Martin (reference 3, p. 112).

cave downward beyond a roof line or cliff top and convex downward near the foot of a cliff or wall.

Figure 20-16 suggests several things about the implications for air pollution of flow around obstacles. First, pollutants entering the air in the cavity will be distributed relatively uniformly throughout the cavity by the turbulence there. Second, a properly located stack will inject pollutants outside the cavity. Third, a streamlined obstacle reduces the height of stack needed to inject pollutants beyond the cavity. Finally, Figs. 20-16 and 20-17 together show that pollutants injected beyond the cavity but within the displacement zone will reach the ground within the wake. The general features shown here associated with buildings will also hold true for terrain features and such obstacles as thick patches of woodland on otherwise open ground.

VI. The Mesoscale—Flow through Rough Terrain

Although several principles governing flow around obstacles, and through rough terrain consisting of many obstacles, are understood and may be explained, the possible combinations of roughness elements and atmospheric responses is nearly infinite. For a particular combination, therefore, research generally centers on use of scale models in wind tunnels (reference 3, p. 221). In this section some of the principles related to wind flow in simple ridge-and-valley terrain are discussed. Mesoscale responses of wind to more complex terrain are virtually unknown, and are likely to be discovered best through scale modeling or through extensive measurements in a particular field site.

TABLE 20-3

Generalized Mesoscale Windflow Patterns Associated with Different Combinations of Wind Direction and Ridgeline Orientation

Wind direction relative to ridgeline	Time of day	Ridgeline orientation	
		East–West	North–South
Parallel	Day	South-facing slope is heated—single helix	Upslope flow on both heated slopes—double helix
	Night	Downslope flow on both slopes—double helix	Downslope flow on both slopes—double helix
Perpendicular	Day	South-facing slope is heated. North wind—stationary eddy fills valley. South wind—eddy suppressed, flow without separation	Upslope flow on both heated slopes—stationary eddy one-half of the valley
	Night	Indefinite flow—extreme stagnation in valley bottom	Indefinite flow—extreme stagnation in valley bottom

Wind behavior in ridge–valley topography depends, broadly speaking, on the relationships between the wind direction and the solar azimuth on one hand and the orientation of the ridge lines (and valleys) on the other. Table 20-3 is an attempt to summarize the net results of several principles operating in ridge–valley topography on the mesoscale. First, an east–west trending valley has only one sunny slope, with heated upslope flow, during the day. (More will be said on such circulations in the next section). A north–south trending valley has slopes about equally heated during the high-sun hours of best heating. Second, windflow parallel to a ridgeline will scour a landscape without the tendency to separated flow encountered in flow perpendicular to a ridgeline. Finally, thermally produced circulations on slopes—upslope under heating, and downslope with cooling—combine with mechanically produced circulations such as eddies to produce distinctive patterns such as eddies and helices.

Figure 20-18 shows sketches of six of the combinations described in Table 20-3. Features shown are the single and double helix, and the stationary eddy filling either half or all of a valley.

The reader should keep several points in mind when trying to apply the information in this section to an actual field site. First, vigorous macroscale circulation will suppress the features shown here, as has been mentioned several times previously in this book. These are mesoscale circulations, and as such may be suppressed by macroscale circulations. Second, under some circumstances one flow pattern may alternate with another during the course of several hours under conditions when neither one has both thermal and mechanical components strongly reinforcing each other. Third, variable macroscale wind speed and direction may change the loca-

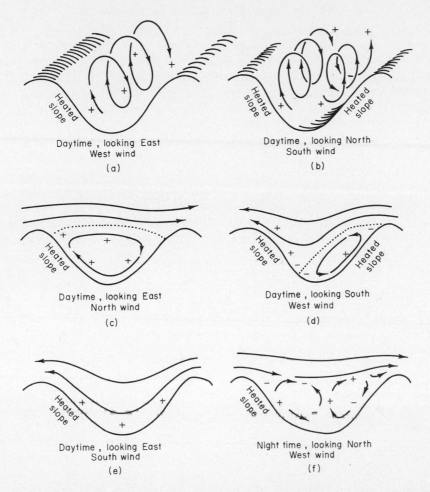

Daytime , looking East
West wind
(a)

Daytime , looking North
South wind
(b)

Daytime , looking East
North wind
(c)

Daytime , looking South
West wind
(d)

Daytime , looking East
South wind
(e)

Night time , looking North
West wind
(f)

Fig. 20-18. Six of the mesoscale flow patterns described in Table 20-3. Localities where thermal and mechanical circulations reinforce each other are marked +; where they oppose each other, −.

tion of the separation surface up- or downslope, unless the presence of a salient edge tends to fix its location. The result, for a particular site on an upper slope, may be a reversal of wind direction as the surface moves across the site.

The applications of Table 20-3 and Fig. 20-18 to air pollution control arise primarily through source siting and zoning. There are probably circumstances in which increased stack height may overcome a problem already in existence (16), and others where the extreme stagnation of crossridge nighttime flow may be forecast and sources shut down.

The details of mesoscale windflow through elements such as the organized patterns of urban skyscrapers is just now being systematically studied (17). As noted above, details of mesoscale wind behavior in complex terrain are not well understood, and perhaps never will be. Wind tunnel modeling is likely to hold the best answers to specific problems in this area of interest.

VII. The Mesoscale—Local Wind Systems

In the previous section mention was made of upslope windflow on heated slopes. So-called "drainage winds" downslope are often observed on rapidly cooling slopes at night, resulting in the thermal components of nighttime flow suggested in Table 20-3. These thermal components are results of differential heating on surfaces of equal pressure, and produce a variety of mesoscale circulations known as *local winds*. The local wind systems are usually taken to be the land–sea breeze, the lake–shore breeze, the mountain-valley wind, and the urban–rural circulation.

The *sea breeze* and the *lake breeze* result from the differential heating of a pressure surface such as *P-P*, across a shoreline. In Figs. 20-19a and 20-19b, the air above the land surface is heated more quickly and deeply in the morning sun because water carries much of the heat absorbed at the surface quickly beneath the surface and dissipates most of the rest in evaporation. Thus the relatively dry surface of the land, with only conduction into the soil, has a distinctly higher surface temperature soon after sunrise. The result, in the absence of a vigorous macroscale circulation, is an onshore flow across the shoreline to replace the air which is rising above the land. Offshore flow aloft completes the system. Many observational and theoretical studies confirm the essential components of this system (18, 19). On a western shoreline the regional wind and the local wind reinforce each other; while on an eastern shore (different meanings apply for lake shores) the opposition of regional and local winds produces (a) a reduced frequency of sea breeze, but (b) sudden and vigorous wind shifts when the sea breeze does occur.

In Fig. 20-19c, the reasons which make the water relatively cool during the day have made it relatively warm at night: ready resupply of large amounts of heat to the surface from beneath the surface. Consequently, an offshore flow—the *land breeze* or *shore breeze*—develops. By day, the influence of the regional wind aloft

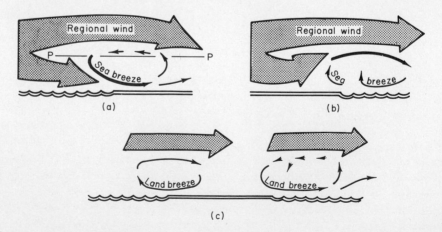

Fig. 20-19. (a) Onshore flow during daylight—the sea breeze and the lake breeze, west-facing shoreline. (b) Same, east-facing shoreline. (c) Offshore flow at night—the land breeze or shore breeze.

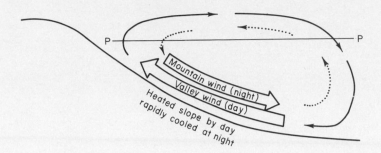

FIG. 20-20. Differential heating on slopes results in the valley wind by day and the mountain, or canyon wind by night.

more readily reaches the surface (because of greater vertical instability) and affects the surface limb of the local wind system. At night, the regional wind remains aloft and the land breeze is less affected. The regional wind is shown in Fig. 29-19 primarily to remind the reader that local wind systems are components superimposed upon regional winds, rather than independent systems unto themselves.

Figure 20-20 shows the nature of the mountain-valley local wind system. On a constant pressure surface, such as P-P, the heating of the air at the up-valley or upslope end, relative to that out over the plain, results in a pressure gradient along the surface. The differential heating results from the fact that the air on the slope is nearer to the major heat exchanger at the surface, which is in turn (according to aspect and time of day) more normal to the solar beam than the horizontal surface out on the plain. East-facing slopes have the advantage in generating more vigorous cells. Once air has been set in motion in this thermally driven pressure field, the resulting upslope limb, the *valley wind*, is completed by an opposing flow aloft and subsidence over the plain. At night, air next to the slope is cooled most readily because of its nearness to the surface. The downslope *mountain wind* flows out to the plain, the circulation cell being completed aloft. If the cold drainage wind leaves the upper slopes through a canyon on the lower slopes, the velocities are much greater and the familiar *canyon winds* of the American Rocky Mountains are the result. A wind pattern which acts similarly, but which is due to other causes, results from leakage of impounded cold air out of a high basin or plateau through a canyon or gorge to nearby lowlands. The Santa Ana of southern California, and the Gorge Winds of the Columbia and the Fraser in the Pacific Northwest are examples of this.

The existence of an urbanized area produces an urban–rural circulation based on differential heating along a pressure surface *P-P*, as in the cases of the other local wind systems just described. For various reasons (20) the city is warmer than the countryside day and night. The city's shape promotes multiple reflections and absorption of shortwave energy between buildings. The city has many artificial heat sources not present in such number and density in the country—furnaces, automobiles, factories, etc. The city is made of materials which quickly absorb the sun's heat and carry it beneath the surface, to be released the following night. As far as mesoscale wind circulation is concerned, the result is a continual flow into the city

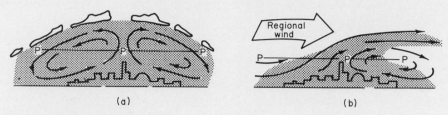

(a)   (b)

FIG. 20-21. (a) Dome-shaped rural–urban circulation with calm regional winds. (b) Plume shaped rural–urban circulation with gentle regional winds.

from its environs, ascent above the city center, and return flow aloft above the suburbs, as suggested in Fig. 20-21a. If the city is roughly circular, the circulation would be dome shaped under light regional winds. Figure 20-21b suggests the "heat plume," which has been inferred from careful observations of temperature in and above several cities under conditions of gentle regional winds. It is quite likely that any major city induces a mesoscale circulation looking like some combination of the dome and the plume under conditions of light regional wind flow. The pattern exists day and night without reversal, but is enhanced at night when differential heating is greatest.

As in the last section, the various circulations suggested here sometimes reinforce and sometimes oppose each other. For example, upslope winds on east-facing near-shore mountains may reinforce the morning sea breeze. For another, canyon winds on a west-facing mountain front may oppose the morning sea breeze across a shore-line at the base of the mountains.

Applications of these principles of mesoscale circulation to air pollution control are most probable in regard to source siting, or to understanding anomolous concentrations in the vicinity of an existing source. The scale of these processes is sufficiently greater than those in the preceding section that plant design (such as stack height) after site selection is probably not an appropriate factor in control. In recent years appreciation of the complex interactions of these mechanisms has grown (21). In addition to the effects of terrain and shorelines, related effects of differential thermal stability and surface roughness enter the processes of pollutant dispersion. Clearly, attention to sequence in assessing climate for air pollution control is particularly important in areas where differential heating produces the local wind systems, which bring diurnal reversals of windflow. These systems tend to produce "sloshing" of pollutants back and forth in a limited volume of air, without real dispersal. More will be said on this subject in a later chapter.

VIII. The Macroscale—Balance of Forces and Horizontal Flow

Macroscale winds and systems are those depicted on ordinary weather maps. They are features whose dimensions are on the order of hundreds or thousands of kilometers horizontally and several kilometers vertically. Horizontal motion in a straight line is due to the resultant of the horizontal components of three real forces: the pressure gradient, gravitation, and friction. Because of the rotation of the earth, straightline motion in an absolute coordinate system (as seen from space) appears to veer to the right (northern hemisphere) when viewed from the relative coordinate system attached to the rotating earth. Thus observers on the rotating coordinate system describe motion as the balance of four forces—pressure gradient, gravity, friction, and the apparent *coriolis force* which explains the veering.

The effect of the coriolis force may be simulated by attaching a paper to a turn-table rotating counterclockwise (northern hemisphere), and then drawing a pencil across the paper along a straight string attached to two points outside the turn-table. The line on the paper, drawn during a unit of elapsed time, will exhibit a veering to the right and a greater veering for a greater velocity (i.e., greater length of line per unit of time). Also, the curvature of the line (amount of veering per unit length) will be greater when the line is drawn nearer the center of the turntable. Following this reasoning, the expression for the horizontal component of the coriolis force is $(2\Omega \sin \phi \cdot U)$, where Ω expresses the rate of rotation of the earth on its axis, and ϕ is the latitude. By definition the coriolis force acts perpendicular to the velocity, and, as seen above, it acts to the right of the flow in the northern hemisphere. It increases toward the poles, and it is proportional to the wind speed.

Unaccelerated (straightline), frictionless, horizontal motion is called *geostrophic flow*. It takes place well above the reach of the earth's frictional drag, along lines parallel to straight isobars (lines of equal pressure) as shown in Fig. 20-22a. It may be seen that the geostrophic wind is that velocity whose coriolis force exactly balances the pressure gradient force (per unit volume). Thus since the pressure gradient force and the coriolis force are equal and opposite, geostrophic flow takes place with low pressure to the left in the northern hemisphere.

Figure 20-22b shows horizontal unaccelerated frictional flow nearer the earth's surface. Here the resultant of the backward-directed friction force and the coriolis force balances the pressure gradient force. With the friction present, the coriolis need not be so large as in geostrophic flow. Thus the wind velocity need not be so great, and is across isobars toward lower pressure. Because of this, flow in the layers experiencing friction tends to equalize pressures by causing this mass transport toward lower pressures.

The magnitude of the frictional force decreases with altitude. Fig. 20-23, called a *hodograph*, is the result and gives the name *spiral layer* to that region in which the

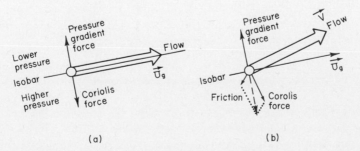

(a)   (b)

Fig. 20-22. (a) Balance of forces producing geostrophic flow. (b) Balance of forces producing frictional flow.

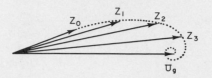

FIG. 20-23. Hodograph of the spiral layer.

wind vector approaches the geostrophic vector with increasing height. Imagining a force diagram like Fig. 20-22b for each level, with its own frictional force, will enable the reader to see the meaning of Fig. 20-23 directly. As may be imagined, the roughness of the earth's surface has considerable influence on the magnitude of the exponent in Eq. (20-1) and on the relationships among the wind vectors in the spiral layer. The difference in magnitude and direction between the vector at level z_o and the geostrophic wind U_g, for example, would be much greater over the land than over the open ocean with its relatively smaller surface frictional force.

Figure 20-24 shows horizontal, frictionless flow along curved isobars. This *gradient flow* is the most commonly observed above the friction layer. As shown, the gradient wind is that vector whose coriolis force balances the vector difference between the pressure gradient force and the centripetal force constraining the motion along a curved path. In the northern hemisphere, counterclockwise flow at a given speed takes place in association with a greater pressure gradient (closer-spaced isobars) and greater curvature than does clockwise flow of the same speed. As will be seen presently, counterclockwise flow is around low pressure centers; clockwise flow is around high pressure centers; and pressure gradients as well as isobar curvature tend to be much greater near lows than near highs, following the rule just given.

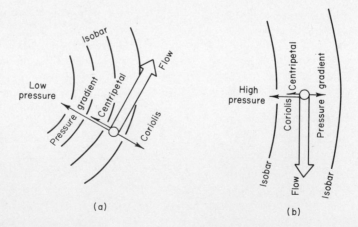

FIG. 20-24. (a) Gradient flow around a low-pressure center. (b) Gradient flow around a high-pressure center. The wind speed is the same in both cases. Notice the differences in pressure gradients and isobar curvature.

TABLE 20-4

Generalized Summary of the Relationship of Wind Flow to Altitude above the Earth's Surface in the Troposphere

Level (approximate) (m)	Description of atmospheric motion
Above 1000	Frictionless, horizontal flow; either geostrophic or gradient
50–1000	Speed and direction change with altitude according to Fig. 20-23
5–50	Direction constant; speed varies according to Eq. (20-1)
0.1–5	Direction constant; speed varies according to the logarithm of altitude according to $(U_z/U_o) \sim \ln(z/z_o)$.[a]

[a] This "logarithmic wind profile" is derived and described in most standard micrometeorology texts (5b). It is not dwelt on here because it represents motion without great interest in air pollution control.

Table 20-4 summarizes the layers of windflow found at increasing altitudes above the earth's surface. The reader must understand that at any instant the actual flow and the vertical dimensions may deviate significantly from these generalizations, but the concepts are valid in the sense that the patterns described appear in average circumstances under what may be called "steady-state" conditions.

IX. The Macroscale—Upper Air Patterns of Pressure and Wind

Figure 20-25 shows large scale flow patterns several kilometers above the northern hemisphere. The lines may be thought of either as lines of equal pressure for some altitude, or as lines of equal altitude for some pressure. At altitudes near 5.5 km (18,000 ft) where pressure is about 500 mbar, flow is generally gradient, along these contours. An area such as the one marked R is an area with higher pressure or with a greater altitude and is called a *ridge*. Similarly, an area such as T is a *trough*.

At any one time, a map of the 500-mbar surface around the northern hemisphere will show several of these *long waves*, and of course their number must be an integer. Though in a given season of the year there seem to be preferences for both the number of waves and their location, both number and location may change within several days, and may be different from one year to another in the same season. Whatever the upper air pattern at one time, however, it has a great deal to do with determining the weather observed at a particular place on the surface.

Since the wave number for a hemisphere must be an integer, and since the integer changes every so often, it follows that at some places on the surface the nature of the pattern overhead must change rather markedly and rapidly, with similar marked and rapid changes in weather. This sort of occurrence marks a very difficult forecast problem for the meteorologist in some—but not all—areas of the world.

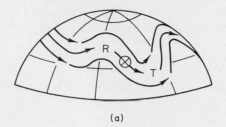

(a)

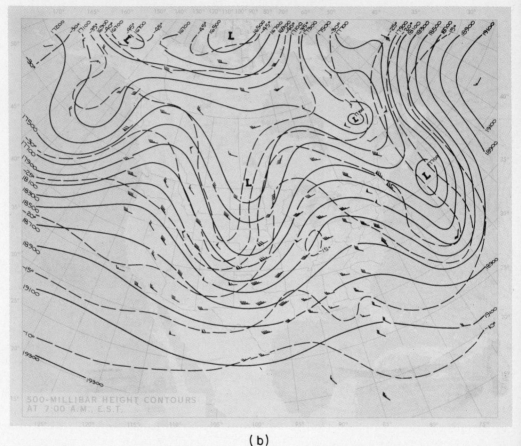

(b)

Fig. 20-25. (a) Generalized 500-mbar map, showing ridges and troughs. (b) Actual 500-mbar map for the United States, April 1, 1971. Solid lines are surface height; dashed lines temperature. For corresponding surface map, see Fig. 20-27b.

In addition to variation in the number of long waves around the hemisphere, the amplitudes and orientations of the waves change. One is likely to find either small amplitude waves or large amplitude waves all the way around the globe at any given time rather than to find the mixture given for illustrative purposes in Fig. 20-25a. Meteorologists refer to a pattern of small amplitude waves as a "high index" pattern, since it represents a large space-averaged value of the westerly component of flow around the hemisphere. "Low index" refers to a pattern of large amplitude waves, and such a pattern generally accompanies surface weather which is very persistent and slow to change.

X. The Macroscale—Storms and Fronts

Near the earth's surface, macroscale weather patterns may be described in very simplified terms as being made up of *cyclones*, or "lows", and *anticyclones*, or "highs." In general the lows are traveling features which are formed in regions of semipermanent low pressure and move downstream through patterns of much less mobile highs. The analogy of these patterns and movements to those of a mountain stream with its traveling eddies often aids understanding of the situation. Lows form in preferred regions, move downstream as they first intensify and then lose their identity, to be followed in the same sequence along roughly the same *storm track* by other lows. The major terrain features of earth—such as mountain ranges—and the distributions of land and water produce a pattern of heat distribution (Chapter 16), which in turn determine the areas of preferred generation of lows, or "storm systems." The bottom topography of a mountain stream acts similarly to produce a sequence of eddies emanating from preferred locations. In general, moving lows intensify (pressure gradients and wind speeds increase) in areas beneath upper trough lines, and become less intense as they move beneath upper ridgelines. The generalizations presented in this section are only generalizations, to which, of course, there are exceptions.

Figure 20-26 depicts, again in very simplified form, the life history of one of the traveling storm systems, lows, or cyclones. This model is most commonly accepted

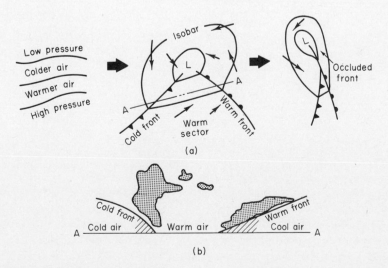

Fig. 20-26. (a) Sequence showing life history of a storm system. (b) Cloud and front cross section through a storm.

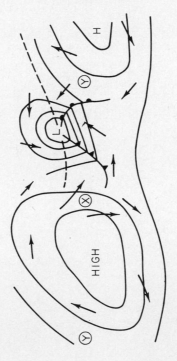

FIG. 20-27. (a) Interrelations among semi-permanent anticyclones and traveling storm system. (b) Surface weather map for the United States, April 1, 1971, showing a traveling storm over the upper midwest, trailing a cold front into Texas. Shaded areas are receiving precipitation. For corresponding 500-mbar map, see Fig. 20-25b.

FIG. 20-27a

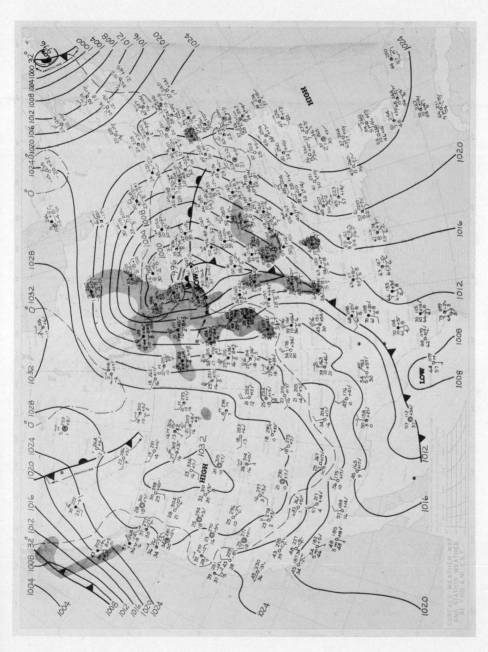

FIG. 20-27b

by operational meteorologists and should be clearly understood by anyone intending to work with a meteorologist on an air pollution problem.

Figure 20-26a shows in plan view several stages in the life of a low as it forms, moves downstream while intensifying, and then dissipates. The *cold front* is the line of intersection, on the surface, of cold air to the west overtaking and flowing beneath warmer air to the east. The *warm front* is the line where warm air is overtaking and flowing over cooler air to the east. The sequence shows that the cold front moves more rapidly across the map than the warm front, and in time the two fronts merge into an *occluded front*. Following the principles in Figs. 20-22, 20-23, and 20-24 the wind directions around the low will be as indicated. Figure 20-26b shows a cross section through *A-A* in the storm's "middle age." The cold and warm fronts, and clouds are indicated. Recalling that, behind a cold front, cold air is moving over a recently heated surface, and that behind a warm front warm air is moving over a recently cooled surface, the reader may appreciate the nature of the cloud forms by referring to the discussion of Fig. 18-13.

Figure 20-27 shows an idealized pattern of the relationships between surface areas of semipermanent high pressure and one of the travelling lows. The low is represented as being middle-aged, and its path across the map will be something like the storm track on the dashed line through its center. The regions marked with a circled *X* in Figs. 20-25a and 20-27 correspond to each other and to the region most likely to experience macroscale atmospheric stagnation, reduced vertical and horizontal ventilation, and thus accumulation of pollutants. As may be seen by the streamlines in Fig. 20-27, this is a region of divergence of windflow at the surface. By continuity of mass, such a divergence at the surface must be accompanied by subsidence from aloft. This subsidence produces not only a suppression of positive (upward) vertical ventilation, but also opposes the production of clouds which might otherwise form through the lifting process *C* in Fig. 18-16. The clear skies in these areas increase the likelihood of radiation inversions being formed at night (Figs. 17-7 and 20-2). If the surface air is maintained at temperatures cooler than the subsiding air, another result of the subsidence at *X* is the formation of a subsidence inversion (Fig. 18-14) over large areas of the lower atmosphere. At sea, the required lower temperatures are maintained by upwelling cold water along western shorelines, as in southern California and Baja California, northwestern Chile, southwestern Africa, and western Australia. On land, lower temperatures near the surface are often maintained by snow cover in such regions as the Appalachian Mountains of southeastern United States.

The presence of both radiation and subsidence inversions, plus the presence of a low-index tendency for persistence of this weather for several days at a time without any major change, all combine to produce the greatest incidence of pollution potential. All the major instances of air pollution disaster have been in areas with the same relative location as indicated by *X* in Figs. 20-25a and 20-27.

If the whole pattern in Fig. 20-27 were to move slowly downstream (to the east), a fixed station originally at *X* would soon be found in the relative location of *Y*— near the location of a warm front. Since in these circumstances, ahead of a warm front, the column of air above the station would be cooler near the surface than aloft,

TABLE 20-5

Sequence of Observations of Temperature, Wind, Sky Conditions, and Visibility at Philadelphia International Airport, January 3–12, 1971

Date	Hour	Weather situation in brief	Temperature (°C) T/T_d	Wind dir/speed[a] (deg/knots)	Sky condition cover/ceiling[b] (10ths/100 ft)	Visibility[c] (miles)
Jan 3	7 A.M.	Under high center	23/22	230/7	00/Unl	4 KH
	7 P.M.		31/28	230/4	00/Unl	7 KH
Jan 4	7 A.M.	Ahead of warm front	32/28	350/6	10/30	5 RKH
	7 P.M.		37/37	100/6	10/4	0.4 RFK
Jan 5	7 A.M.	Post warm front	45/45	100/20	10/30	3 RF
	7 P.M.	In warm sector	35/25	270/11	10/120	15
Jan 6	7 A.M.	Post cold front	29/19	290/10	04/Unl	15
	7 P.M.		27/10	290/10	07/220	15
Jan 7	7 A.M.	NE of high center	22/10	270/8	08/Unl	12
	7 P.M.		21/7	260/10	00/Unl	12
Jan 8	7 A.M.	E of high center	16/7	280/11	07/Unl	15
	7 P.M.		24/12	320/7	10/120	15
Jan 9	7 A.M.	SE of high center	24/12	050/8	10/60	10
	7 P.M.		28/19	050/8	09/25	10
Jan 10	7 A.M.	Pre-warm front	24/18	140/5	07/38	6 K
	7 P.M.	Warm front passage	35/30	220/10	04/Unl	5 KH
Jan 11	7 A.M.	Warm sector	29/29	230/3	07/120	5 GFK
	7 P.M.		33/32	200/7	04/Unl	4 KH
Jan 12	7 A.M.	Cold front passage	31/30	260/6	03/Unl	6 K
	7 P.M.	Post cold front	32/19	360/12	08/140	10

[a] Wind direction is direction from which the wind blows, with 360 being north and 090 being east. Wind speed in knots: 10 mph = 8.7 knots = 16.1 km/hour.

[b] If less than one half (05 10ths) of lower sky is cloud covered, and there are no high clouds above, ceiling is "unlimited." If more than half of the lower sky is cloud covered, the cloud base is given in hundreds of feet above the station. Thus 09/25 means 9/10 of lower sky has clouds with bases at 2500 ft above the station.

[c] Visibility is defined in the next chapter. Values less than 8 miles require an explanation of the reason: R = rain; K = smoke; H = haze; F = fog; GF = ground fog.

a *frontal inversion* would exist, with continued trapping of pollutants. As the macro-scale pattern continued to move downstream, the station would in all probability experience the passage of a cold front, and with it a vigorous ventilation and replacement of the stagnant air mass. The wake of a cold front is generally considered the region of least stagnation, but on its heels may well come the arrival of another high pressure area with its characteristic subsidence conditions. The cycle would repeat itself.

Table 20-5 contains the sequence of observations, from the Philadelphia International Airport, during a 10-day period in which two storm systems, with their fronts, passed through the northeastern United States. Careful examination of these sequences together with Fig. 20-27 will disclose agreement between the two. In particular, notice the variation of visibility during the sequence. This rough indicator of air quality will be discussed in more detail in a later section.

References

1. Smith, M., Meteorological Monograph 1, No. 4, pp. 30–35. American Meteorological Society, Boston, Massachusetts, 1951.
2. Geiger, R., "The Climate near the Ground," p. 104. 3rd edition, 494 pp. Harvard Univ. Press, Cambridge, Massachusetts, 1959.
3. Slade, D. (ed.), "Meteorology and Atomic Energy," 445 pp. U.S. Atomic Energy Commission, Washington, D.C., 1968.
4. After Munn, R., and Stewart, I., *J. Air Pollut. Contr. Ass.* 17(2), 98–101 (1967).
5a. For example, Slade, D., *op. cit.*; Sellers, W., "Physical Climatology," 272 pp. Univ. of Chicago Press, Chicago, Illinois, 1965.
5b. Haltiner, G. J., and Martin, F. L., "Dynamical and Physical Meteorology," 470 pp. McGraw-Hill, New York, 1957.
6. Roberts, J., Croke, E., and Kennedy, A., *in* "Proceedings of Symposium on Multiple-Source Urban Diffusion Models" (A. C. Stern, ed.). APCO Publ. AP-86, U.S. Environmental Protection Agency, Research Triangle Park, North Carolina, 1970, pp. 6–1 to 6–72.
7. Thompson, R., *Quart. J. Roy. Meteorol. Soc.* 97, 93–98 (1971).
8. Panofsky, H., *Amer. Sci.* 57(2), 269–285 (1969).
9. Turner, D., *J. Appl. Meteorol.* 3, 83–91 (1964).
10. Clarke, J., *J. Air Pollut. Contr. Ass.* 14, 347–352 (1964).
11. Miller, M., and Holzworth, G., *J. Air Pollut. Contr. Ass.* 17(1), 46–50 (1967).
12. Bowne, N., *J. Air Pollut. Contr. Ass.* 19, 370–374 (1969).
13. Reiquam, H., *Atm. Environ.* 4, 233–247 (1970).
14. Reiquam, H., *Atm. Environ.* 5, 57–64 (1971).
15. Moses, H., *in* "Proceedings of Symposium on Multiple-Source Urban Diffusion Models," *op. cit.*, pp. 14–13 to 14–18.
16. Scorer, R., "Air Pollution," 151 pp. Pergamon, Oxford, 1968.
17. For example, Bowne, N., and Ball, J., *J. Appl. Meteor.* 9(6), 862–873 (1970).
18. Defant, F., *in* "Compendium of Meteorology," 1334 pp. American Meteorological Society, Boston, 1951.
19. Schroeder, M., *Bull. Amer. Meteorol. Soc.* 48, 802–808 (1967).
20. Lowry, W., *Sci. Amer.* August 1967, p. 51.
21. For example, Hewson, E. W., and Olsson, L., *J. Air Pollut. Contr. Ass.* 17, 757–761 (1967).

Suggested Reading

Byers, H. R., "General Meteorology," 540 pp. McGraw-Hill, New York, 1959.
Hess, S. L., "Introduction to Theoretical Meteorology," 362 pp. Holt, New York, 1959.

Munn, R. E., "Descriptive Micrometeorology," 245 pp. Academic Press, New York, 1966.

Pasquill, F., "Atmospheric Diffusion," 297 pp. Van Nostrand-Reinhold, Princeton, New Jersey, 1962.

Sutton, O. G., "Micrometeorology," 333 pp. McGraw-Hill, New York, 1953.

Questions

1. (a) The anemometer at a plant indicates a 10 min mean wind speed of 10 m sec^{-1} at the 25 m site where the instrument is exposed. Estimate the mean wind speed at the height of the plant's stack: 65 m.
   (b) List the assumptions that were made in the course of making the above estimate.
   (c) Place an upper and lower limit on your estimate, representing the extremes of thermal stability.

2. (a) Assuming the chart record in Fig. 20-4 is of "typical" turbulence, classify the thermal stability of the lower atmosphere during the hour 06–07, using nomenclature of Table 20-1.
   (b) The half hour 0730–0800.
   (c) The half hour 0830–0900.

3. If the instrument of Question 1 had been located above the one in Question 2, estimate the so-called vertical intensity of turbulence (σ_w/U) at 25 m, where σ_w is the standard deviation of the vertical eddy components, for the three times in Question 2.

4. Is the exhaust of a single automobile, moving in a straight line in calm air at a constant speed, modeled by the Gaussian plume? Explain.

5. A pollutant source is located in a valley with an inversion lid at ridgetop. The valley is 0.5 km wide at the location of the source and opens at a 20° angle downwind.
   (a) If the mean wind speed is 4 m sec^{-1} at 1 km downwind and if the concentration of the pollutant is uniform downwind, what is the wind speed at 0.5 km downwind?
   (b) If, instead, the concentration is doubled between 1 km and 2 km downwind of the source due to a second source at 1 km, what is the strength of the second source relative to the first?
   Note: Assume all valley cross sections normal to the centerline have the same slope on the walls and the same depth to ridgetop.

6. A plume from a 50 m stack is Gaussian, and the wind field is uniform and constant.
   (a) Under class A stability (see Appendix D) how far from the ground level centerline 0.2 km downwind must one go to find a mean concentration equal to 0.1 that of the centerline concentration at the same distance downwind?
   (b) In class D stability what are the values of x and y that exhibit the same centrations as in (a)?

7. Two puffs leave a ground level source at the major airport in Table 20-2: one at 0600, the other at noon.
   (a) Estimate the puff diameter and puff volume for each at the end of its first hour of travel, with class D stability.
   (b) What is the location (direction and distance) of the second puff relative to the first at 1500?

8. Conditions are as follows at a pollution source (i) the pressure is 1000 mbar, (ii) the temperature sounding is as given in Question 6, Chapter 18, (iii) the mean wind speed is 5 m sec^{-1} at stack height, (iv) the exit velocity of the stack effluent is 8 m sec^{-1}, at temperature 40°C from a stack with radius 4 m.
   (a) Calculate the plume rise by Eq. (20-8).
   (b) By Eq. (20-9).

9. Explain how the Miller–Holzworth area source model uses the principle of relative addition. Does the Reiquam model of Eq. (20-10) use the principle?

10. Using hourly wind speed and direction observations covering a 5 year period for a major airport near you, prepare the following:
    (a) A wind rose for all January hours midnight to 0400. Wind roses for the same hours in July and for the hours 1200–1600 in January and July.

(b) Thirty (30) trajectories for each of the same four combinations of month and time. Select the 30 starting dates from each combination by choosing a day at random from each 5 day period in the five January and five July records.

(c) For each combination of month and time, compare the trajectory endpoint obtained using the most frequent speed–direction combination, assuming this lasts constantly for 5 hours, with the endpoint defined by the mean of all x and all y coordinates of the 30 actual trajectories of the period.

11. Discuss the probable effects of the orientation of a major mountain valley (i.e., of the azimuth of the centerline) on the effectiveness of the local mountain–valley wind system in dispersing pollutants generated in the valley.

12. Prepare sketches and a discussion to show that the upper air maps of (a) the height of a constant pressure surface and (b) the pressure at a constant height will both show essentially the same ridge–trough and High–Low features.

Chapter 21

EFFECTS OF POLLUTANTS ON THE ATMOSPHERE

Although interactions between pollutants and the atmosphere have been mentioned previously to illustrate some point or another about atmospheric behavior, there remain other sets of interactions—syndromes one might call them—which deserve examination. Some have been recognized and studied for many years. Others have only recently emerged to the point where they are dimly seen as problems to be reckoned with in the near future. It would not be possible to discuss all recognized problems of interactions between pollutants and the atmosphere in a volume such as this; but it is possible to suggest the span of syndromes from the microscale to the global, and to point out which of the principles discussed in previous chapters are at work in the operation of each of the syndromes mentioned. That is the aim of this chapter.

I. Seeing through the Atmosphere

For people who live in flat country, or who have grown up in a place that has had dirty air all their lives, the impact of limited visual range through the atmosphere is small. For those who live in places devoid of major topographic features, but who have seen air become distinctly less transparent in the past few years—people on the Texas Gulf coast, for example—the problem is now apparent as an aesthetic one. For those in areas where mountain views are a major part of the pleasure of life, the amount of outcry from residents is often proportional to their number and to the recency of decline in visual range. Nearly a whole generation of southern Californians has grown up without being able to see clearly the mountains at their back doors. In the Pacific Northwest and in parts of the New England states, the impact is more recent and the outcry more vigorous. The segment of our society most in touch with the real magnitude and time scale of the problem of vision through the atmosphere is the fraternity of airline pilots. The availability of electronic technology to make seeing unnecessary during flight does not reduce their negative reactions to the impact of reduced visibility.

For very few people is seeing through the atmosphere more than an aesthetic matter, but to anyone who has ever really enjoyed a good long view through clear air, the fouling of our air is a source of considerable sorrow and frustration. To all, no matter what their personal history, place of residence, or level of income, the loss of visual range ought to be perceived as "nature trying to tell us something before it is too late."

In meteorological and climatic records, *visibility* is the standard measure of the transparency of the atmosphere in the visible spectrum. The term is defined according to the needs of the industry that prompted its definition—the airline and aviation industry. Visibility is "the greatest horizontal visual range which is attained or surpassed throughout half the horizon circle, not necessarily continuous." *Visual range* is simply the maximum distance one may see in a given direction at a given time. Thus the measure that is reported and recorded—visibility—is a kind of average visual range for all directions. Official observers arrive at a reportable figure through the use of a set of known landmarks, of prescribed characteristics, at known distances and directions from their duty post. Visibility is determined with considerable subjectivity.

The term *meteorological range* is more directly relatable to air quality in a physical and numerical sense. Following is a development of its definition (1).

If the luminance of an object B is the intensity of the beam of visible light leaving the object in the direction of an observer, its difference from the luminance of a second, adjacent object is $(B - B')$. Because light is scattered and otherwise removed from the sight path by air molecules and contaminants, and also because light is scattered into the path when the objects are at some distance L from the observer, the apparent luminances will be different: B_L and $(B_L - B_L')$. By defining an *extinction coefficient* b one may write an empirically correct relationship for luminance which has the form of Beer's Law (Chapter 17):

$$(B_L - B_L') = (B - B') \exp(-bL) \qquad (21\text{-}1)$$

Since objects in the environment are "visible" partly because they contrast with their surroundings, define the close-range contrast ("inherent contrast") as $C_0 = (B - B')/B'$ and the distant contrast ("apparent contrast") as $C_L = (B_L - B_L')/B_L'$. Given these definitions, plus the assumption that the primary object is seen against the sky, and the assumption that the sky's luminance is independent of distance, then $B'/B_L' = 1$, and Eq. (21-1) becomes

$$C_L = C_0 \exp(-bL) \qquad (21\text{-}2)$$

This expresses the exponential fashion in which visual contrast deteriorates with distance due to light scattered out of and into the intervening visual path.

If the object seen against the sky is in fact perfectly black, then $B = 0$, and the contrast $C_0 = (B - B')/B' = -1$, leading from Eq. (21-2) to

$$C_L = -\exp\ (-bL)\quad \text{or}\quad e^{+bL} = |\,1/C_L\,| \tag{21-3}$$

for a black target seen against the horizon sky. The assumption of a "standard" performance for the human eye, in the form of a threshold contrast value for C_L equal to 0.02 (i.e., 2%), leads from Eq. (21-3) to the definition of the *meteorological range* L_m:

$$\exp\ (bL_m) = |\,1/0.02\,|\quad \text{or}\quad L_m = \ln(50)/b = 3.9/b \tag{21-4}$$

The units of the extinction coefficient are the same as those of its physical analog, the absorption coefficient of Beer's Law, which is $(\text{length})^{-1}$. Thus, for example, a meteorological range of 10 km would occur with $b = 3.9 \times 10^{-4}\ \text{m}^{-1}$, and a range of 1 km would occur when $b = 3.9 \times 10^{-3}\ \text{m}^{-1}$. The latter would be found in very dirty air, with roughly the same depletion powers as a cloud in which $k_s = 0.005$ (see Fig. 17-4).

What determines the magnitude of b? What is it a function of? First of all, when light is scattered by particles the size of which is very much smaller than the wavelengths of the light (i.e., much smaller than 0.5 μ), the process is called *Rayleigh scattering* and the value of b would be the order of 1.5×10^{-5}. The molecules of pure air produce such an effect. The extinction of light by air pollutants, however, is due to particles whose diameters are the same order of magnitude as the wavelength of light. This *Mie scattering* produces an effective coefficient b_s, which is related to the particle radius r, the number of particles involved N, and the "scattering area ratio", κ, which depends on the materials which make up the particles, in the following manner:

$$b_s = \pi r^2 N \kappa \quad \text{or} \quad b_s = \pi \sum_{i=1}^{n} r_i^2 N_i \kappa_i \tag{21-5}$$

In the first case, all particles have the same radius. In the second, there are N_i particles in the ith size interval, of which there are n intervals making up the size spectrum (2).

The area ratio κ is a function of the refractive index m, and the ratio of the circumference of the spherical particle and the wavelength being scattered $\alpha = 2\pi r/\lambda$. Figure 21-1 shows the value of κ as a function of $2\alpha\ (m - 1)$.

With the assumption of monochromatic radiation, and aerosols which are all spheres of the same material and diameter, several effects may be examined by combining Eqs. (21-4) and (21-5). First note the mass of aerosols in a unit volume is $(4/3)\pi r^3 \rho N$, where ρ is the particle density and the other symbols are defined above. Then, designating the density as χ, form the product:

$$L_m \chi = (3.9/\pi r^2 N \kappa)(1.33\ \pi r^3 \rho N) = 5.2\ \rho r/\kappa = 2.6\ \rho d/\kappa \tag{21-6}$$

For a water aerosol, $\rho = 1$, $m = 1.33$, and $d = 0.8\ \mu$. With $\lambda = 0.5\ \mu$ then from

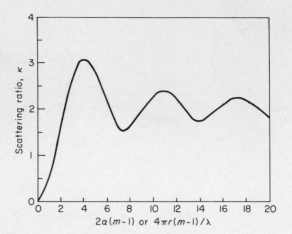

FIG. 21-1. Approximate values of the scattering area ratio κ for nonabsorbing spheres as a function of $2\alpha(m - 1)$. From Robinson (22).

Figure 21-1, $\kappa = 2.9$ approximately. Thus for this particular aerosol $L_m\chi = 0.72$ gm·m^{-2}. The physical interpretation is that in a sight path of 1 m^2 cross section, the meteorological range will be the length of the path containing 0.72 gm of water.

A measure of the amount of depletion due to scattering for a unit mass of spherical aerosol is

$$(b_s/1.33 \, \pi r^3 \rho N) = (\pi r^2 N\kappa)/(1.33 \, \pi r^3 \rho N) = 3 \, \kappa/4 \, \rho r = 1.5 \, \kappa/\rho d \quad (21\text{-}7)$$

In Fig. 21-2, the effect of aerosol size on the meteorological range L_m and on the scatter per unit mass, 1.5 $\kappa/\rho d$, are shown for two aerosols: oil droplets and water droplets. For both aerosols, it is clear that spheres of diameter between 0.5 and 1.0 μ are much more effective in reducing visibility than either smaller or larger sizes. The implications for control technology are very important.

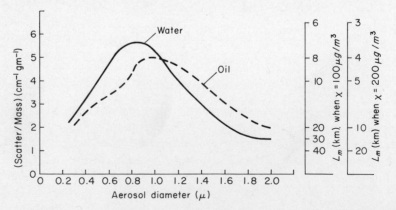

FIG. 21-2. The effect of aerosol diameter on scatter per unit mass and on meteorological range, for oil ($\rho = 0.9$; $m = 1.5$) and for water ($\rho = 1$; $m = 1.33$).

TABLE 21-1

Relationships of Visibility and Relative Humidity at Coastal Stations[a]

Relative humidity (%)	Average visibility (km), Los Angeles	Frequency (%) of visibility <7 km, Leeuwarden, Netherlands	
		Maritime air	Continental air
100–99	<0.2	95	100
98–97	0.5	89	97
96–93	0.9	62	88
92–89	1.6	32	72
88–83	2.5	19	57
82–77	3.1	11	47
76–69	5.6	4	45
68–61	7.2	1	35
60–50	7.1	0	20

[a] From Robinson (2), after Neiburger and Wurtele, and Buma.

Although natural aerosols are not of a single diameter, and sunlight is not monochromatic, the features of Fig. 21-2 are still qualitatively correct, as may be demonstrated in similar calculations by use of the spectral version of Eq. (21-5). Still another effect of an aerosol–environment interaction is that which results from relative humidity. Figure 21-1 shows the wide range of values of κ for submicron particles or droplets, and Fig. 19-2 suggests the major effect of relative humidity on droplet size when nuclei are hygroscopic, as in the case of sea salt. Thus when relative humidity is near saturation it may be expected to affect κ, and thus b_s, and thus L_m, or visibility. Table 21-1 shows the rapid deterioration of visibility in maritime air as the relative humidity exceeds the value of 75%, where sea salt

TABLE 21-2

Contributions to Final Extinction Due to Various Portions of the Sight Path in a Uniformly Illuminated and Polluted Atmosphere[a]

Portion of path to L_m (%)	0–20	20–40	40–60	60–80	80–100
Percent of total extinction	56	27	9	6	2

[a] After Robinson (2).

crystals start to become droplets. The data for continental air show a different rate of decrease in visibility, presumably because the nuclei are industrial rather than maritime.

The direct interpretation of either visual range or visibility as a measure of particulate loading in the atmosphere is of very shaky validity. As already noted, several things about the mixture of pollutants must be known. In addition, the value of b_s along the sight path might be quite variable in the presence of one or several strong point sources whose plumes intersect the path. Finally, differential illumination of the path, as for example under scattered clouds, may also affect the assumptions made in the derivation of Eq. (21-4). Whatever the nature of the

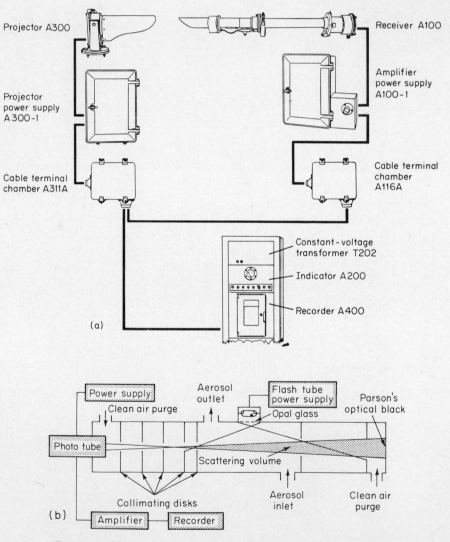

FIG. 21-3. (a) A transmissometer. (b) An integrating nephelometer.

nonuniformity along the path, its importance in the determination of the visual range is greater the nearer it is to the observer, as shown in Table 21-2.

In order to avoid the necessity of space averaging of b_s with the human eye as sensor, various instruments have been developed to examine much shorter paths selected for study. The transmissometer (Fig. 21-3a) senses a fixed path of about 100 m, usually along an airport runway, and converts the intensity of light received at one end of the path, compared with a known light intensity within the system, to a value of L_m (3). It is intended to sense mainly fog along the runway, providing a continuous record of L_m for airport use. The assumptions involved in the calibration make the instrument of value only for specific air pollution problems, and bulk and path length of the instrument necessitate its use as a fixed installation.

The integrating nephelometer senses a 2-m path length and, by means of the measured intensity of light scattered to the side from a sample of ambient air, yields a reading of L_m directly (4). The instrument is small, portable (Fig. 21-3b), and capable of being continuously recorded. It has been used atop an automobile to sense b_s along a path, on the ground, known to include variations in b_s. It has been mounted on light aircraft to sense vertical as well as horizontal variations of b_s in various combinations of thermal stratification and pollution sources (5). Valuable though it is, the nephelometer's record is interpretable directly as L_m only based on a number of assumptions leading to a relationship such as Eq. (21-4). One of the assumptions is that the attenuation due to absorption, expressed in the coefficient b_a, is negligible compared with the attenuation due to scattering, ex-

Fig. 21-4. The Voltz sun photometer.

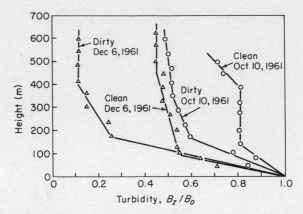

FIG. 21-5. Turbidity profiles for clean and polluted air, Cincinnati, Ohio. "Clean" soundings are upwind of city; "polluted" soundings are above city center. After Robinson (22) from McCormick and Baulch (7).

pressed by b_s. Though relatively little is known about b_a, in some cases it is of the same order of magnitude as b_s (2, 6) and must therefore be included in a more general form of calibration equation for an instrument such as the integrating nephelometer:

$$L_m = 3.9/(b_s + b_a) \tag{21-8}$$

In particular, based on a comparison of ocular estimates of L_m with values from the nephelometer sensing air containing substantial concentrations of carbon particles from open burning of woody, organic materials, $b_a \doteq 0.44 \, b_s$. As noted by Robinson (2), "for smokes of obvious color . . . absorption becomes an important factor in the light extinction calculation."

A hand-held "Voltz sun photometer" (Fig. 21-4) has been used with increasing frequency to assess the turbidity of the atmosphere between the sun and the instrument (2). It collimates the solar beam through optical filters, which make an essentially monochromatic beam ($\lambda = 0.5 \, \mu$) impinge on a photocell detector connected for direct reading to a microammeter. The collimation procedure also provides a measure of the relative optical path length. The nomographs with the instrument lead from the reading on the microammeter to a value for the *turbidity coefficient*, B, which is closely akin to $(b_s + b_a)$, but not related to the luminance B in Eq. (21-1).* Through a number of assumptions about the pollutant aerosols producing the turbidity, McCormick and Baulch (7) have concluded that the value of the turbidity coefficient at ground level B_o is linearly related to the total mass of the particulate loading in the air column above the observation point.

Figure 21-5, taken from McCormick and Baulch, shows two pairs of soundings of B_z, the turbidity of the air layer above altitude z. The soundings are normalized with respect to the value of B_o. October 10, 1961 was less polluted over Cincinnati than December 6, 1961, according to these data. But each of the four soundings tends to exhibit one slope up to some intermediate level and then essentially another

---

* In particular, $B \sim [-\log_{10}(S_n/S)][z\sec\theta]^{-1}$, whereas, $b \sim [-\ln(S_n/S)][z\sec\theta]^{-1}$.

constant slope above that. It can be shown* that the concentration of pollutants is essentially constant from the surface up to the level of the inflection point in the sounding, and then essentially constant, but less, above that level. This kind of vertical homogeneity of pollutant concentration is a necessary assumption in the box models of dispersion for pollutants from area sources (see discussion of Fig. 20-12).

Visual range need not be restricted to the horizontal. Given the inherent contrasts and the assumptions about uniform turbidity and illumination in connection with Eqs. (21-2) and (21-4), the ideas can be extended to the concept of the *slant visual range*, L_s. Figure 21-6a shows the geometry of the slant visual range in three dimensions. In a mass of uniformly polluted air, in which the visual range is L_s, an observer at 0 can see throughout the spherical volume with radius L_s. In particular he can see a circular section of an intersecting surface whose nearest distance to him is AL_s. The radius of the circular section is GL_s. The rapid increase in the radius of the horizon circle shortly after the observer has approached to within L_s of the surface (Fig. 21-6b) accounts for the frequent reports of obscured objects "suddenly looming up" before an approaching observer. It lies behind Edinger's vivid description (8) of the pilot's passage through the top of the Los Angeles smog layer:

> As one rises up through the polluted air in a light aircraft he is first aware that he is nearing the top of the layer when the sky above begins to turn blue rapidly. He begins to see a fuzzy horizon encircling him, blue above, beige or brown below. He may think this is the top but it is not. He is still submerged. The horizon becomes less fuzzy. Then suddenly it is sharp and straight. He has risen out of the murky marine air [above point B in Fig. 18-14b] and entered the desert above.

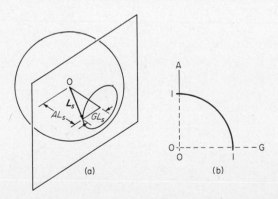

$$(a) \qquad\qquad\qquad (b)$$

FIG. 21-6. (a) Geometry of the slant visual range in three dimensions. (b) Relationship of relative distance to a surface, A, and the relative radius of the surface horizon circle, G.

* In a layer of constant slope for a sounding in Fig. 21-5, the relationship may be expressed as $d(B_z/B_0)/dz = -S$. Since $B \sim N$ in a layer by Eq. (21-5), this gives $(N_2/N_0) - (N_1/N_0) \sim -S(dz)$, where level 2 is above level 1. From this when $dz = 1$, $N_1 - N_2 \sim SN_0$. Since $N_1 - N_2 \sim \chi$, the statement above follows. Thus for a given slope in Fig. (21-5), doubling B_0 or N_0 doubles χ; or increasing the slope by a given amount (i.e., increasing the departure from vertical) within a given sounding increases χ in the layer a like amount.

Likewise, these remarks apply to the visual range afforded a ground-based observer in mountainous terrain. In nonuniformly polluted, or nonuniformly illuminated air, of course, the relationships become much more complex.

II. Mesoscale Topography, Pollutants, and Stagnation

In the discussion of Eq. (20-4), it was noted that in theory the downwind concentration of pollutants from a source of strength Q cannot fall below (Q/ULW), where U is the mean wind speed, L is the depth of vertical mixing, and W the width of a valley. If, instead of a point source upstream in the valley there is an area source of strength Q' (mass/area-time) and of area WS_c, then $Q'WS_c$ replaces Q. Thus, from this area source, which may be an industrial area or a city, the downwind concentration in the valley cannot fall below $(Q'S_c/UL)$. If vertical mixing to height L is restricted because of a very stable layer—often an inversion—and pollutants accumulate in the valley, the presence of the accumulated pollutants often acts, in several ways, to prolong the very conditions which led to stagnation of the air and the accumulation of pollutants. The syndrome is thus a self-perpetuating one.

In the first place, higher concentrations of particulate pollutants in the enclosed valley make the top of the polluted layer more reflective of sunlight, thus reducing the amount of shortwave radiant energy penetrating to the valley walls and floor. The heating of the valley walls and floor, and thus the enclosed air mass, would lead to decreased stability of the air, but the pollutants counteract this tendency. Second, the overabundance of particles available as condensation nuclei in the valley air increases the likelihood that droplets, once formed, will not grow beyond the droplet size into drops. The resulting fog, which forms in the colder air beneath the inversion, more effectively reflects sunlight from its top, further reducing the chances of warming in the valley. Finally, the prolonged absence of warming in the valley may cause increased combustion of particulate-producing fuels for domestic heating. This, of course, further magnifies the problem.

III. Air Pollutants and Urban Climate

In the discussion of Fig. 20-21, it was noted that the existence of a city, for various reasons, causes the climate in that place to be different from the climate which would exist if the city were not there. There seems to be little doubt that this "urban climate"—various changes superimposed on the regional climate—exists and is attributable to the city (9–13). The magnitudes of the various changes, and certainly the processes that produce them, are far from definitely established. In Chapter 20 brief mention was made of some of the effects of the city's shape, its surface materials, and its artificial heat sources. In this section, mention will be made of some of the suspected effects whose causes involve air-quality directly.

Landsberg (9, 10) gives as a general result a 15% reduction in the annual receipt of solar radiation on a horizontal surface in a mid-latitude city. Robinson (2) shows by simple calculations that this is quite a reasonable expectation. For example, for a path length of 0.3 km through a polluted urban atmosphere in which the meteorological range is 3 km, Eq. (21-4) gives $b = 3.9/3 = 1.3$, with which Eq. (17-8) (Beer's Law) gives

$$S_U/S_o = \exp(-1.3)(0.3) = 0.68$$

where S_U is the urban shortwave flux on a surface and S_o is the shortwave flux above the polluted layer, on a surface of the same orientation. Robinson points out that this calculated 32% reduction in S_o in the direct beam may be partly made up for by the forward scattering of the pollutants, making perhaps a 25% actual net reduction, for the particular path length and turbidity specified. Comparable calculations for other turbidities (i.e., values of L_m), times of day, latitudes, and seasons would give information for specific dates, hours, and localities.

While the reductions above are for the entire shortwave flux, the reduction is apparently quite selective, being considerably greater for ultraviolet and violet than for the blue-to-red part of the spectrum. Landsberg makes the generalization (10) that UV is reduced typically 15% in summer and 30% in winter in mid-latitudes. He cites a 90% reduction in UV in Paris, France (9) as does Robinson (2) in Pasadena, California, under rather ordinary circumstances. Although the amount of reduction of S_o or the UV portion of it is not established in general, the selectivity by wavelength is. The reason for this may be seen in the contents of Table 21-3. For a value of $m = 1.4$ representative of urban air pollutants, values of the function $2\alpha (m - 1)$ are shown, for use in reference to Fig. 21-1. In the figure values of the function increasing to 4.0 result in values of increasing κ, and thus increasing b and decreasing transmission. Just above 4.0, the function shows decreasing values of κ. The net result is that which has been observed: for the smaller aerosol particles in the urban air (r less than 0.3μ), ultraviolet ($\lambda \leq 0.3 \mu$) is depleted distinctly more than longer wavelengths of the visible spectrum. For the same kinds of reasons, polluted air, or air full of salt crystals, produces redder sunrises and sunsets.

Some writers (10, 14) suggest that the particulate aerosols of polluted air pro-

TABLE 21-3

For an Assumed Value of $m = 1.4$, Values of the Function $2\alpha(m - 1)$, or $4\pi r(m - 1)/\lambda$ as a Function of Particle Radius r and Radiation Wavelength λ

Wavelength (μ)	"Color"	Particle radius (μ)				
		0.1	0.2	0.3	0.4	0.5
0.3	UV	1.66	3.4	5.1	6.7	8.4
0.4	Violet	1.26	2.5	3.8	5.1	6.0
0.5	Blue–green	1.0	2.0	3.0	4.0	5.1
0.6	Red	0.84	1.66	2.5	3.4	4.2
0.7	Near IR	0.72	1.44	2.2	2.9	3.6

TABLE 21-4

Probabilities of Low Visibility Due to Fog in Paris, France, and Environs[a]

Type of visibility reduction due to fog	Probability[b] (%)					
	Summer			Winter		
	City	Suburb	Country	City	Suburb	Country
Light fog ($\frac{1}{4}$ to 1 mile)	4.9	4.9	0.6	35.0	21.9	6.0
Moderate (300 ft to $\frac{1}{4}$ mile)	0.3	0.3	0.2	4.9	4.3	2.8
Dense (less than 300 ft)	0	0.1	0	0.8	1.4	0.5

[a] After Landsberg (9) from Besson.
[b] Besson's data are for 9 AM.

duce increases in precipitation-related climatic variables by acting as condensation and freezing nuclei (see Chapter 19, Sections I and III). Measures of total precipitation, cloudiness, days with precipitation of a certain amount, frequency of thunderstorms, and frequency of hail have been shown to increase (9, 11, 14, 15) by roughly 10% in times and places which suggest they may be responding to air pollution. These same writers, however, note that many of the observed effects may be as readily attributed to increased vertical velocities, with resulting increased drop residence time and precipitation rate (Chapter 19) due to the mechanical effects of urban roughness elements and the thermal effects of urban heat sources (Chapter 20). The riddle has yet to be unraveled.

About the frequency of fog, however, there seems less doubt that the increases in urban areas are directly related to air quality. Naturally, reduced visibility in cities is likewise related to the air quality directly and to fog indirectly. Robinson (2) cites the greater number and greater solubility of particulate urban air pollutants as assuring the observed increase in frequency, density, and persistence of fog in urban areas. Landsberg offers data due to Besson, showing probabilities of fog in three classes in the vicinity of Paris, as shown in Table 21-4. Such data are so rare there is little chance of telling if these probabilities are at all typical of mid-latitude cities. Lowry has only a speculation to offer (12) about the reason for a greater likelihood of suburban fog than would be expected from an interpolation between city and country: greater numbers of nuclei and greater amounts of urban-generated water vapor encountered by inflowing air before very much heating of the air has taken place.

Landsberg also offers as typical the data in Table 21-5, on low visibility due to fog and smoke at Detroit, Michigan. These occurrences were all with wind speeds less than 5 knots, but do not take account of any information on mixing height. Robinson (2) notes that L_m should be directly related to mixing height and to wind speed [Eqs. (20-4) and (21-4)], and he offers some evidence that this is actually observed.

Effects of air quality upon urban temperatures have only been suggested.

TABLE 21-5

Comparative Frequency of Visibility Being a Mile or Less, at an Urban and a Rural Airport Site near Detroit, Michigan[a]

| | Observed frequencies (hr/month) | | | | | |
| | Due to fog | | | Due to smoke | | |
Month	Urban	Rural	Difference	Urban	Rural	Difference
Jan.	12	10	+2	5	1	+4
Feb.	8	6	+2	6	2	+4
Mar.	6	3	+3	7	1	+6
Apr.	6	5	+1	3	0	+3
May	4	3	+1	2	0	+2
June	2	4	−2	2	0	+2
July	1	3	−2	0	0	0
Aug.	3	5	−2	1	0	+1
Sept.	8	13	−5	1	0	+1
Oct.	17	16	+1	7	0	+7
Nov.	11	7	+4	7	1	+6
Dec.	22	8	+14	8	1	+7
Total	100	83	+17	49	6	+43

[a] After Landsberg (9).

Whereas it appears definitely established that cities are warmer than their surroundings, more so at night and in winter (9, 10, 13), the portion of the difference due to air pollution is not known. Basically, polluted air should act thermostatically, as clouds do, in reducing the diurnal range of temperature without significantly changing the mean temperature. Such an effect is shown in the change from curve 1 to curve 2 in Fig. 21-7. If an increase in the mean temperature, due to a city's artificial heat sources, were then superimposed on the thermostatic effect due to clouds and pollutants, the result would look like curve 3 in Fig. 21-7. The relationships between curves 1 and 3 are qualitatively the same as observed in the diurnal temperature variations of many cities compared with their surroundings (9, 12).

Urban–rural temperature differences are typically +4°C at night, −0.5°C at midday, and +0.7°C for the annual mean. Nocturnal differences as large as +6 or +7°C are not uncommon when winds are light and skies are clear.

IV. Air Pollutants and Weather Modification

To atmospheric scientists, the term "weather modification" has come to mean the intentional activity of men to alter the kind or amount of precipitation falling on a certain "target" area or region. Not simply "rainmaking," these activities are also aimed at hail suppression and fog dispersion (11). A few tentative trials of hurricane modification have also been made (17). All of these activities have

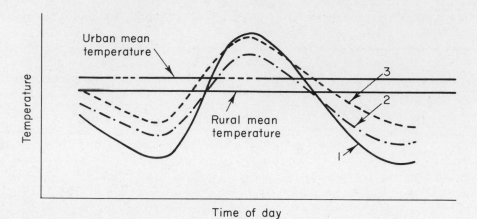

FIG. 21-7. Comparative diurnal temperature variation in the city and environs, as the result of effects due to clouds, pollutants, and artificial heat sources in the city.

involved the injection into the atmosphere of specific chemical substances to act as nuclei (Chapter 19) in the various stages of the formation of precipitation—air pollutants in every sense. They have all been intentional, but their successes have been mixed, despite the apparent validity of the physical theories on which they are founded. From the controversies surrounding the evaluations of these techniques have come various questions about the advisability of their operational use, even if they could be perfected. Ecological systems analysis, which insists on examination of both advantages and disadvantages throughout a system, leaves us asking "Even if we can do it, should we? (11)"

The several ways in which air pollutants are suspected of modifying urban climates, discussed in the last section, are cases of "inadvertant weather modification," to use the current phrase. Many of the mechanisms and results have to do with precipitation processes, but others do not. All having to do with precipitation seem to affect downwind, rural areas as well as, and in some cases more than, the urban areas which generate them. For example Huff and Changnon (18) conclude that the cities of St. Louis, Missouri, and Chicago–Gary, on the Illinois–Indiana border, produce increases in various measures of precipitation in areas perhaps 30–35 km downwind from the principal urban-industrial parts of the cities. Some of the increases are ascribed partially to pollution. Much doubt has been expressed about the amount of increase beyond 10% downwind from Chicago–Gary (19, 20), but increases up to 10% seem not in doubt.

As another example, Hobbs et al. have mapped a measure of the change in mean annual precipitation between two 20-year periods (1927–1946 and 1947–1966) over the Pacific Northwest [Washington, Oregon, Idaho, and British Columbia (21)]. They propose there is a causal connection between the facts that many industrial sources of pollutants were constructed near the end of the first period; and that there were outstanding increases in precipitation (absent elsewhere in the region) on areas roughly downwind of many of these sources. Because not all new sources have increases downwind, and because there are increases not associated with any known source which came into existence around 1946–1947, the evidence must be considered intriguing but shaky (21a).

With reports such as these, suggesting weather modification has been "discovered" after the fact, it is not surprising that concern is expressed about the possible effects of such relatively new (in North America at least) technology as cooling towers for power generating plants. In the absence of definitive studies, either theoretical or at operating sites, only the broadest and most tentative conclusions about this problem are available at this writing (22). The cooling towers, built to dissipate the unused two thirds of the heat generated by a nuclear plant, inject this heat into the atmosphere at a rate between 5 and 10% that of a large city, and on this basis might be expected to produce detectable effects on precipitation processes. Unlike a city, however, cooling towers, which dissipate heat by evaporating water, inject "wet heat" in the form of warm vapor rather than particle-laden plumes of warmer, dried air (23). Dry towers, of course, behave otherwise, but their relatively lower efficiency makes them distinctly less popular and less of a concern.

Huff *et al.* (22) have concluded that lower, forced draft towers (as opposed to taller natural draft towers) may have marked effects on fogging and snowfall under relatively rare meteorological circumstances. With the density of plants and towers contemplated today, no major, detectable weather modification appears likely from cooling towers. If these installations become more numerous, however, today's tentative conclusions may well have to be changed.

V. Air Pollutants and Global Climatic Change

In and downwind of specific, known urban-industrial areas there are changes in weather and climate, of reasonably established kind and magnitude, which are ascribable in part to air pollutants released in the areas. The mechanisms which link the pollutants and some of the changes are in doubt, though physical theory provides some bases for speculation and experimentation. On the hemispheric and global scales, however, a "signal" of man-made climatic change has not yet been separated from the "noise" of natural climatic change (11). Figure 21-8 shows the 230-year records of temperature and annual precipitation reconstructed by Landsberg for northeastern North America. The variations before mid-nineteenth century are almost certainly natural, and show the "noise" on which any man-made signal would be imposed. The last half-century of the records, which includes most of today's anthropogenic (man-made) effects, looks very little different.

Table 21-6 suggests qualitatively why global temperatures are not likely to respond in any simple way to changes in gaseous and particulate pollution. The table shows whether increases $(+)$, or decreases $(-)$ in pollutants or clouds, taken by themselves, are likely to produce increases, decreases, or no change in the mean temperature of the earth's surface, T_e, in the mean radiative temperature of the earth–atmosphere system, T_π, or in the ratio of the mean temperatures of atmosphere and earth (T_a/T_e). This last parameter is taken as an indicator of the vigor

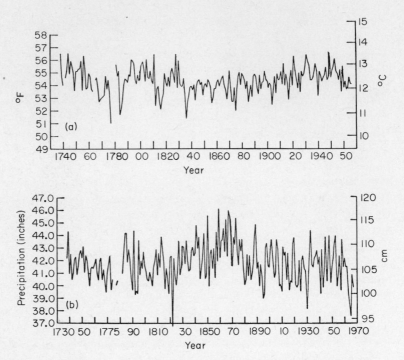

FIG. 21-8. (a) Annual mean temperature at Philadelphia, 1738–1967. (b) Annual precipitation total at Philadelphia, 1738–1967. From Landsberg (11).

of large scale convective mixing, since mixing (lapse rate) is enhanced as T_a grows smaller relative to T_ϵ.

Equations for the three temperature parameters in terms of the atmosphere's shortwave reflectivity r_{as}, shortwave transmissivity τ_{as}, and longwave transmissivity, τ_{aL}, are obtained from the three-layered model of the global radiation balance in Chapter 17. Recalling that (a) $S' = (1 - r_{as})(S/4)$; that (b) $(r_{as} + \tau_{as} + a_{as}) = 1$; and that (c) $(\tau_{aL} + a_{aL}) = 1$, the three equations obtained from the model are

From Eq. (17-16): $\sigma T_\epsilon^4 = (1 - r_{as})(S/4)\left[(1 + \tau_{as})/(1 + \tau_{aL})\right]$ (21-9)

From Eq. (17-15): $(T_a/T_\epsilon)^4$
$$= [1 - (\tau_{as}/F_\epsilon)(S/4)(1 - r_{as})]/(1 - \tau_{aL}) \qquad (21\text{-}10)$$

From Eq. (17-13): $\sigma T_\pi^4 = (1 - r_{as})(S/4)$ (21-11)

The remarks in Table 21-6 are taken in part from recent papers (24–26). For example, Robinson (24) remarks that gaseous pollution, in the absence of changes in cloudiness, can become a "destabilizing condition" with positive feedback (processes I and II). In this model, increases in ocean temperatures due to the greenhouse effect would produce greater evaporation of water and greater vaporization of dissolved CO_2 from the oceans. Along with these changes, however,

TABLE 21-6

Summary of Qualitative Effects of Gaseous Pollutants (CO_2, H_2O), Particulate Pollutants, and Cloudiness on Three Parameters of Global Temperature ᵃ

Pollutants and processes	Increase (+), decrease (−) or absence of change (0) in ...						Remarks
	r_{as}	τ_{as}	τ_{aL}	T_ϵ	(T_a/T_ϵ)	T_π	
(I) Increases in CO_2 and/or H_2O; no change in cloud	0	0	−	+	−	0	These two processes represent positive feedback in which an increase in gaseous pollution leads to further increase in gaseous pollution. A decrease in T_a might alter radio propagation.
(II) Increases in CO_2 and H_2O due to increase in T_ϵ	0	0	−	+	−	0	
(III) Increase in cloud due to increased T_ϵ, H_2O and convective mixing [due to decreased (T_a/T_ϵ)]	+	−	−	$\begin{cases} -(\tau_{as}) \\ -(\tau_{as}) \\ +(\tau_{aL}) \end{cases}$	$\begin{cases} +(\tau_{as}) \\ +(\tau_{as}) \\ -(\tau_{aL}) \end{cases}$	−	Counteracts effects of the two processes above on T_ϵ and (T_a/T_ϵ) but lowers T_π
(IV) Increases in particulate pollution producing							
(a) Increased shortwave scatter	+	0	0	−	+	−	Shortwave effects tend to reduce T_ϵ and thus CO_2, H_2O and cloud.
(b) Increased shortwave absorption (reduced τ_{as})	0	−	0	−	+	0	Reduced convective mixing further reduces cloud.
(c) Increased longwave absorption (reduced τ_{aL})	0	0	−	+	−	0	

ᵃ For derivation and physical meanings of T_ϵ, (T_a/T_ϵ), and T_π, see text.

would come increased convective mixing (decrease in T_a/T_ϵ) and the production of more cloudiness (process III). This would tend to counteract the effects of processes I and II, so that the net changes in thermal parameters would not be simply determined. Robinson notes in addition that cooling of the upper mesosphere and lower thermosphere might well change patterns of radio propagation.

Bryson and Wendland (25) argue that between the late 1800's and 1950 increases in CO_2 led to increases in T_ϵ (processes I and II), but that in the past two decades the observed decrease in T_ϵ has resulted from process IV overbalancing I and II as particulate aerosols have increased rapidly.* They also argue that in the particular region of Northwest India, process IV has led to an increase in subsidence over the Rajasthan Desert (see remarks in Table 21-6), and that this has led to suppression of clouds and rain in another cycle with positive feedback.

Atwater (26) cautions that the interactions among the values of the surface albedo of the earth (taken to be zero in Table 21-6), the scattering coefficient of the aerosol b_s, and the absorption coefficient of the aerosol b_a, may produce such unexpected results as surface warming from increased near-surface particulate loadings.

The surface albedo of the earth, of course, is not zero. Though it is small over most of the surface, it is quite sizeable over ice and snow-covered areas. The extent of this area, and its effect on the mean surface albedo, is a complex function of T_ϵ, atmospheric water vapor content, and convective mixing. The extent of this area may also be a function of the rate of dustfall on existing glaciers (24), so that consideration of this factor in the determination of surface albedo constitutes another complex aspect of the problem not even dealt with in Table 21-6, nor in any existing models of global heat balance and climatic change.

Robinson (24) considers briefly the rate of meridional (equator-to-pole) mixing of heat and water vapor through macroscale storm systems, which is in turn a function of the distribution of the heat and water vapor being mixed. This horizontal mixing on the hemispheric scale is still another complex aspect of the problem not dealt with in Table 21-6. Robinson also considers the possibilities of using intricate mathematical simulation models of macroscale atmospheric behavior to study changes in global climate resulting from hypothetical changes, due to air pollution, in boundary conditions and parameters. The conclusion is that no model which does not include provision for the essential features of cloud production, migration, and dissipation will provide dependable results. At the time of writing, no existing models include such provisions.

The recent controversy over the predicted effects of the supersonic transport airplane (SST) involved not only processes I, II, and III of Table 21-6, but also the concept that increased water vapor injected into the lower stratosphere might lead to a chemical chain reaction that would reduce the concentrations there of ozone, which protects surface life from damage by absorbing ultraviolet radiation (27).

Finally, several points can be made within the scope of this text. First, simplistic

---

* In the same volume (25) Mitchell argues the increase in aerosols has been due to volcanic activity, not to anthropogenic pollution.

reasoning about global effects of air pollution is misleading. Second, natural variations in basic climatic parameters (e.g., Fig. 21-8) and natural concentrations of air pollutants are distinctly greater than those currently being attributed to man's activities (11, 24, 25). Third, our knowledge of the complex earth–atmosphere system and of the values of the critical parameters and variables is so scant as to make any definitive statements impossible for at least a decade hence. Fourth, despite the first three statements, climatic change could be produced regionally by such regional effects as urbanization. This last assertion is based on the fact that the atmosphere moves, behaves, and produces regional climates as a result not of average values of such parameters as T_ϵ and (T_a/T_ϵ) and CO_2 content, but as a result of their gradients or distributions across the face of the earth. It is conceivable that man's non-random activities, such as urbanization and industrialization, may have global effects greater than the effects of nature's larger but more random variations. Probing that assertion seems to be among mankind's most crucial tasks for the future.

References

1. Middleton, W. E. K. "Vision through the Atmosphere," 250 pp. University of Toronto Press, Toronto, 1952.
2. Robinson, E. *in* "Air Pollution" (A. C. Stern, ed.), Vol. 1, 694 pp. Academic Press, New York, 1968.
3. U. S. Department of Commerce. "Instruction Manual, Transmissometer System." Publication ESSA-DC-WB1024, Washington, D.C., August 1966.
4. Ahlquist, N., and Charlson, R. *J. Air Pollut. Contr. Assoc.* **17,** 467–469 (1967).
5. Adams, D., and Koppe, R. *J. Air Pollut. Contr. Assoc.* **19,** 410–415 (1969).
6. Nelson, W., Boubel, R., and Lowry, W. A comparison of several techniques for estimation of aerosol loading and visibility. *in* "Proceedings of the Second International Clean Air Congress," p. 406. Academic Press, New York, 1971.
7. McCormick, R., and Baulch, D. Reference 9, Chapter 17.
8. Edinger, J. "Watching for the Wind," p. 106. Doubleday Anchor Science Study Series, New York, 1967.
9. Landsberg, H. The climate of towns *in* "Man's Role in Changing the Face of the Earth" (W. L. Thomas, ed.), 1193 pp. Univ. Chicago Press, Chicago, Illinois, 1956.
10. Landsberg, H. City air—better or worse?, *in* "Air over Cities." U.S. Public Health Service: Washington, 1962.
11. Landsberg, H. *Science* **170,** 1265–1274, Dec. 18, 1970.
12. Lowry, W. Reference 4, Chapter 17.
13. Lowry, W. *Sci. Amer.* p. 15, Aug. 1967.
14. Changnon, S. *Bull. Amer. Meteorol. Soc.* **49,** 4–11 (1968).
15. Changnon, S. *Bull. Amer. Meteorol. Soc.* **50,** 411–421 (1969).
16. Changnon, S. *in* "Urban Climates," 390 pp. Technical Note 108, World Meteorological Organization: Geneva, Switzerland, 1970.
17. Gentry, R. *Science* **168,** 473 (1970).
18. Huff, F., and Changnon, S. Urban effects on daily rainfall distribution. Paper presented to the Second National Conference on Weather Modification, Santa Barbara, California, April 1970.
19. Changnon, S. *Bull. Amer. Meteorol. Soc.* **51,** 337–342 (1970).
20. Changnon, S. *J. Appl. Meteorol.* **10,** 165–168 (1971).
21. Hobbs, P., Radke, L., and Shumway, S. *J. Atm. Sci.* **27,** 81–89 (1970).

21a. Elliott, W. P., and Ramsey, F. *J. Atm. Sci.* **27**, 1215–16 (1970); Hobbs, P., Radke, L., and Shumway, S. *J. Atm. Sci.* **27**, 1216–17 (1970).
22. Huff, F., Beebe, R., Jones, D., Morgan, G., and Semonin, R. Preliminary report: effect of cooling tower effluents on atmospheric conditions in northeastern Illinois. Circular 100, Illinois State Water Survey: Urbana, 39 pp, 1971.
23. Woodson, R. *Sci. Amer.* p. 70, May 1971.
24. Robinson, G. "Long-Term Effects of Air Pollution—a Survey." Publication CEM 4029-400, Center for the Environment and Man: Hartford, Connecticut, June 1970.
25. Bryson, R., and Wendland, W. Climatic effects of atmospheric pollution, *in* "Global Effects of Environmental Pollution" (S. Singer, ed.), 218 p. D. Reidel: Dordrecht, Holland. 1970.
26. Atwater, M. *Science* **170**, 64–66, 1970.
27. Nuessle, V., and Holcomb, R. *Science* **168**, 1562, 1970.

Suggested Reading

Brooks, C. E. P. "Climate through the Ages," 395 pp. Benn, London, 1949.
Clairborne, R. "Climate, Man and History," 444 pp. Norton, New York, 1970.
Council on Environmental Quality. Environmental Quality: The First Annual Report to the Congress. U.S. Government Printing Office: Washington. 1970. 326 pp.
Study of Critical Environmental Problems (SCEP) "Man's Impact on the Global Environment," 319 pp. MIT Press, Cambridge, Massachusetts, 1970.

Questions

1. When might the visibility and the meteorological range actually have the same value? *Hint*: First list the assumptions made in arriving at each.
2. Prepare a graph or a table showing how sensitive the meteorological range is to the definition of "the standard performance of the human eye."
3. Does a colored automobile "fog light" (usually yellow) penetrate fog better than an ordinary "white" headlamp? Assume fog to be made up of droplets as in Fig. 19-4. Does the same answer apply if the question is about "smog lights" referring to particulate aerosols?
4. Given the following industrial effluent ($m = 1.4$) made up of spheres, calculate the smallest particle diameter down to which a filter must be effective if (a) the total extinction of sunlight ($0.5\ \mu$) must be reduced by 80% and (b) the filter removes 90% of all particles in the sizes it filters and none in other sizes.

Size range (μ)	<0.5	0.5–0.8	0.8–1.2	1.2–1.5	1.5–1.8	1.8–2.4	2.4–3.0
Relative mass (%) per unit volume	5	2	10	18	40	20	5

5. Over a city at latitude 40°, at noon on the shortest day of the year, the meteorological range in a polluted layer is 4 km. What is the depth of the polluted layer if the sun's intensity is reduced by 50% in passing through the layer?
6. Explain the assertion (Section V) that the increased particulate pollution in the atmosphere over a desert becomes a self-perpetuating phenomenon (i.e., with positive feedback.)
7. Air flows across isobars toward the center of a traveling cyclone, or Low, yet the cyclone does not necessarily "fill up." Where does the inflowing air go?

Chapter 22

STUDIES OF AIR POLLUTION CLIMATOLOGY

Why "air pollution climatology?" Why not "air pollution meteorology?" Except for in-depth studies of specific, important, acute episodes of meteorologically based air pollution damage, after the fact, e.g. (1), research into the role of the atmosphere in air pollution problems* should be statistical and probabilistic. That is, its aim should be to develop methods for evaluating and then reducing the odds in favor of damaging episodes, based on examination of a sample of past cases of damaging or potentially damaging episodes and based on study of the probabilistic nature of the atmosphere's behavior as it relates to air pollution. The atmosphere is not deterministic in any way we could ever predict accurately. It is probabilistic. It will never again exhibit precisely the same combination of conditions which accompanied any of the famous air pollution episodes. Thus it is the climatological examination of past examples, to extract the relevant general principles, which appears to be potentially the most fruitful avenue of approach. Detailed studies of episodes are badly needed, therefore, as parts of samples, but not as ends in themselves.

The methods of research are essentially limitless (2), but unfortunately the full diversity of tools is seldom used in operational studies of actual problems in air

---

* Topics such as scavenging by precipitation, visibility, and radiation balance of the upper atmosphere are generally referred to by atmospheric scientists as atmospheric physics. These are not the kinds of problems being spoken of in this chapter.

pollution climatology. The tools are statistical, graphic, physical, and conceptual. All are potentially important if research is to be truly complete. Statistical tools permit numerical examination of relationships, and are of varied and familiar kinds. Graphic tools such as maps and charts often produce insight and understanding at a glance, which might come only laboriously by purely numerical methods. Physical modeling, in wind tunnels as an example, often produce valuable answers for novel and complex combinations of elements for which not even first approximations of understanding are otherwise available. Finally, conceptual methods, such as system modeling and the testing of hypotheses, are perhaps the essential tools of any complete research program. Too often, for example, the avoidance of an honest attempt to disprove an attractive conclusion by further testing of an hypothesis leads to the properly ridiculed "lying with statistics (3)."

I. Two Conceptual Models

Earlier, in the introduction to Chapter 17, it was noted that the atmosphere plays roles in conditioning the release of air pollutants and in conditioning the effects of air pollutants on receptors. The third role of the atmosphere as transporter and diffuser may be seen through two simple conceptual models. First, the atmosphere, in either a local or a global context, may be viewed as a wash basin with an inlet tap and a drain. Pollutants enter through the tap, representing sources with varied strengths. They exit through the drain, representing natural cleansing processes with a fixed upper limit of capacity. If the rate of entry does not exceed the upper limit of the drain, the basin will empty with only the time in doubt. If the rate of entry does exceed the upper limit of the drain, the basin will overflow with only the time in doubt. Globally there seems no way of increasing the size of the drain, but we do not yet know its size very well. Locally the ebb and flow of macroscale weather systems to some extent enlarges and then constricts the size of the drain in much better understood fashion. Everyday experiences tell us the drain capacity is frequently exceeded locally, as when visibility and other factors in the syndrome become aggravated. Less well known is the evidence that the global

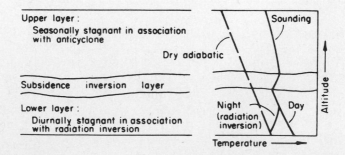

FIG. 22-1. Schematic model relating diurnal and seasonal potential of the atmosphere for dispersion and dilution of pollutants.

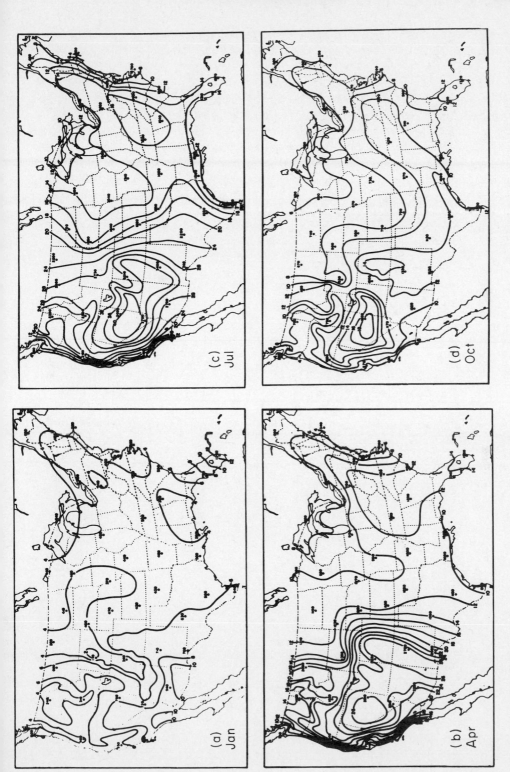

FIG. 22-2. Mean maximum mixing height above the surface (100's of meters) of the continental United States. (a) January, (b) April, (c) July, and (d) October. From Holzworth (5).

drain size is probably being exceeded for some pollutants (4). No matter what the numerical values involved, the nature of the problem is inescapably one of rates, in the sense of the tap–basin–drain model.

The second simple, conceptual model deals with the linkages between the size and time scales of local and regional atmospheric cleansing processes—the drain size. In the model, the atmosphere consists of two layers, one above the other, as in Fig. 22-1. The thickness of the bottom layer is closely related to the Maximum Mixing Height (MMH), which is in turn governed in part by the presence or absence of macroscale subsidence (Chapter 18). Holzworth has mapped the mean MMH in different seasons across the continental United States (5), and his maps appear as Fig. 22-2. Except in the Southwest, and in winter at all localities, the thickness of the lower layer is generally between 1 and 2 km. The upper layer extends from there on up to around 10 km, the top of the troposphere.

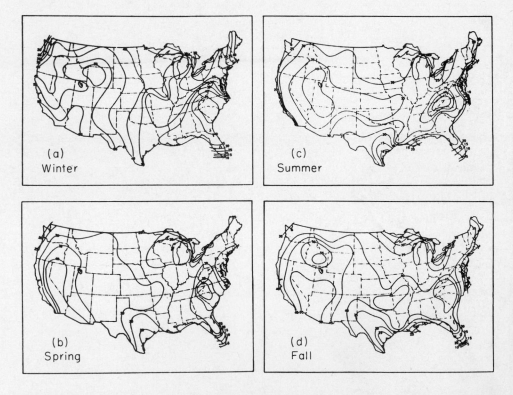

Fig. 22-3. Frequency of near-surface inversions (base at 150 m or below) as a percent of total hours across the continental United States. (a) Winter, (b) spring, (c) summer, and (d) fall. From Hosler (6).

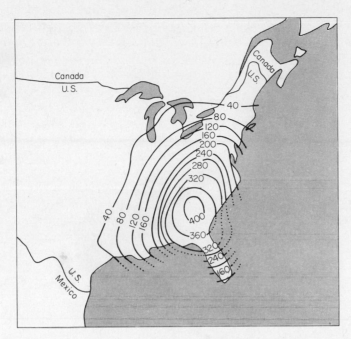

Fig. 22-4. Total number of extreme stagnation days during 1936–1965 east of the Rocky Mountains. From Korshover (8).

The two layers may be either stagnant or in motion. If they are coupled (connected), both exhibit the same sort of motion or lack of it. If they are uncoupled, their motion may or may not be the same. Coupling occurs when macroscale subsidence is absent; decoupling when it is present. This is so because, as discussed in Chapter 18 (Section VI), the subsidence inversion strongly suppresses vertical motion, and thus restricts the sharing of both momentum and pollutants between layers. The lower layer tends to alternate *diurnally* between the stagnation accompanying radiation inversions (Chapter 17, Section VII) and the turbulent motion brought by midday heating. Hosler has mapped the frequency (percent of total hours) of near-surface inversions, by season, across the United States (6); his maps are shown in Fig. 22-3. The western mountains (not coastal) and the Appalachians exhibit the highest frequencies in all seasons. Over most of the country, such inversions, with their temporary stagnation of the lower layer, occur during part of 50% of all nights, and during 90% or more of all nights over large areas in the western states (7).

The upper layer tends to alternate *seasonally* between motion and the relative stagnation associated with the presence of semipermanent anticyclones (Chapter 20). Korshover has mapped the frequency of upper level stagnation due to anticyclones east of the Rocky Mountains (8). The total number of such days in the 30-year period 1936–1965 is shown in Fig. 22-4. The months of April–June and

September–October exhibit the greatest frequency. Korshover's maps show only the cases of extreme stagnation; but cases of intermediate severity follow the same seasonal patterns as his study shows. In addition to the center of stagnation over the southern Appalachians, a center appears in winter over the northern Great Basin (Idaho, Wyoming) (7).

In summary, the second simple model, as shown in Fig. 22-1, suggests that the local potential for dispersion and dilution of air pollutants is the net result of the interactions of generally diurnal processes in the lower layer and generally seasonal processes in the upper layer. The least severe potential for accumulation of pollutants is when both layers are coupled and in motion. The most severe potential [which has occurred in all cases pointed to as classic air pollution episodes (see Fig. 11-3)] consists of the stagnation of both layers over the period of several days, when the diurnal processes in the lower layer were suppressed (Chapter 21, Section II). Other combinations produce intermediate severity of potential. Among other things, it should now be clear to the reader that air pollution climatology is many times more complicated than the often-heard comment that only in times of an inversion does an air pollution problem exist.

II. Studies in Air Pollution Potential

Air pollution potential has been defined as "the inability of the atmosphere to disperse and dilute pollutants which may be emitted into it." Thus high potential may exist even in the absence of pollutants.

While the two-layer model addresses the problems of variable dispersion rate

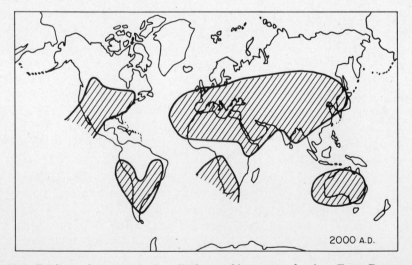

FIG. 22-5. Regions where severe air pollution problems may develop. From Pettersen (9).

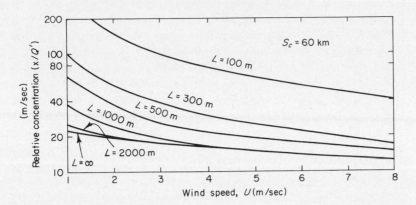

FIG. 22-6. Average relative concentration (χ/Q') over a 60-km city as a function of thickness of the mixed layer L and mean wind speed in the mixed layer U. After Miller and Holzworth (10).

(drain size) on the local and regional scales, we must consider the maximum potential cleansing rate (drain size) as being fixed on the global scale. Some of the consequences of exceeding that rate were discussed in Chapter 21, Section V. Pettersen has published a map (9) suggesting which regions on earth will first experience severe air pollution problems if the implications of the two conceptual models just outlined are ignored. His map is shown in Fig. 22-5.

In a sense Figs. 22-2 through 22-5 represent studies of air pollution potential, presenting information on atmospheric behavior which could produce problems if combined with excessive emissions. Although Pettersen's map is the result of synthesizing several individual variables, it is nonnumerical. Holzworth, however, has proceeded from the Miller–Holzworth city model (10) to a numerical statement about the mean relative concentration (χ/Q) to be expected over a city of a given size (i.e., a given downwind dimension S) if it were to be built in a given climate. The climate for these purposes is characterized by the daily combinations of mixing height L, and mean wind speed in the mixed layer, U. Figure 22-6 shows the relative concentration as a function of L and U for a city with a downwind dimension $S = 60$ km. Figure 22-7 shows Holzworth's results (11) obtained from the mixing height–wind speed climates of those 62 stations in the continental United States which twice daily obtain wind and temperature soundings from radiosonde ascents (Figs. 18–6 and 18–7).

By calculating relative concentration on each morning and each afternoon, Holzworth developed probability curves (12) from which statements such as the one in Fig. 22-7 could be made. Interpretation of a value of (χ/Q') from the figure might be as follows: For any combination of L and U which gives $(\chi/Q') = 100$ sec·m⁻¹ over a 10-km city, an average area source strength, Q', of 0.1 μg m⁻² sec⁻¹ for a particular pollutant will result in a mean concentration throughout the city of $(\chi/Q')(Q') = (100)(0.1) = 10$ μg m⁻³. In Fig. 22-7, relative concentrations for a

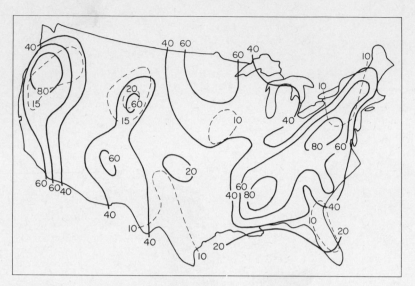

Fig. 22-7. Relative concentration (χ/Q') exceeded in a 10-km city on 10% of all mornings (solid lines) and 10% of all afternoons (dashed lines). After Holzworth (11).

100-km city would be roughly 10 times the values shown for a 10-km city. In general, relative concentrations in the Holzworth model are roughly proportional to city dimension S_c [see formulation of $(Q'S_c/UL)$ in Chapter 21, Section II].* With proper interpretation, the relative concentration calculated by Holzworth is a valuable index, which synthesizes several pertinent variables into a single number whose space and time variations may be analyzed. Various other such indexes have been developed and discussed in the technical literature, of which three will serve as examples.

Robinson (13) proposed what became known as the Stanford Index of air pollution potential, in the form

$$I_1 = (\Delta\theta)^2/(Z_i \cdot \Delta Z + 3) \qquad (22\text{-}1)$$

where $\Delta\theta$ is the difference in potential temperature (°C) between the top and the bottom of a subsidence inversion layer (see A, B in Fig. 18-14), Z_i is the height (100 m) of the inversion base above the surface, and ΔZ is the thickness (100 m) of the inversion layer. The number 3 avoids the necessity of dividing by zero when the inversion is based on the surface. Robinson and his colleagues have used the

* With calm winds, a simple box model indicates average concentrations over a city should increase in proportion to the persistence of the stagnation and inversely with the mixing height:

$$\text{Concentration} = \frac{\text{(Total emission)}}{\text{(Volume enclosed)}} = \frac{[Q't(\text{city area})]}{[L(\text{city area})]} = \frac{Q't}{L}$$

TABLE 22-1

Median and Tenth Decile Limits of Length of Stagnation Episodes by Months at Salem, Oregon, as Defined by I_2 in Eq. (22-2) [a]

	Jan.	Feb.	Mar.	Apr.	May	Jun.	Jul.	Aug.	Sep.	Oct.	Nov.	Dec.
Median	2.5	2.5	2.6	2.9	2.6	2.8	4.4	4.6	3.6	3.1	2.7	2.7
10th Decile	11	5.5	5.6	6.2	6.2	9.5	11	12	8.4	8.5	8.0	7.2

[a] From Lowry and Reiquam (14).

index both as a tool for climatic analysis and as a predictor of air quality in Los Angeles, where the subsidence inversion is a common and important feature of the air pollution problem.

Lowry and Reiquam (14) described an index of potential, which like Robinson's, uses data from routine radiosonde flights as input. Using data from Salem, Oregon, they developed statistics for a decade of the daily index:

$$I_2 = 14 + (T_{9,\text{AM}} - T_{s,\text{AM}}) + (T_{9,\text{PM}} - T_{s,\text{PM}}) \tag{22-2}$$

where T is temperature in degrees Centigrade, subscripts 9 and s refer to the 900 mbar and surface pressure levels, and AM and PM refer to morning and afternoon radiosonde flights. The factor 14 is chosen subjectively so that the index is near zero on a day with a zero net accumulation of pollutants due to vertical ventilation. To study stagnation episodes, I_2 is summed for consecutive days until terminated by one or more of the following conditions: (a) a negative value of I_2 (strong vertical

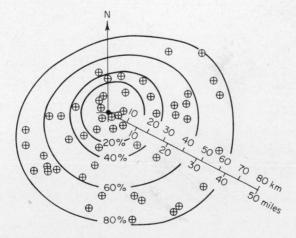

FIG. 22-8. Theoretical fractional containment of pollutants, emitted at the origin, during low-wind episodes at Allentown, Pennsylvania, 1965–1970. ⊕ = trajectory endpoint. For explanation of episode, see text.

ventilation), (b) a large daily resultant wind vector (horizontal ventilation), or (c) a daily precipitation total of 0.13 cm (0.05 in.) or greater (washout of pollutants). Their presentation of Salem's climate in terms of the lengths of such stagnation episodes is shown in Table 22-1. They claim its merits are (i) sensitivity to local and temporal changes, (ii) ease of calculation and comprehension by control agencies and planners, and (iii) flexibility of summation, resetting criteria (episode ending rules), and threshold [14 in Eq. (22-2)].

Weedfall and Linsky (15) propose a scheme for assessing the potential for three kinds of air pollution problems on the mesoscale. Problems arising from (i) incomplete combustion (SO_2, smoke, etc.), (ii) photochemical pollution (oxidants, PAN, etc.), and (iii) fog-reactive gases (e.g., SO_2) are examined in conjunction with such data as hourly wind speeds, mesoscale terrain factors (the depth/width ratio of a valley), and the frequencies in Fig. 22-3.

A mesoscale version of Pettersen's macroscale concept (Fig. 22-5) is shown in Fig. 22-8. The envelopes were developed from three-hourly wind tabulations at the Allentown, Pennsylvania, Airport. For all episodes (during 1965–1970) in which the record showed 4 kt or less persisting for 24 hours, a trajectory was drawn from a common origin, and the "nodes" of the trajectory were recorded on a master sheet. For episodes longer than 24 hours, several overlapping 24-hour periods each produced a trajectory. The isopleths in the figure enclose 20%, 40%, 60%, and 80% of all the 672 nodes recorded. The assertion is that at the end of any such episode at Allentown, approximately the indicated fraction of all persistent emissions during the episode from a given source would be contained within the corresponding isopleth when the origin is placed at the source. For reference, Philadelphia is about 50 miles (80 km) to the SSE, and New York is about 80 miles (130 km) east. The area of potential dispersion—a circle of radius (4 kt) (24 hours)—contains 12.7 times the area enclosed by the 80% isopleth, indicating a distinct tendency, during low wind-speed episodes, for the "sloshing" (Chapter 20, Section VII) produced by local wind systems, probably mountain–valley winds at Allentown. To emphasize

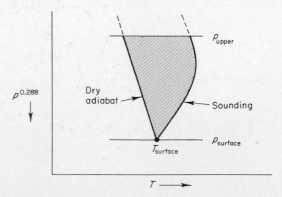

FIG. 22-9. Area representing the energy required to raise a unit mass of air from the surface to an upper level. After Pavelka (17).

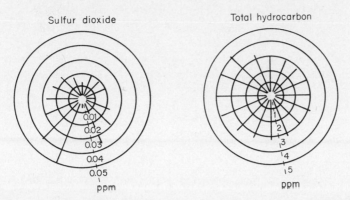

FIG. 22-10. Mean concentration of pollutant as a function of wind direction, Cincinnati, Ohio. From McCormick (7).

the idea, the figure shows all trajectory endpoints falling within the 80% isopleth—61% of all the endpoints generated in the study.

III. Studies of Interactions between the Atmosphere and Pollutants

Typical of the early attempts to relate atmospheric variables and crude measures of air quality is the study by Markee (16) in which 24-hour averages of particulate

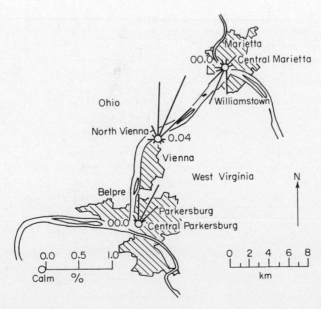

FIG. 22-11. Frequency of wind direction when SO_2 concentration exceeds 0.1 ppm (262 $\mu g/m^3$) near Parkersburg, West Virginia. From Munn (2, p. 111).

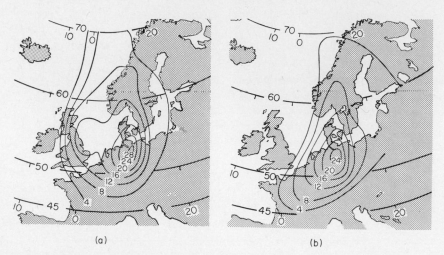

Fig. 22-12. Mean concentration of total sulfur ($\mu g/m^3$) (a) during a simulated air pollution episode in winter over northwest Europe; (b) same except all sulfur emissions in the United Kingdom were assumed to be eliminated. From Reiquam (21).

concentration were compared with basic weather observations. In such simple, direct attempts at defining relationships the absence of clear statistical correlation underscores the complexity of the interactions between pollutants and the atmosphere. Among the more successful attempts at simple linear correlation was described by Pavelka (17). During summer and fall at Medford, Oregon, he found a correlation ($r = +0.74$) between the mean particulate concentration for the 8-hour period 0330–1130LST and the area shown in Fig. 22-9 for the 0300LST radiosonde flight from Medford. In his study, the area is a measure of the energy required to lift a unit mass of air from the surface (about 970 mbar at Medford) to the 850-mbar level.

Correlation of air quality with wind-vector quantities has been variously attempted. For example, McCormick shows roses (7) of mean pollutant concentration in central Cincinnati, Ohio, as a function only of wind direction. Figure 22-10 contains two of his roses. The intriguing roses shown by Munn (2) depict frequency of wind direction at times when SO_2 concentration exceeded 0.1 ppm during 1965–1966 near Parkersburg, West Virginia (Fig. 22-11). Munn cautiously states "The diagram suggests, but of course does not prove, that a major source of SO_2 is situated between the sampling stations."

In a well-known early study, Pooler (18) used wind speed and direction frequencies together with estimated monthly mean area emission rates of SO_2 for Nashville, Tennessee, as inputs to a Gaussian model (see Chapter 20, Section III) to calculate mean monthly concentrations of SO_2 at points on a regional grid. The following is Wanta's (19) summary of Pooler's success—one-half of observed concentrations at 123 stations were within a factor of 1.25 of predicted; less than 5% of observed differed by a factor greater than 2 of predicted. This is typical of the successes of Gaussian models used in various ways, according to Wanta's tabulation of models.

Although Pooler did not examine questions about the results of changing emission patterns in Nashville, his model would have permitted it. Such an investigation was undertaken in Chicago by Golden and Mongan (20), however, as they examined the theoretical results of eliminating all SO_2 emissions from electric power generating plants in the area.* Because emissions from those sources are from elevated stack sources, while most of the other SO_2 comes from domestic heating and is injected nearer ground level, the elimination of today's 78% of total emissions attributable to power plants would produce only an estimated 11% reduction in annual mean concentration in the city's most polluted areas. Reductions in annual mean concentration near 50% in suburban areas, and reductions of 14% in total annual SO_2 dosage to the area's population were other estimates reported.

Employing the same basic concept of the "box model" as described previously for mesoscale application (Chapter 20, Section IV), Reiquam (21) calculated concentrations of sulfur over northwestern Europe, based on estimated source strengths in the region and on data simulating macroscale atmospheric behavior during a "typical" 10-day episode of light southwesterly winds during winter. Figure 22-12a shows Reiquam's results, which he judges "similar, both quantitatively and qualitatively," to results obtained for observed concentrations by Scandinavian investigators. Figure 22-12b illustrates other of Reiquam's results which are conceptually similar to those above from Chicago. It shows the concentrations of sulfur calculated on the assumption that all sulfur emissions were eliminated in the United Kingdom.

IV. Operational Forecasting of Air Pollution Potential

It is one thing to study the potential for pollutant accumulation before sources are present, or to study past examples of pollutant–atmosphere interactions. It is

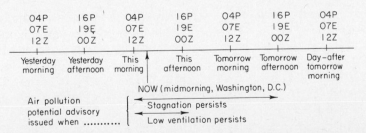

FIG. 22-13. The time structure of the National Air Pollution Potential Forecast Program.

* Golden and Mongan used the Air Quality Display Model (AQDM) developed for general, nonsite-specific studies by the U.S. Air Pollution Control Office, Environmental Protection Agency. One of several "package models," the AQDM is Gaussian and accepts meteorological and source inventory data to produce various estimates of air quality.

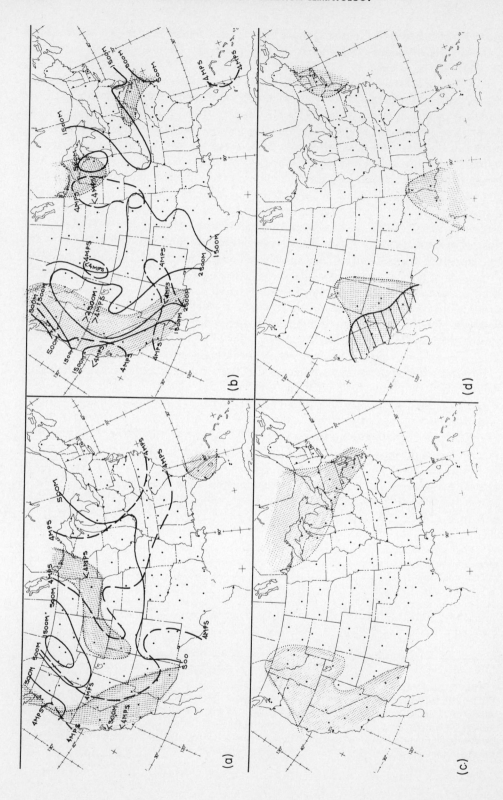

quite another thing to forecast the occurrence of atmospheric conditions conducive to accumulations, regionally or locally, on a daily operational basis. A program to provide such an "early warning" capability was undertaken jointly in the early 1960's by the United States Departments of Commerce (Weather Bureau) and Health, Education, and Welfare (Air Pollution Control Administration). It has grown from small, tentative experiments, made in the eastern states without forecasts actually being issued to the public, to today's daily issuance of air pollution potential advisories on teletype circuits to the entire nation.

The program consists of (22, 23) three coordinated efforts, each operating daily. First, a tally is made of the values of five parameters* observed at, or predicted for, the U.S. radiosonde stations in the network of Fig. 18-6. From these tallies, areas exhibiting tendency to stagnation are delimited. Second, within stagnation areas, estimates are made of the maximum mixing height (MMH, Fig. 18-12) and the average wind speed within the mixed layer. The product of the mixing height L and the transport wind speed U gives the ventilation factor, UL, which is inversely proportional to the relative pollutant concentration [Eq. (20-4) and Chapter 21, Section II]. Third, special low-level radiosonde flights are made within designated urban areas of existing pollution problems† to supplement the national radiosonde network. With slower ascent rates, these special sondes released by personnel of Environmental Meteorological Support Units (EMSU's) are capable of greater height resolution of inversion layers, and thus of greater precision in estimating UL.

Figure 22-13 shows the time dimensions of the forecast program and the intricate coordination required to produce the national advisories and the local advisories made by EMSUs. Radiosonde flights are made at 12-hourly intervals—Midnight and noon, Greenwich times (00Z and 12Z). Eastern Standard (E) and Pacific Standard (P) times are shown for the continental United States. Predictions of macroscale atmospheric changes out to 24 hours are made by computer at the National Meteorological Center at 00Z. Special soundings are made by EMSUs at or near sunrise. Using these observations and forecasts, midmorning advisories are issued (NOW in Fig. 22-13) based on the predicted persistence of stagnation conditions (first step, above) from "this morning" through "tomorrow afternoon"

* Wind speed 1.5 km (5000 ft) above station; 12-hour temperature change at 1.5 km; 6-hour precipitation amount; absolute vorticity (a measure of macroscale "spin" in the atmosphere) at 500 mbar above the station; and the 12-hour change in absolute vorticity at 500 mbar.

† At the time of writing programs existed in these areas—Washington, Philadelphia, New York, St. Louis, Chicago, Cleveland, Denver, and Louisville. The following are scheduled for initiation during 1972—Los Angeles, Pittsburgh, Houston, Seattle, Boston, and San Francisco. In 1974: Portland (Oregon), Charleston (West Virginia), and Detroit.

FIG. 22-14. Sample transmission accompanying the Pollution Potential Advisory of March 21, 1970. Stippled areas are stagnation areas. (a) This morning (observed): solid lines are observed values of L, and dashed lines are observed values of U. (b) This afternoon (forecasted); lines are forecasted values of L and U. (c) Tomorrow morning (forecasted). (d) Tomorrow afternoon (forecasted); the hatched area meets the criteria for persistence over 36 hours from "this morning" through "tomorrow afternoon." From Gross (22).

and of low ventilation (second step) from "this morning" through "this afternoon" within stagnation areas. Figure 22-14 shows the four-panel package transmitted daily to all forecast offices of the National Weather Service, together with the advisory for the day.

The threshold, or critical values, of the various indicators used on the way to the daily decisions about issuance of advisories are the result of examining past examples of atmospheric behavior as they relate to air pollution—climatology. Hence the inclusion of this section in a chapter on climatology. Efforts to improve the accuracy of the advisories, and efforts to enhance effective use of advisories by user agencies (air pollution control and public health) continue. Whether this coordination of supplier and users of the potential advisories actually prevents future air pollution

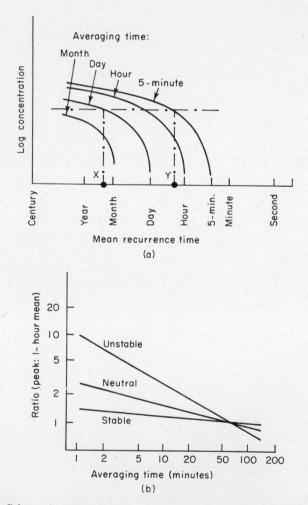

FIG. 22-15. (a) Schematic relationships among mean concentration, averaging time, and recurrence time for a hypothetical pollutant. [After Wanta (25).] (b) Schematic relationships among atmospheric stability, sampling time, and "peak-to-mean ratio." [After Turner (26).]

disasters remains to be seen, but the existence of the ambitious program constitutes major progress made in less than a decade in the United States.

V. Undertaking Future Studies of Air Pollution Climatology

Following the philosophy set forth in the introduction to this chapter, future studies of the atmosphere's role in air pollution problems should aim to evaluate, then reduce the odds in favor of damage from polluted air resources. Hilst (24) states the objective in language expressive of the viewpoint of an industrial manager: ". . . to exploit the natural capacity of the atmosphere to dispose of waste materials without incurring undue risks of loss and harm because of temporary or long-term alteration of the quality and composition of the air and the underlying surface and vegetation."

These two statements probably appeal to quite divergent tastes and viewpoints, but they have at least two things in common. First, they are both speaking of the avoidance of exceeding the "drain size" of the atmosphere, either locally or globally (Section I). Second, they both implicitly contain a principle which will lead to more effective research. The principle admonishes the researcher to *ask of the data the questions he actually wants answers to, and then analyze the data with these questions in mind.* Too often, researchers take the analyses others have made, for other purposes, and change their questions to fit the ready-made answers. Clearly, if followed too regularly this can lead to incorrect, or at best inappropriate, answers to actual operational questions. Several examples follow.

Hilst, in emphasizing the point about asking the right questions (24), uses as an example the 24-hour average concentrations of various pollutants obtained by the National Air Sampling Network (NASN) stations. While useful for many purposes, these data provide no information about variability within a day at the station or variability spatially around the station. Since equal damage or harm can result from a dosage involving short-term or long-term exposures (2, p. 19), and since threshold dosages vary widely with the nature of receptors, some questions simply cannot be properly answered by the 24-hour averages.

Several relationships concerning changes of measured values with changing sampling time are shown in Fig. 22-15. For example, while a certain 24-hour average value of the pollutant may recur only about once in 3 months (Point X in Fig. 22-5a) it is likely to recur about every 6 hours (Point Y) in a sequence of 5-minute samples of concentration. Moreover, this 5-minute average may represent a value between 1.2 and 4 times the hourly average, depending on the atmosphere's thermal stability at the time of measurement (Fig. 22-15b).

Another example is the standard wind rose analysis. A wind rose (Appendix E) may be a very useful tool for determining local placements of buildings with respect to pollutant sources or for determining alignment of an airport runway, but it is not of much use in answering a question such as "where will pollutants from this source move during the next 12 hours if they are released under conditions of low wind speed?" (Fig. 22-8). Detailed and complete though they may be, the answers given in Figs. 22-2 and 22-3 simply do not respond to the questions answered by Fig. 22-7,

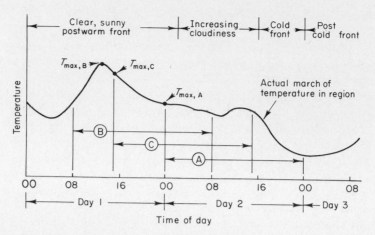

FIG. 22-16. Schematic relationship between hypothetical regional march of temperature and the maximum temperatures recorded at stations with different observation schedules. Station A observes hourly, reporting on a midnight-to-midnight day; Station B observes daily at 8 AM; Station C daily at 5 PM. T_{max} as recorded at each station on day 2 is shown on the temperature record.

though the same data went into the construction of all three. Finally, even the great utility of a presentation such as Fig. 22-7 breaks down when the question is "What is the probability of experiencing a three-day episode of high relative concentrations at various points in the United States?" This question too can be answered by the same data as went into these figures, but the question must first be asked explicitly.

What of the availability of climatic data for analysis? National governments have increasingly good networks of stations within which observations are made and data collected in archives under procedures recommended as standard by the World Meteorological Organization, an agency of the United Nations.* In the United States the variety of data and the forms in which they are available in published form have been catalogued (27). Various collections of regional climatic data have been published (28), and many specialized records and summaries are available upon inquiry.† McCormick has summarized briefly the better known sources of air quality data (7).

For some studies, having data summaries available may not be enough. There are still errors to be avoided by knowing details of the observation site and the observation schedule. For example, observations are usually made at major airports at a nonstandard location well above ground level on a building surrounded by paved areas. The same remark holds for many urban observations. Under some conditions of stagnation, such locations may introduce differences in wind speed or temperature which would be important for careful analyses. Another factor sometimes overlooked

* *Address:* The Secretary-General, World Meteorological Organization, 1211 Geneva 20, Switzerland.

† *Address:* Director, National Climatic Center, Federal Building, Asheville, North Carolina 28801

is the variation in observation time among the cooperative observers who supply most of the once-a-day data published by the U.S. government. Figure 22-16 suggests the kind of error which might arise, for example, in the preparation of a map of maximum temperatures for a region on a particular day. The error arises because data are published under the date on which the instruments are read, while the maximum temperature thermometer "remembers" the maximum value that occurred since the last reading. By no means the only error to be avoided by a thorough knowledge of the observation program, this one is mentioned in detail to emphasize the point.

For many questions about air pollution climatology, answers are possible only from special observations made expressly for the purpose of answering that question. Other answers may come from routine observations made by industrial or other private or nonfederal agencies. In any case, instrumentation is required for the observational program.* Meteorological instrumentation has been mentioned and displayed in earlier chapters, as discussion centered on the variables measured. The reader is here referred to the excellent treatments by Moses *et al.* (29), by Hewson (30), and by Turner (31) of available instrumentation and its proper calibration, installation, and utilization in air pollution climatology. Sellers (32) presents an excellent section on radiation instruments, as does Gates (33). Broader discussions of climatological data collection and analysis, not specifically concerned with air pollution problems, have been published by Middleton and Spilhaus (34), by Platt and Griffiths (35), and by Day and Sternes (36).

Lowry (37) includes a discussion of the rationale with which an environmental scientist approaches his task of meteorological data gathering. According to the discussion, the basic requirements for proper data collection are (a) knowledge of the nature of the environment, (b) a scheme for characterizing the environment in terms appropriate to the question being answered, (c) knowledge of instruments available and their interactions with the environment, and (d) a plan for obtaining measurements which is compatible with the scheme for characterization of the environment. In the end, the discussion comes down to the principle mentioned earlier, that the most effective applied researcher (in this case, data collector) keeps clearly in mind the specific question he is attempting to answer and the plan set out at the beginning for answering it. The experience of too many research programs shows that this direct, almost platitudinous concept is all too often overlooked.

References

1. Fensterstock, J., and Fankhauser, R. "Thanksgiving 1966 Air Pollution Episode in the Eastern United States," 45 pp. National Air Pollution Control Office Publication AP-45, July 1968.
2. Munn, R. E. "Biometeorological Methods," 336 pp. Academic Press, New York, 1970.
3. Platt, R. *Science* **146**, 347–353 (1964).
4. Robinson, G. Reference (24), Chapter 21.
5. Holzworth, G. *Monthly Weather Rev.* **92**, 235–242 (1964).
6. Hosler, C. R. *Monthly Weather Rev.* **89**, 319–339 (1961).

* As Munn (2, p. 66 *et seq.*) points out, good visual observations and simple questionnaires applied as "instrumentation" at the beginning of a program may greatly enhance and clarify the remainder of the program which follows.

7. McCormick, R. *in* "Air Pollution" (A. C. Stern, ed.), Vol. 1, 694 pp. Academic Press, New York, 1968.
8. Korshover, J. "Climatology of Stagnating Anticyclones East of the Rocky Mountains, 1936–1965," 15 pp. Public Health Service Publication, 999-AP-34, 1967.
9. Pettersen, S. *Bull. Amer. Meteorol. Soc.* 47(12), 950–963 (1966).
10. Miller, M., and Holzworth, G. Reference (11), Chapter 20.
11. Holzworth, G. Meteorological potential for urban air pollution in the contiguous United States. Paper presented to the Second International Clean Air Congress, Washington, D.C., December 1970.
12. Holzworth, G. *J. Appl. Meteorol.* **6**, 1039–1044 (1967).
13. Robinson, E. *Bull. Amer. Meteorol. Soc.* **33**, 247–250 (1952).
14. Lowry, W., and Reiquam, H. An index for analysis of the buildup of air pollution potential. Paper presented to the Annual Meeting, Air Pollution Control Association, St. Paul, Minnesota, June 1968.
15. Weedfall, R., and Linsky, B. *J. Air Pollut. Contr. Assoc.* **19**, 511–513 (1969).
16. Markee, E. *J. Amer. Indust. Hyg. Assoc.* **20**, 50–55 (1959).
17. Pavelka, B. M.S. Thesis, Oregon State University, Corvallis, June 1966.
18. Pooler, F. *Intern. J. Air Water Pollut.* **4**, 199 (1961).
19. Wanta, R. *in* "Air Pollution" (A. C. Stern, ed.), Vol. 1, *op. cit.*
20. Golden, J., and Mongan, T. *Science* **171**, 381–383 (1971).
21. Reiquam, H. *Science* **170**, 318–320 (1970).
22. Gross, E. "The National Air Pollution Potential Forecast Program," 28 pp. Technical Memorandum WBTM-NMC-47, U.S. Department of Commerce, Washington, May 1970.
23. Kirschner, B. "Proceedings of the Second International Clean Air Congress," pp. 987–993, Academic Press, New York, 1971.
24. Hilst, G. *in* "Air Pollution" (A. C. Stern, ed.), Vol. 1. *op. cit.*
25. Wanta, R. *in* "Air Pollution" (A. C. Stern, ed.), Vol. 1. *op. cit.*
26. Turner, D. B. Pollutant concentration variation (4-page mimeo). Collected materials published as training course manuals in air pollution meteorology by the U.S. Department of Health, Education and Welfare.
27. "Selective Guide to Climatic Data Sources." Key to Meteorological Records Documentation No. 4.11, U.S. Dept. of Commerce: Washington, D.C., 1969.
28. For example, "Climatological Handbook, Columbia Basin States." Pacific Northwest River Basins Commission, Vancouver, Washington, 1969.
29. Moses, H., and others. *in* "Meteorology and Atomic Energy." Reference (3), Chapter 20.
30. Hewson, E. W. *in* "Air Pollution" (A. C. Stern, ed.), Vol. II, 684 pp. Academic Press, New York, 1968.
31. Turner, D. B. "Meteorological Instruments," 26 pp. mimeo. Collected materials published as training course manuals in air pollution meteorology by the U.S. Department of Health, Education and Welfare.
32. Sellers, W. Chapter 6 in Reference (1), Chapter 16.
33. Gates, D. "Energy Exchange in the Biosphere," 151 pp. Harper, New York, 1962.
34. Middleton, W., and Spilhaus, A. "Meteorological Instruments," 286 pp. Univ. of Toronto Press, Toronto, 1953.
35. Platt, R., and Griffiths, J. "Environmental Measurement and Interpretation," 235 pp. Reinhold, New York, 1964.
36. Day, J., and Sternes, G. Chapters 2–4 in Reference (3), Chapter 19.
37. Lowry, W. Chapter 9 in Reference (4), Chapter 17.

Questions

1. What "operational" conclusions can be drawn, for either the local or the global scale, from the tap–basin–drain model of Section I?
2. Does Fig. 22-6 agree with the concept in Chapter 21, Section II? Should it?

3. According to Fig. 22-7, what would be the maximum permissible emission rate for SO_2 averaged over the area of your city if reasonable ambient air quality standards are to be met? For particulate aerosols less than 0.5 μ?

4. The AM sounding at a station is as follows, but the PM sounding has been modified until the surface temperature is 15°C. (a) Calculate I_1 from Eq. (22-1) and (b) calculate I_2 from Eq. (22-2).

Pressure (mbar)	1000	950	900	850	800	750
Temperature (°C)	10	8	7	10	12	10

5. With the January and July trajectories from Chapter 20, Question 10, (60 from each month) construct a diagram for each of the two months similar to Fig. 22-8. Your trajectories will each have only 5 "nodes" instead of 24 and will not be restricted to periods of low wind speeds. Nevertheless, do your diagrams suggest any regional zoning policies for your region?

6. The study of Golden and Mongan (20) drew a reply by J. J. MacKenzie in *Science* **172**, 792–793. Did he interpret Golden and Mongan correctly? Did he prepare an adequate rebuttal? What would you have answered if you were Golden or Mongan?

7. Explain qualitatively why the effects of atmospheric stability on the peak-to-mean ratio should be as they are in Fig. 22-15b.

Part IV

THE CONTROL OF AIR POLLUTION

Chapter 23

SOURCES OF AIR POLLUTION

I. General

The sources of air pollution are nearly as numerous as the grains of sand. In fact, the grains of sand themselves are air pollutants when the wind entrains them and they become airborne. We would class them as a natural air pollutant which infers that such pollution has always been with us. Natural sources of air pollution are defined as sources not caused by man in his activities.

Consider the case where man has removed the ground cover by his activities and left a layer of exposed soil. Later the wind picks up some of this soil and transports it a considerable distance to deposit it at another point where it affects other men. Would this be classed as a "natural pollutant" or a "man-made pollutant?" We might call it "natural" pollution if the time span between when man removed the ground cover and when the material became airborne was long enough. How long would be long enough? The answers to such questions are not as simple as they first appear. This is one of the reasons why pollution problems require careful study and analysis before a decision to control them at a certain level can be made.

A. Natural Sources

An erupting volcano emits particulate matter. Pollutant gases such as SO_2, H_2S, methane, etc. are also emitted. The emission from an eruption may be of such magnitude as to cause harm to the environment for a considerable distance from the volcanic source. Clouds of volcanic particulates and gases have remained airborne for very long periods of time.

Accidental fires in forests and on the prairies are usually classed as natural sources even though they may have been originally ignited by man or his activities. There are many cases where man intentionally sets fires in forest lands to burn off the residue, but lightning setting off a fire in a large section of forest land could only be classed as "natural." A large uncontrolled forest fire, as shown in Fig. 23-1, is a frightening thing to behold. Such a fire emits large quantities of pollutants in the form of smoke, unburned hydrocarbons, carbon monoxide, oxides of nitrogen, and ash. Forest fires in the Pacific Northwest section of the United States have been observed to emit a plume which causes visibility reduction and reduced amounts of sunlight as far away as 350 km from the actual fire.

Dust storms that entrain large amounts of particulate matter are a common natural source of air pollution in many parts of the world. Even a relatively small dust storm can result in suspended particulate readings one or two orders of magnitude above ambient air-quality standards. Visibility reduction during major dust storms is frequently the cause of severe highway accidents and can even effect air travel. The particulate transferred by dust storms from the desert to urban areas

FIG. 23-1. Uncontrolled forest fire. From Information and Education Section of the Oregon Department of Forestry.

causes problems to housewives, industry, and automobiles. The materials removed by the air cleaner of an automobile are primarily "natural pollutants" such as road dust and similar entrained material.

The oceans of the world are an important natural source of pollutant material. The ocean is continually emitting aerosols to the atmosphere, in the form of salt particles, which are corrosive to metals and paints. The action of the waves on the rocks reduces them to sand which may eventually become airborne. Even the shells washed up on the beach are eroded by wave and tidal action until they are reduced to such a small size that they too may become airborne.

An extensive source of natural pollutants is the plants and trees of the earth. Even though these green plants play a large part in the conversion of carbon dioxide to oxygen through photosynthesis, they are still the major source of hydrocarbons on the planet (1). The familiar blue haze over forested areas is nearly all from the atmospheric reactions of the volatile organics given off by the trees of the forest. Another air pollutant problem, that can be attributed to plant life, is the pollens which cause so much respiratory distress and allergic reaction among humans.

Other natural sources such as alkaline and salt water lakes are usually quite local as far as their effect on the environment. Sulfurous gases from hot springs also fall into this category in that the odor is extremely strong when close to the source but disappears a few kilometers away.

B. Man-Made Sources

1. Industrial Sources

The reliance of modern man upon industry to produce his needs has resulted in transferring the pollution sources from the individual to industry. A soap factory will probably not emit as much pollution as did the sum total of all the home soap-cooking kettles it replaces, but the factory is a source that all who consume the soap can point to and demand it be cleaned up.

A great deal of industrial pollution comes from manufacturing products from raw materials—(1) iron from ore, (2) lumber from trees, (3) gasoline from crude oil, and (4) stone from quarries. Each of these manufacturing processes produces a product along with several waste products which we term pollutants. Occasionally part or all of the polluting material can be recovered and converted into a useable product.

Industrial pollution also is emitted by industries that convert products to other products—(1) automobile bodies from steel, (2) furniture from lumber, (3) paint from solids and solvents, and (4) asphaltic paving from rock and oil.

Industrial sources are stationary and each emits relatively consistent qualities and quantities of pollutants. A paper mill, for example, will be in the same place tomorrow that it is today, emitting the same quantity of the same kinds of pollutants unless a major process change is made. Control of industrial sources can usually be accomplished by applying known technology. The most effective regulatory control is that which is applied uniformly within all segments of industries in a given region, e.g., "Emission from all asphalt plant dryers in this region shall not exceed 230 mg of particulate matter per standard dry cubic meter of air."

2. Utilities

The utilities in our modern society are so much a part of our lives that it is hard to imagine how we survived without them. An electric power plant generates electricity to heat and light our home in addition to additional power for the television, refrigerator, and electric toothbrush. When our homes were heated with wood fires, and home-made candles were used for light, there was no television and food was stored in a cellar, the total of the air pollution generated by all the individual sources probably exceeded that of the modern generating station supplying the energy. It is easy for the citizens to point out the utility as an air pollution source without connecting their own use of the power to the pollution from the utility. The electric utility counters this by erecting billboards which proclaim, "Heat your home with pollution free electric heat." Figure 23-2 illustrates one source of such "pollution free" electric heat.

Fig. 23-2. Smoking power plant.

Utilities are in the business of converting energy from one form to another and transporting that energy. If a large steam generating plant, producing 2000 MW, burns a million kilograms per hour of 4% ash coal it must somehow dispose of 40,000 kg per hour of ash. Some will be removed from the furnaces by the ash handling systems but some will go up the stack with the flue gases. If 50% of the ash enters the stack and the fly ash collection system is 95% efficient, 1000 kg/hr of ash will be emitted to the atmosphere. The gaseous emissions would include 341,000 kg of oxides of sulfur per day and 185,000 kg of oxides of nitrogen per day. If this is judged as excessive pollution, the management decision can be to (1) purchase lower ash or lower sulfur coal, (2) change the furnace so more ash goes to the ash pit and less goes up the stack, or (3) install more efficient air pollution control equipment. In any case, the cost of operation will be increased and this increase will be passed on to the consumer.

Another type of utility that is a serious air pollution source is that which handles the wastes of modern society. An overloaded, poorly designed, or poorly operated sewage treatment plant can cause an air pollution problem which will arouse citizens to demand immediate action. A burning dump is a sure source of public complaint even though it may be explained to this same public that it is the "cheapest" way to dispose of their solid waste. The public has shown their willingness to ban burning dumps and pay the additional cost of adequate waste disposal facilities to have a "pollution free environment."

3. Personal

Even though society has moved toward centralized industries and utilities we still have many personal sources of air pollution for which we alone can answer—(1) automobiles, (2) home furnaces, (3) home fireplaces, (4) backyard barbeque grills, and (5) open burning of refuse and leaves. Figure 23-3 illustrates the personal emissions of a typical United States family.

The energy release and air pollution emissions from personal sources in the United States is greater than that from industry and utilities combined. In any major city in the United States, the mass of pollutants emitted by the vast numbers of private automobiles exceeds that from all other sources.

Control of these personal sources of pollution takes the form of (1) regulation (fireplaces may only be used when atmospheric mixing is favorable), (2) change of lifestyle (sell the automobile and ride public transportation), (3) change from polluting source to less polluting source (convert the furnace to natural gas), or (4) change the form of pollution (instead of burning leaves haul them to the city dump). Whatever method is used for control of pollution from personal sources it will probably be difficult and unpopular to enforce. It is difficult to get a citizen to believe that his new, highly advertised, shiny, unpaid for, automobile is as serious a pollution problem as the smoking factory stack he can see on the horizon. It is likewise a very ineffective argument to point out that the workers at that factory put more total air pollution into the air each day just driving their automobiles to and from work than does the factory with its visible plume.

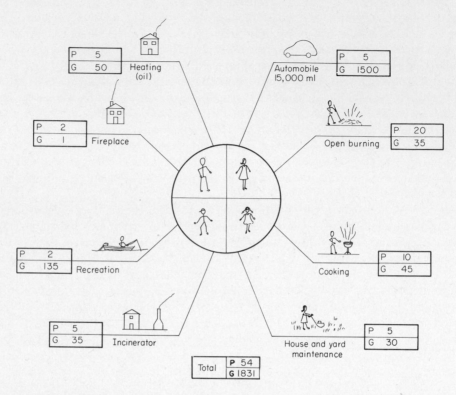

FIG. 23-3. Estimated personal emissions from United States family of four persons. (P, Particulate in kg/year; G, gases in kg/year.)

II. Combustion

Combustion is the most widely used, and yet one of the least understood, chemical reactions at man's disposal. Combustion is defined as the rapid union of a substance with oxygen accompanied by the evolution of light and heat (2).

Man uses combustion primarily for heat by changing the potential chemical energy of the fuel to thermal energy. He does this in a fossil fuel fired power plant, a home furnace, or an automobile engine. Man also uses combustion as a means of destruction for his unwanted materials. He reduces the volume of a solid waste by burning the combustibles in an incinerator. He subjects undesirable combustible gases, such as odors, to a high temperature in an afterburner system to convert them to less objectionable gases.

The simple combustion equations are very familiar:

$$C + O_2 \rightarrow CO_2 \tag{23-1}$$

$$2\,H_2 + O_2 \rightarrow 2\,H_2O \tag{23-2}$$

They produce the products, carbon dioxide and water, which are odorless and invisible.

The problems with the combustion reaction occur because in going from the start to finish there are also produced many other products, most of which are termed "air pollutants." These can be: carbon monoxide, oxides of sulfur, oxides of nitrogen, smoke, flyash, metals, metal oxides, metal salts, aldehydes, ketones, acids, polynuclear hydrocarbons, and many others. Only in the past few years have combustion engineers become concerned about these relatively small quantities of materials emitted from the combustion process. An Automotive Engineer, for example, was not overly concerned about the 1% carbon monoxide in the exhaust of the gasoline engine. If he could get this 1% to burn to carbon dioxide inside the combustion chamber, he could expect an increase in gasoline mileage of something less than $\frac{1}{2}$ of 1%. This 1% carbon monoxide, however, is 10,000 ppm by volume and a number of such magnitude cannot be ignored by an engineer dealing with air pollution problems.

Combustion is an extremely complicated reaction but is generally accepted to be a free radical chain reaction. Several reasons exist to support the free radical mechanism—(1) simple calculations of heats of dissociation and formation for the molecules involved do not agree with the experimental values obtained for heats of combustion, (2) a great variety of end products may be found in the exhaust from a combustion reaction. Many complicated organic molecules have been identified in the effluent from a system burning pure methane with pure oxygen, and (3) inhibitors, such as tetraethyl lead, can greatly change the rate of reaction (3).

A useful concept when visualizing a combustion process is to think of it in terms of the three T's, time, temperature, and turbulence. Time for combustion to occur is necessary. A combustion process that is just about ready to get going and suddenly has its reactants discharged to a chilled environment will not go to completion and will emit excessive pollutants.

A high enough temperature must exist for the combustion reaction to be initiated. Combustion is an exothermic reaction (it gives off heat) but it also requires energy to be initiated. This is illustrated in Fig. 23-4.

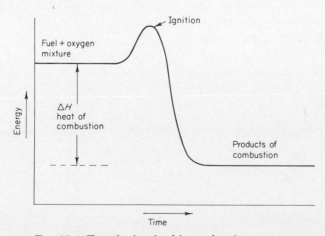

FIG. 23-4. Energies involved in combustion processes.

Turbulence is necessary to assure that the reacting fuel and oxygen molecules in the combustion process are in intimate contact at the proper instant that the temperature is high enough to cause the reaction to begin.

The physical state of the fuel for a combustion process dictates the type of system to be used for burning. A fuel may be composed of either, or both, volatile material and fixed carbon. The volatile material burns as a gas and exhibits a visible flame while the fixed carbon burns without a visible flame in a solid form. If a fuel is in the gaseous state, such as natural gas, it is very reactive and can be fired with a simple burner.

If a fuel is in a liquid state, such as fuel oil, most must be vaporized to the gaseous state before combustion occurs. This vaporization can be accomplished by supplying heat from an outside source but usually the liquid fuel is first atomized and then the finely divided fuel particles are sprayed into a hot combustion chamber to accomplish the gasification.

With a solid fuel, such as coal or wood, a series of steps are involved in combustion. These steps occur in a definite order and the combustion device must be designed with these steps in mind. Figure 23-5 shows what happens to a typical solid fuel during the combustion process.

The cycle of operation of the combustion source is very important as far as emissions are concerned. A steady process, such as a large steam boiler, operates with a fairly uniform load and a continuous fuel flow. The effluent gases along with any air pollutants are discharged steadily and continually from the stack. An automobile engine, on the other hand, is a series of intermittent sources. The emissions from the automotive engine will be vastly different than those from the boiler both in terms of quantity and quality. An eight cylinder automotive engine operating at 2500 rpm has to have 10,000 separate combustion processes started and completed each minute of its operation. Each of these lasts about 1/100th of a second from beginning to end.

The emissions from combustion processes may be predicted to some extent if the variables of the process are completely defined. Figure 23-6 indicates how the

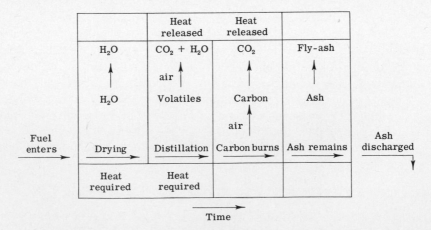

Fɪɢ. 23-5. Solid fuel combustion schematic.

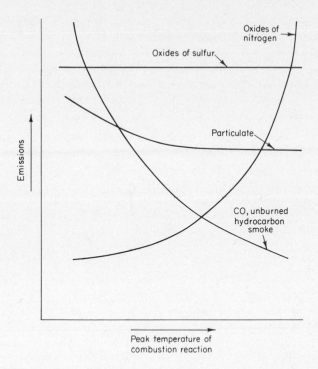

Fig. 23-6. Combustion emissions as a function of peak combustion temperatures.

emissions from a combustion source would be expected to vary with the temperature of the reaction. No absolute values are shown as these will vary greatly with fuel type, independent variables of the combustion process, etc.

A comparison of typical emissions from various common combustion sources may be seen in Table 23-1.

III. Stationary Sources

Emissions from industrial processes are varied and often complex (4). These emissions can be controlled by applying the best available technology. The emissions may vary slightly from one facility to another using apparently similar equipment and processes, but in spite of this slight variation, similar control technology is usually applied (5). For example, a method used to control the emissions from steel mill "X" may be applied to control similar emissions at plant "Y." It should not be necessary for plant "Y" to spend excessive amounts for research and development if plant "X" has a system that is operating satisfactorily. The solution to the problem is often to look for a similar industrial process, with similar emissions, and find the type of control system used.

A. Chemical and Allied Products

The emissions from a chemical process can be related to the specific process. A plant manufacturing a resin might be expected to emit not only the resin being

TABLE 23-1

Comparison of Combustion Pollutants[a]

Contaminant	Power plant emission (gm/kg fuel)			Refuse burning emission (gm/kg refuse)		Automotive emission (gm/kg fuel)	
	Coal	Oil	Gas	Open burning	Multiple chamber	Gasoline	Diesel
Carbon monoxide	Nil	Nil	Nil	50.0	Nil	165.0	Nil
Oxides of sulfur (SO$_2$)	(0.84)x	(0.86)x	(0.70)x	1.5	1.0	0.8	7.5
Oxides of nitrogen (NO$_2$)	0.43	0.68	0.16	2.0	1.0	16.5	16.5
Aldehydes and ketones	Nil	0.003	0.001	3.0	0.5	0.8	1.6
Total hydrocarbons	0.43	0.05	0.005	7.5	0.5	33.0	30.0
Total particulate	(0.322)y	(0.12)y	Nil	11	11	0.05	18.0

[a] $x = \%$ sulfur in fuel; $y = \%$ ash in fuel.

manufactured, but some of the raw material and also some other products which may or may not resemble the resin. A plant manufacturing sulfuric acid can be expected to emit sulfuric acid fumes and SO$_2$. A plant manufacturing soap products could be expected to emit a variety of odors. Depending on the process, the emissions could be any one of, or a combination of, dust, aerosols, fumes, or gases. The emissions may or may not be odorous or toxic. Some of the primary emissions might be innocuous but later react in the atmosphere to form an undesirable, secondary pollutant. A flow chart and material balance sheet for the particular process is a great aid in understanding and analyzing any process and its emissions (6).

In any listing of the importance of emissions from a particular process for an area, several factors must be considered—(1) the percentage of the total emissions of the area that the particular process emits, (2) the degree of toxicity of the emissions, and (3) the obvious characteristics of the source (which can relate to either sight or smell).

B. Resins and Plastics

Resins are solid or semi-solid, water insoluble, organic substances with little or no tendency to crystallize. They are the basic components of plastics and are also used for coatings on paper, particle board, and other surfaces that require a decorative, protective, or special purpose finish. The common characteristic of resins is that heat is used in their manufacture and application, and gases are exhausted from these processes. Some of the gases that are economically recoverable may be condensed but a large portion are lost to the atmosphere. One operation, coating a porous paper with a resin to form battery separators, was emitting to the atmosphere about 85% of the resin purchased. This resin left the stacks of the plant as a blue haze and the odor was routinely detected over two kilometers away. Since most resins and their by-products have low odor thresholds, disagreeable odors is the most common complaint against any operation using them.

C. Varnish and Paints

In the manufacture of varnish, heat is necessary for formulation and purification. The same may be true of operations preparing paints, shellac, inks, and other protective or decorative coatings. The compounds emitted to the atmosphere are gases, some with extremely low odor thresholds. Acrolein with an odor threshold of about 4000 $\mu g/m^3$ and reduced sulfur compounds with odor thresholds of 2 $\mu g/m^3$ are both possible emissions from varnish cooking operations. The atmospheric emissions from varnish cooking appear to have little or no recovery value while some of the solvents used in paint preparation are routinely condensed for recovery and return to the process. If a paint finish is baked to harden the surface by removal of organic solvents, the solvents must either be recovered, destroyed, or emitted to the atmosphere. The latter course, emission to the atmosphere, is undesirable and may be prohibited by the air pollution control agency.

D. Acid Manufacture

Acids are used as basic raw materials for many chemical processes and manufacturing operations (7). Figures 23-7a and 23-7b illustrate an acid plant with its flow diagram. Sulfuric acid is one of the major inorganic chemicals of our modern industry. The atmospheric discharges from a sulfuric acid plant can be expected to contain gases including SO_2 and aerosol mists, containing SO_3 and H_2SO_4, in the submicron to 10-μ size range. The aerosol mists are particularly damaging to paint, vegetation, metal, and synthetic fibers (8).

Other processes producing acids such as nitric, acetic, phosphoric, etc., can be expected to produce acid mists from the processes themselves as well as various toxic and nontoxic gases. The particular process must be thoroughly studied to obtain a complete listing of all the specific emissions.

E. Soap and Detergents

Soaps are made by reacting fats or oils with a base. Soaps are produced in a number of grades and types. They may be liquid, solid, a granule or powder. The air pollution problems of soap manufacture are primarily odors from the greases, fats, and oils although particulate emissions do occur during drying and handling operations. Detergents are manufactured from base stocks similar to those used in petroleum refineries so the air pollution problems are similar to refinery air pollution problems.

F. Phosphate Fertilizers

Phosphate fertilizers are prepared by benefication of phosphate rock to remove its impurities followed by drying and grinding. The H_3PO_4 in the rock may then be reacted with sulfuric acid to produce the normal superphosphate fertilizer (6).

Fig. 23-7. (a) Phosphoric acid manufacturing plant.

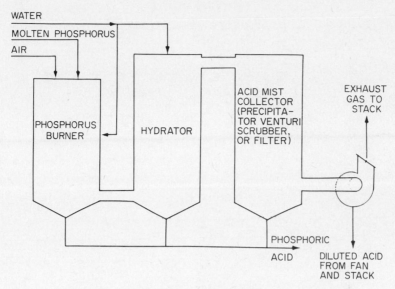

FIG. 23-7. (b) Flow diagram for phosphoric acid plant.

Over 200 plants are operating in the United States producing over a billion kilograms of phosphate fertilizer per year. Figure 23-8 is a flow diagram for a normal superphosphate plant which notes the pollutants emitted. The particulate and gaseous fluoride emissions cause greatest concern near phosphate fertilizer plants.

G. Other Inorganic Chemicals

Production of the large quantities of inorganic chemicals necessary for modern industrial processes can result in air pollutant emissions as undesirable by-products. Table 23-2 lists some of the more common inorganic chemicals produced along with the major atmospheric emissions from the specific process (6).

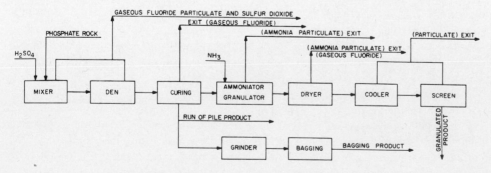

FIG. 23-8. Flow diagram of normal superphosphate plant.

TABLE 23-2

Miscellaneous Inorganic Chemicals and Associated Air Pollution Emissions

Inorganic chemical produced	Major air pollution emissions
Calcium oxide (lime)	Lime dust
Sodium carbonate (soda ash)	Ammonia—soda ash dust
Sodium hydroxide (caustic soda)	Ammonia—caustic dust and mist
Ammonium nitrate	Ammonia—nitric oxides
Chlorine	Chlorine gas
Bromine	Chlorine gas

H. Petroleum and Coal

Petroleum and coal supply the majority of the energy in all industrial countries. This fact gives one an idea of the vast quantities of materials handled and also hints at the magnitude of the air pollution problems associated with obtaining the resource, transporting it, refining it, and transporting it again. The emission problems from burning fossil fuel have been previously covered.

1. Petroleum

Petroleum products are obtained from crude oil. In the process of getting the crude oil from the ground to the refinery many possibilities for emission of hydrocarbon and reduced sulfur gaseous emissions occur. In many cases, these operations take place in relatively remote regions and only affect those employed by the industry so that little or no control is attempted because of lack of complaints.

As shown in Fig. 23-9 a petroleum refinery is a potential source of large tonnages of atmospheric emissions. All refineries are odorous, the degree being a matter of the housekeeping practices around the refinery. Since refineries are essentially closed processes, emissions are not normally considered a part of the operation. Refineries do need pressure relief systems and vents, and emissions are possible from them. Most refineries are under very strict control measures for economic as well as regulatory reasons. The recovery of one or two percent of a refinery throughput which was previously lost to the atmosphere can easily pay for the cost of the control equipment. The expense of the catalyst charge in some crackers and regenerators absolutely requires that the best possible control equipment be used to prevent emission to the atmosphere.

Potential air pollutants from a petroleum refinery could include (1) hydrocarbons from all systems, leaks, loading, and sampling, (2) sulfur oxides from boilers, treaters, and regenerators, (3) carbon monoxide from regenerators and incinerators, (4) nitrogen oxides from combustion sources and regenerators, (5) odors from air and steam blowing, condensers, drains, and vessels, and (6) particulate matter from boilers, crackers, regenerators, coking and incinerators.

Loading facilities must be designed to recover all vapors generated during filling of tank trucks or tanker ships. Otherwise these vapors will be lost to the atmosphere.

Since they may be both odorous and photochemically reactive, serious air pollution problems could result. The collected vapors must be returned to the process or disposed of by some means.

2. Coal

The air pollution problems associated with combustion of coal are of major concern. These problems generally occur away from the coal mine. The problems of atmospheric emissions due to mining, cleaning, handling, and transportation of coal from the mine to the user are of lesser significance as far as the overall air pollution problems are concerned. Whenever coal is handled, particulate emission becomes a problem. The emissions can be either coal dust or inorganic inclusions. Control of these emissions can be relatively expensive, but coal storage and transfer facilities are often located near residential areas.

I. Primary Metals Industry

Metallurgical equipment has long been an obvious source of air pollution. The effluents from metallurgical furnaces are submicron size dusts and fumes and hence are highly visible. The emissions from associated coke ovens are not only visible but may be odorous as well.

1. Ferrous Metals

Iron and steel industries have been concerned with emissions from their furnaces and cupolas since the industry started. The recent pressures for control have forced the companies to such a low level of permissible emissions that some of the older operations have been closed rather than spend the money to comply. The companies

Fig. 23-9. Petroleum refinery.

TABLE 23-3

Changes in Steel-Making Processes in the United States

Year	Production by specific process (%)				
	Bessemer	Open hearth	Electric	Basic oxygen furnace	Total
1920	21	78	1	0	100
1940	6	92	2	0	100
1960	2	89	7	2	100
1970	1	36	14	48	100

controlling these operations have not gone out of business but rather have opened a new, controlled plant to replace each old plant. Table 23-3 illustrates the emissions of the steel-making processes.

Air polluting emissions from steel-making furnaces include metal oxides, smoke, fumes, and dusts to make up the visible aerosol plume. They also may include gases, both organic and inorganic. If steel scrap is melted, the charge may contain appreciable amounts of oil, grease, and other combustibles that further add to the organic gas and smoke loadings. If the ore used has appreciable fluoride concentrations the emission of both gaseous and particulate fluorides can be a serious problem (10).

Emissions from foundry cupolas are relatively small but still significant. An uncontrolled 2 m cupola can be expected to emit up to 50 kg of dust, fumes, smoke, and oil vapor per hour. Carbon monoxide, oxides of nitrogen, and organic gases may also be expected. Control is possible but the cost of the control may be expensive in the case of the small foundry which only has one or two heats per week.

2. Nonferrous Metals

Around the turn of the century, one of the most obvious effects of industry on the environment was the complete destruction of vegetation downwind from copper, lead, and zinc smelters. This was caused by the smelting of the metallic-sulfide ores. As the metal was released in the smelting process, huge quantities of sulfur were oxidized to SO_2, which was toxic to much vegetation fumigated by the plume. Present smelting systems go to great expense to prevent the uncontrolled release of SO_2 but in many areas the recovery of the ecosystem will take years and possible centuries.

Early aluminum reduction plants were responsible for air pollution because of the fluoride emissions from their operations. Fluoride emissions can cause severe damage to vegetation and to animals feeding on such vegetation. The end result is an area surrounding the plant devoid of vegetation. Such scenes are reminiscent of those downwind from some of the uncontrolled copper smelters. New aluminum reduction plants are going to considerable expense to control fluoride emissions. Some of the older plants are finding the cost of control will exceed the original capital investment in the entire facility. Where the problem is serious, control agencies have developed extensive sampling networks to continually monitor the plant of concern.

Emissions from other nonferrous metal facilities are primarily metal fumes or metal oxides of extremely small diameter. Zinc oxide fumes vary from 0.03 to 0.3 μ and are toxic. Lead and lead oxide fumes are extremely toxic and have been extensively studied. Arsenic, cadmium, bismuth, and other trace metals can be emissions from many metallurgical processes.

J. Stone and Clay Products

The industries which produce and handle various stone products emit considerable amounts of particulate matter at every stage of the operation. These particulates may include fine mineral dusts of the size to cause damage to the lungs (1–8 μ). The threshold limit values for such dusts have been set quite low to prevent disabling diseases for the workman.

In the production of clay, talc, cement, chalk, etc., an emission of particulate matter will usually accompany each process (11). These processes may involve grinding, drying, sieving, etc. which can be enclosed and controlled to prevent the emission of the particles from the plant. In many cases the recovered particles can be returned to the process for a net economic gain for the recovery equipment.

During the manufacture of glass, considerable dust, averaging about 300 μ in size, will be emitted. Some dusts may also be emitted from the handling of the raw materials involved. Control of this dust to prevent a nusiance problem outside the plant is a necessity. When glass is blown or formed into the finished product smoke and gases can be released from the contact of the molten glass with lubricated molds. These emissions are quite dense but of a relatively short duration.

K. Forest Products Industry

1. Wood Processing

Trees are classed as a renewable resource which are being utilized in most portions of the world on a sustained yield basis. A properly managed forest will produce wood for lumber, fiber, and chemicals forever. Harvesting this resource can generate considerable dust and other particulates. Transportation over unpaved roads causes excessive dust generation. The cultural practice of burning the residue left after timber harvest, called slash burning, is still practiced in some areas and is a major source of smoke, gaseous, and particulate air pollution in the localities downwind from the fire.

Processing the harvested timber into the finished product may involve sawing, peeling, planing, sanding, and drying operations which can release considerable amounts of wood fiber and lesser amounts of gaseous material to the atmosphere. Control of wood fiber emissions from the pneumatic transport and storage systems can be a major problem, of considerable expense, for a plywood mill or a particle board plant.

2. Pulp and Paper

Pulp and paper manufacture is increasing in the world at an exponential rate. The demand for paper will continue as new uses are found for this product. Since

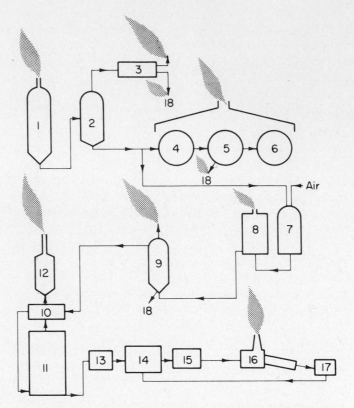

FIG. 23-10. Schematic diagram of Kraft pulping process (6).

most paper is manufactured from wood or wood residue, it provides an excellent use for this renewable resource.

The most widely used pulping process is the Kraft process, as shown in Fig. 23-10, which results in recovery and regeneration of the chemicals (12). This occurs in the recovery furnace which operates with both oxidizing and reducing zones. Emissions from such recovery furnaces are particulate matter, very odorous reduced sulfur compounds, and oxides of sulfur. If extensive, and expensive, control is not exercised over the Kraft pulp process the odors and aerosol emissions will effect a wide area Odor complaints from over 100 km have been reported. A properly controlled and operated Kraft plant will handle huge amounts of material and produce millions of kilograms per day of finished products with little or no complaints regarding odor or particulate emissions.

L. Noxious Trades

As the name implies, these operations, if uncontrolled, can be a serious air pollution problem. The main problem is the odors associated with the process. Examples of such industries are tanning works, rendering plants, and many of the food processing plants such as fish meal plants. In most cases the emission of particulates and gases from such plants are not of concern, only the odors. The concept of requiring these industries to locate away from the centers of commerce, or residences, no longer is acceptable as a means of control.

IV. Mobile Sources

A mobile source of air pollution can be defined as one capable of moving from one place to another under its own power. According to this definition an automobile is a mobile source while a portable asphalt batching plant is not. Generally, mobile sources infers transportation but sources such as construction equipment, gasoline powered lawn mowers, and gasoline powered tools are included in this category.

Mobile sources therefore consist of many different types of vehicles, powered by engines using different cycles, fueled by a variety of products, and emitting varying amounts of both simple and complex pollutants. Table 23-4 includes the more common mobile sources (13).

A. Gasoline-Powered Automobiles

The predominant mobile air pollution source in all industrialized countries of the world is the automobile, powered by a 4-stroke cycle (Otto Cycle) engine and using gasoline as the fuel. In the United States, over 85 million automobiles were in use in 1970. If the 15 million gasoline-powered trucks and buses and the 2 million motorcycles are included, the United States total exceeds 100 million vehicles. The engine used to power these millions of vehicles has been said to be the most highly engineered machine of the century. When one considers the present reliability, cost, and life expectancy of the internal combustion engine it is not difficult to see why it has remained so popular. A modern V-8 engine in traveling 100,000 km will have about 5×10^8 power cycles.

The emissions from a gasoline powered vehicle come from many sources. Figure 23-11 illustrates what might be expected from an uncontrolled (1960 model) automobile and a controlled (1976 model) automobile if it complies with the 1976 federal standards (14).

TABLE 23-4

Emissions from Mobile Sources

Power plant type	Fuel	Major emissions	Vehicle type
Otto Cycle	Gasoline	HC, CO, NO_x	Auto, truck, bus, aircraft, marine, motorcycle, tractor
2-Stroke Cycle	Gasoline	HC, CO, NO_x, Particulate	Motorcycle, outboard motor
Diesel	Diesel oil	NO_x, Particulate	Auto, truck, bus, railroad, marine, tractor
Gas turbine (jet)	Turbine	NO_x, Particulate	Aircraft, marine, railroad
Steam	Oil, coal	NO_x, SO_x, Particulate	Marine

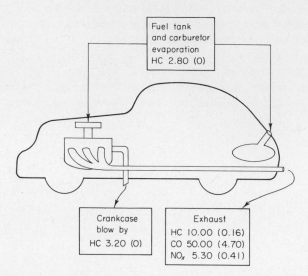

FIG. 23-11. Emissions from controlled and uncontrolled automobiles, gm/mile. (Controlled automobile at 1976 E.P.A. standards.)

The crankcase emissions have been effectively controlled since 1963 by positive crankcase ventilation systems which take the gases from the crankcase, through a flow control valve, and into the intake manifold. The gases then enter the combustion chamber with the fuel–air mixture where they are burned.

Evaporative emissions from the fuel tank and carburator have been controlled on all 1971 and later model automobiles sold in the United States. This has been accomplished by either a vapor recovery system which uses the crankcase of the engine for the storage of the hydrocarbon vapors or through an adsorption and regeneration system using a canister of activated carbon to trap the vapors and hold them until such time as a fresh air purge through the canister carries the vapors to the induction system for burning in the combustion chamber.

The exhaust emissions from gasoline-powered vehicles are the most difficult to control. These emissions are influenced by such factors as gasoline formulation, air–fuel ratio (which is usually richer than stoichiometric), ignition timing, compression ratio, engine speed and load, engine deposits, engine condition, coolant temperature, and combustion chamber configuration. Consideration of control methods must be based upon elimination or destruction of unburned hydrocarbons, carbon monoxide, and oxides of nitrogen. Methods used to control one pollutant may actually increase the emission of another requiring even more extensive controls.

Control of exhaust emissions for unburned hydrocarbons and carbon monoxide has followed three routes.

1. Fuel modification as to volatility, hydrocarbon types, or additive content. Some of the fuels currently being used are liquified petroleum gas (LPG), liquified natural gas (LNG), compressed natural gas (CNG), and unleaded gasoline. Supply

of some of these fuels is very limited. Other fuel problems involving storage, distribution, and power requirements have to be considered.

2. Minimization of pollutants from the combustion chamber. This approach consists of actually designing the engine with improved fuel–air distribution systems, ignition timing, fuel–air ratios, coolant and mixture temperatures, and engine speeds for minimum emissions. Engines using this system may idle rougher and faster and have slightly lowered power output at some operating conditions.

3. Further oxidation of the pollutants outside of the combustion chamber. This oxidation may either be by normal combustion or by catalytic oxidation. These systems require the addition of air into the exhaust manifold at a point downstream from the exhaust valve. An air pump is employed to provide this air. The pump requires periodic servicing and draws appreciable horsepower from the engine. It is possible that future automobiles may be required to combine all three systems mentioned in order to achieve the required reductions of carbon monoxide and unburned hydrocarbons. Some representative exhaust-emission-control systems are shown in Fig. 23-12.

Systems are currently being developed to control emissions of oxides of nitrogen. Some of the approaches being considered are exhaust gas recirculation and catalytic reaction.

B. Diesel Cycle

The diesel (compression ignition) cycle is regulated by fuel flow only, air flow remaining constant with engine speed. Because the diesel engine is normally operated well on the lean side of the stoichiometric mixture (40:1 or more), emission of unburned hydrocarbons and carbon monoxide is minimized. The actual emissions from a diesel engine are (1) oxides of nitrogen just as for spark ignition engines; (2) particulate matter which at times can be excessive; (3) organic compounds many of which cause irritation to the eyes and upper respiratory system; and (4) oxides of sulfur from the use of sulfur containing fuels. A smoking diesel engine indicates that more fuel is being injected into the cylinder than is being burned and some of the fuel is only partially burned resulting in the emission of unburned carbon.

Control of diesel-engine-powered vehicles is presently accomplished by fuel modification to obtain reduced sulfur content and cleaner burning, and by proper tuning of the engine using restricted fuel settings to prevent overfueling. Recirculation of exhaust gases or injection of water is being considered as a means of lowering NO_x emissions as are chemical and catalytic converters.

C. Gas Turbines and Jet Engines

The modified Brayton cycle is used for both gas turbines and jet engines. The turbine is designed to produce a usable torque at the output shaft while the jet engine allows most of the hot gases to expand into the atmosphere producing usable thrust. Emissions from both turbines and jets are similar, as are their control methods. The emissions are primarily unburned hydrocarbons, unburned carbon

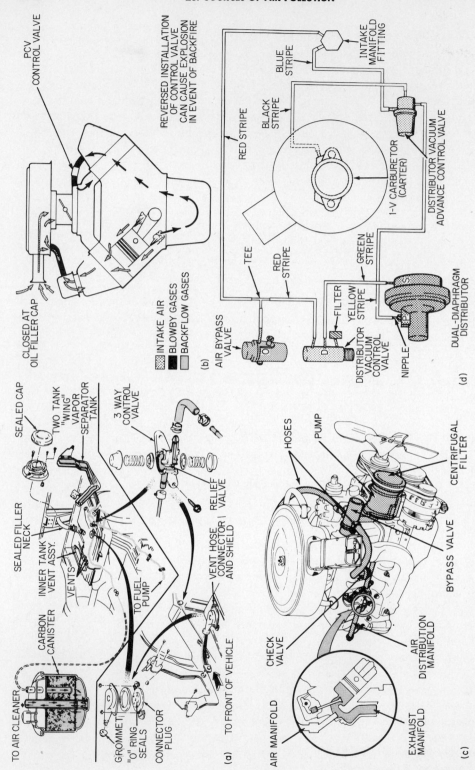

FIG. 23-12. Representative exhaust-emission-control systems.

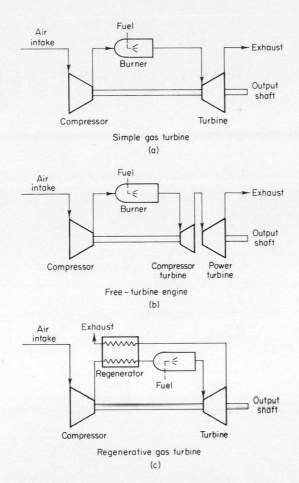

Fig. 23-13. Schematic diagrams of gas turbines.

which results in the visible exhaust, and oxides of nitrogen. Control of the unburned hydrocarbons and the unburned carbon may be accomplished by redesigning the fuel spray nozzles and reducing cooling air to the combustion chambers to permit more complete combustion. United States airlines are currently converting their jet fleets to lower emission engines using these control methods. NO_x emissions may be minimized by reduction of the maximum temperature in the primary zone of the combustors.

The gas turbine engine for automotive use could either be a simple turbine, a regenerative turbine, a free turbine, or any combination. Figure 23-13 shows these basic types which have been successfully tried in automotive use.

A modification of the gas turbine cycle is the free piston engine which uses free, reciprocating pistons as both a compressor and a gasifier to provide hot, high pressure gas to the power turbine. The free piston engine does not appear to be feasible as an alternative automotive power source because of its high weight, bulk, high cost, and need for an auxiliary air source for starting.

D. Alternatives

The atmosphere of the world cannot continue to accept greater and greater amounts of emissions from mobile sources as our transportation systems expand. The present emissions from all transportation sources in the United States exceed 100 billion kg of carbon monoxide per year, 30 billion kg per year of unburned hydrocarbons, and 20 billion kg of oxides of nitrogen. If presently used power sources cannot be modified to bring their emissions to acceptable levels we must develop alternative power sources or alternative transportation systems. All alternatives should be considered simultaneously to achieve the desired result, an acceptable transportation system with a minimum of air pollution.

The Wankel-type engine, which uses a rotary piston arrangement instead of the usual reciprocating piston, has been under intensive development in many countries for several years. Its combustion chamber configuration of high surface to volume ratio results in high hydrocarbon emissions, but these may be easily burned in an exhaust gas reactor. This approach appears to offer great possibilities for meeting required exhaust emission standards.

The stratified-charge engine is an unthrottled, spark-ignition engine using fuel injection in such a manner as to achieve selective stratification of the air–fuel ratio in the combustion chamber. The air–fuel ratio is ignitable only at the spark plug and fuel lean in the other portions of the combustion chamber. The engine is somewhat similar to the diesel engine in that a full charge of air enters the cylinder on the intake stroke and the power output is controlled by varying the amount of fuel introduced into the cylinder. Stratified-charge engines have been operated experimentally for several years and show promise as low emission engines. The hydrocarbon emission levels from this engine are quite variable, the CO levels low, and the NO_x emission quite high.

The Stirling engine is an external combustion engine which adds and removes energy to the working fluid (usually air) through heat exchangers. Since the combustion is continuous and external to the engine cylinders, emissions are much lower than in the conventional internal combustion engine. At the present time the Stirling engine would be heavier and more expensive than the present internal combustion engine.

Another external combustion engine that has been widely supported as a low emission power source is the Rankine cycle steam engine. Many different types of expanders can be used to convert the energy in the working fluid into rotary motion at a drive shaft. Expanders that have been tried or proposed are reciprocating piston engines, turbines, helical expanders, and all possible combinations of these. The advantage of the steam engine is that the combustion is continuous and takes place in a combustor with no moving parts. The result is a much lower release of air pollutants but still not completely zero emissions. Present technology is capable of producing a satisfactory steam driven car, truck, or bus, but costs, operating problems, warm-up time, and weight and size must be considered in the total evaluation of the system. A simple Rankine cycle steam system is shown diagrammatically in Fig. 23-14.

Electric drive systems have been tried as means of achieving propulsion without

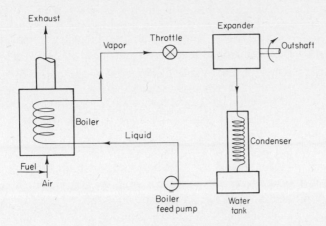

FIG. 23-14. Rankine cycle system.

harmful emissions. Battery operated vehicles are of low power and limited range and require frequent recharging. In power shortage areas this could be a severe additional load on the electrical system. Sulfuric acid, hydrogen, and oxygen emissions from millions of electric vehicles using lead–acid storage batteries for an energy source would be appreciable. Other types of batteries offer some promise but their present cost would be prohibitive for automotive use.

Hybrid systems consisting of two or more energy-conversion processes may offer the greatest promise for lower emission automobiles. A constant speed and load internal combustion engine driving a generator with a small battery for load surges could be made to emit less hydrocarbons, CO, and NO_x than a standard automobile engine but the cost would be much more. Other hybrid systems which have been proposed are steam-electric, turbine-electric, and Stirling-electric.

Probably the ultimate answer to the problem of emissions from millions of private automobiles is an alternative transportation system. It must be remembered, however, that even rail systems and bus systems do emit some air pollution. Rail systems are expensive and lack flexibility. A quick calculation of the number of passengers carried per minute past a single point on a freeway in private automobiles will illustrate the difficulties of a rail system to replace the automobile. Busses offer much greater flexibility at lower cost than rail systems but in order to operate efficiently and effectively they would require separate roadway systems and loading stations apart from automobile traffic.

V. Radioactive Air Pollution

Natural radioactivity in the earth's atmosphere, whether from soils and rocks containing radioactive minerals or from cosmic sources, is relatively small compared to the amount of radioactivity released by man. Furthermore if the pollution is considered on a weight basis, the amount of radioactive material released to the atmosphere during routine nuclear processes is several orders of magnitude smaller than that produced by the world's combustion processes (15).

Of primary concern to environmental groups is the release of pollutants from nuclear power generating stations. The number of such plants is increasing at a rapid rate and they are usually located near urban centers. The mining, concentration, separation, reprocessing, and storage facilities may all be potentially greater polluters but they are located far from urban areas so that concern regarding them is minimized.

With the exception of cooling tower or cooling pond water vapor, which must also be considered as a pollutant if it condenses and affects visibility, essentially all pollutants released to the air from nuclear facilities are radioactive in nature. Most of the radioactive pollutants are directly or indirectly related to the nuclear fuel processing, handling, use, or recovery. The portions of the fuel cycle, other than the use in the reactor itself, present the greatest potential source of air pollution, with the chemical reprocessing plants being the most significant. Figure 23-15 illustrates a typical nuclear fuel cycle. Radioactive air pollutants may be emitted from every step shown. These pollutants may be gases, aerosols, or large particles.

The reactor is surrounded by a containment vessel, or building, designed to contain all products released by the reactor during an accident as well as those

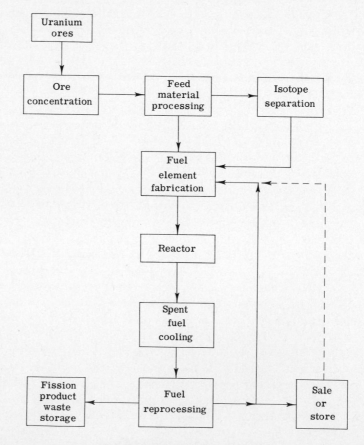

FIG. 23-15. Nuclear fuel cycle.

normally released during routine operations. The containment vessel is maintained at less than atmospheric pressure so that any pollutants released will not escape through leaks. Major gas-cleaning equipment is located at the building exhaust to remove the pollutants before the ventilation air is released to the atmosphere. If the level of radioactivity within the containment vessel becomes so high that the gas cleaning equipment cannot maintain the required maximum allowable exhaust level, the containment vessel can be isolated so that no air flows in or out of the structure. Later, when the radioactivity level has decayed sufficiently, the isolation system is reopened and normal operations may be resumed.

Within the reactor, if rupture of the cladding occurs at fuel temperatures below melting, noble gases and iodine will comprise the major pollutant release. If localized melting of the fuel has occurred, volatile solids such as cesium, tellurium, and ruthenium may be released along with the noble gases and iodine. If a serious accident were to heat the fuel to the extent that it volatilized, the bulk of the fission products would be released including some refractory materials such as rare earths and strontium. A nuclear excursion or loss of coolant flow could be the cause of such an accident.

References

1. Went, F. W. "Dispersion and Disposal of Organic Materials in the Atmosphere," Preprint Series 31, Desert Research Institute, University of Nevada, 1966.
2. Fryling, G. R. (Ed.) "Combustion Engineering," pp. 21–23. Combustion Engineering, New York, 1966.
3. Popovich, M., and Hering, C. "Fuels and Lubricants." 312 pp. Wiley, New York, 1959.
4. Strauss, W. "Industrial Gas Cleaning." 471 pp. Pergamon Press, Oxford, 1966.
5. Danielson, J. A. (Ed.) "Air Pollution Engineering Manual," 99-AP-40. U.S. Government Printing Office, Washington, D.C., 1967.
6. Stern, A. C. (Ed.). "Air Pollution," 2nd ed. Vol. III. Academic Press, New York, 1968.
7. "Atmospheric Emissions from Nitric Acid Manufacturing Processes." 99-AP-27. U.S. Department of Health, Education, and Welfare, Cincinnati, 1966.
8. "Atmospheric Emissions from Sulfuric Acid Manufacturing Processes." 99-AP-13. U.S. Department of Health, Education, and Welfare, Cincinnati, 1965.
9. "Atmospheric Emissions from Petroleum Refineries." U.S. Department of Health, Education, and Welfare, Cincinnati, 1969.
10. Schueneman, J. J., High, M. D., and Bye, W. E. "Air Pollution Aspects of the Iron and Steel Industry." U.S. Department of Health, Education, and Welfare, Cincinnati, 1963.
11. Kreichelt, T. E., Kemnitz, D. A., and Cuffe, S. T. "Atmospheric Emissions from the Manufacture of Portland Cement." U.S. Department of Health, Education, and Welfare, Cincinnati, 1967.
12. Kenline, P. A., and Hales, J. M. "Air Pollution and the Kraft Pulping Industry." U.S. Department of Health, Education, and Welfare, Cincinnati, 1963.
13. "Motor Vehicles, Air Pollution, and Health, A Report of the Surgeon General to the U.S. Congress." U.S. Department of Health, Education, and Welfare, Washington, D.C., 1962.
14. "Control Techniques for Carbon Monoxide, Nitrogen Oxide, and Hydrocarbon Emissions from Mobile Sources." U.S. Department of Health, Education, and Welfare, Washington, D.C., 1970.
15. Slade, D. H. (Ed.). "Meteorology and Atomic Energy 1968." U.S. Atomic Energy Commission, Washington, D.C., 1969.

Suggested Reading

"Air Conservation," 2nd ed., Report of the Air Conservation Commission of The American Association for the Advancement of Science, AAAS publication No. 80, Washington, D.C., 1965.
"Air Pollution." World Health Organization Monograph, No. 46, Columbia University Press, New York, 1961.
Lund, H. F., ed. "Industrial Pollution Control Handbook." McGraw-Hill, New York, 1971.

Questions

1. Calculate the heat generated by dissociation and formation as one molecular weight of methane, CH_4, burns to carbon dioxide and water. How does this heating value compare to the tabular heating value for methane?
2. Los Angeles County has banned the use of private, backyard incinerators. Would you expect a noticeable increase in air quality as a result of this action?
3. Show a free radical reaction which results in ethane in the effluent of a combustion process burning pure methane with pure oxygen.
4. A power plant burns oil that is 4% ash and 3% sulfur. At 50% excess air what particulate (mg/m^3) and SO_x (mg/m^3) would you expect?
5. Los Angeles County has very tight controls over petroleum refineries. Suppose these refineries produce 100 million liters of products per day and required air pollution control devices recover all of the 2% which was previously lost. What are the savings in dollars per year at an average product cost of 3¢ per liter? How does this compare to the estimate that the refineries spent $400 million dollars for control equipment over a ten-year period?
6. Suppose a 40,000 liter gasoline tank is filled with liquid gasoline with an average vapor pressure of 20 mm Hg. At 50% saturation, what weight of gasoline would escape to the atmosphere during filling?
7. If a major freeway with four lanes of traffic in one direction passes four cars per second at 100 km/hr during the rush period, and each car carries two people, how often would a commuter train of 5 cars carrying 100 passengers per car have to be operated to handle the same load? Assume the train would also operate at 100 km/hr.
8. An automobile traveling 50 km/hr emits 1% CO from the exhaust. If the exhaust rate is 80 m^3 per minute, what is the CO emission in grams per km?
9. List the following in increasing amounts from the exhaust of an idling automobile: O_2, NO_x, SO_x, N_2, UBHC, CO_2, and CO.

Chapter 24

EMISSION INVENTORY

An emission inventory is a listing of the amount of pollutants from all sources entering the air in a given time period. The boundaries of the area are fixed (1, 2).

The tables of emission inventory are very useful to control agencies as well as planning and zoning agencies. They can point out the major sources whose control can lead to a considerable percentage reduction in the area. They can be used with appropriate mathematical models to determine the degree of overall control necessary to meet ambient air-quality standards. They can be used to indicate the type of sampling network and the locations of individual sampling stations if the areas chosen are small enough. For example, if an area has a very small use of sulfur bearing fuels it would not be an optimum use of public funds to set up an extensive SO_2 monitoring network in the area. Emission inventories can be used for publicity and political purposes, "If natural gas cannot meet the demands of our area we will have to burn more high-sulfur fuel and the SO_2 emissions will increase 8 tons per year."*

The method used to develop the emission inventory does have some elements of error but the other two alternatives are both expensive and also subject to their own errors. The first alternative would be to continually monitor every major source in the area. The second method would be to continually monitor the pollutants in the

* This chapter will depart from the style of using metric units because English units are presented in all referenced tables and examples.

ambient air at many points and apply appropriate diffusion equations to calculate the emissions. In practice, the best system would be a combination of all three, knowledgeably applied.

I. Inventory Techniques

To develop an emission inventory for an area one must (1) list the types of sources for the area, such as cupolas, automobiles, home fireplaces, etc., (2) determine the type of air pollutant emissions from each of the listed sources, such as particulates, SO_2, CO, etc., (3) examine the literature to find valid emission factors for each of the pollutants of concern, such as, "particulate emissions for open burning of tree limbs and brush are 22 pounds per ton of residue consumed," (4) through an actual count, or by means of some estimating technique, determine the number and size of specific sources in the area. The number of steel making furnaces can be counted but the number of home fireplaces will probably have to be estimated, and (5) multiply the appropriate numbers from (3) and (4) to obtain the total emissions and then sum the similar emissions to obtain the total for the area.

Using a typical example will illustrate the procedure. Suppose we wish to determine the amount of carbon monoxide from oil furnaces, emitted per day, during the

TABLE 24-1

Nationwide Emissions of Particulates, 1968 (3)

Source	Emissions, 10^6 tons/year	Percent of total
Transportation	1.2	4.3
Motor vehicles	0.8	2.8
Gasoline	0.5	1.8
Diesel	0.3	1.0
Aircraft	N[a]	N
Railroads	0.2	0.7
Vessels	0.1	0.4
Nonhighway use of motor fuels	0.1	0.4
Fuel combustion in stationary sources	8.9	31.4
Coal	8.2	29.0
Fuel oil	0.3	1.0
Natural gas	0.2	0.7
Wood	0.2	0.7
Industrial processes	7.5	26.5
Solid waste disposal	1.1	3.9
Miscellaneous	9.6	33.9
Forest fires	6.7	23.7
Structural fires	0.1	0.4
Coal refuse burning	0.4	1.4
Agricultural burning	2.4	8.4
Total	28.3	100.0

[a] N = Negligible.

heating season, in a small city of 50,000 population:

1. The source is oil furnaces within the boundary area of the city.
2. The pollutant of concern is carbon monoxide.
3. Emission factors for carbon monoxide are listed in various ways (4) (2 pounds per 1000 gallons of fuel oil, 50 gm per day per burner, $1\frac{1}{2}\%$ by volume of exhaust gas, etc.). For this example use 2 pounds per 1000 gallons of fuel oil.
4. Fuel oil sales figures, obtained from the local dealers association, average 10,000 gallons per day.

5. $$\frac{2 \text{ pounds CO}}{1000 \text{ gallons}} \times \frac{10,000 \text{ gallons}}{\text{day}} = \frac{20 \text{ pounds CO}}{\text{day}}$$

An example of a completed emission inventory for the United States is illustrated in Table 24-1.

II. Emission Factors

Valid emission factors for each source of pollution are the key to the emission inventory. It is not uncommon to find emission factors differing by 50%, depending upon the researcher, variables at time of emission measurement, etc. Since it is possible to reduce the estimating errors in the inventory to $\pm 10\%$ by proper statistical sampling techniques, an emission factor error of 50% can be overwhelming. It must also be realized that an uncontrolled source will emit pollutants of at least 10 times the amount of those released from one operating properly with air pollution control equipment installed.

Actual emission data are available from many handbooks, government publications, and literature searches of appropriate research papers and journals. It is always wise to verify the data if at all possible as to validity of the source and reasonableness of the final number. Some emission factors, which have been in use for years, were only rough estimates proposed by someone years ago to establish the order of magnitude of the particular source.

Emission factors must be also critically examined to determine the tests from which they were obtained. For example, carbon monoxide from an automobile will vary with load, engine speed, displacement, ambient temperature, coolant temperature, ignition timing, carburetor adjustment, engine condition, etc. However, in order to evaluate the overall emission of carbon monoxide to an area we must settle on an average value that we can multiply by the number of cars, or miles driven per year, to determine the total carbon monoxide released to the area.

III. Data Gathering

To compile the emission inventory requires a determination of the number and types of units of interest in the study area. It would be of interest, for example, to know the number of automobiles in the area and the mileage each was driven per year. This figure would require considerable time and expense to obtain. Instead, it can be closely approximated by determining the gallons of gasoline sold in the area during the year. Since a tax is collected on all gasoline sold for highway use, these figures can be obtained from the tax collection office.

Data regarding emissions is available from many sources. Sometimes the same item may be checked by asking two or more agencies for the same information. An example of this would be to check the gallons of gasoline sold in a county by asking both the tax office and the gasoline dealers association. Sources of information for an emission inventory include (1) city, county, and state planning commissions, (2) city, county, and state chambers of commerce, (3) city, county, and state industrial development commissions, (4) census bureaus, (5) national associations such as the coal associations, (6) local associations such as the County Coal Dealers Association, (7) individual dealers or distributors of oil, gasoline, coal, etc., (8) local utility companies, (9) local fire and building departments, (10) data gathered by air pollution control agencies through surveys, sampling, etc., (11) traffic maps, and (12) insurance maps.

IV. Data Reduction and Compilation

The final emission inventory can be prepared on a computer if one is available. This will enable the information to be stored on magnetic tape so that it can be updated rapidly and economically as new data or new sources appear. The computer program can be written so that changes can easily be made. There will be times when major changes occur and the inventory must be completely changed. Imagine the change that would take place when natural gas first becomes available in a commercial-residential area which previously used oil and coal for space heating.

To determine emission data, as well as determining the effect that fuel changes would make, it is necessary to use the appropriate thermal conversion factor from one fuel to another. Table 24-2 lists these factors for fuels in common use.

A major change in the emissions for an area will occur if control equipment is

TABLE 24-2

Thermal Conversion Factors for Fuels

Fuel	Btu (gross)
Bituminous coal	26,200,000 per ton
Anthracite coal	25,400,000 per ton
Wood (solid)	20,960,000 per cord (128 ft³)
Wood (hogged)	18,300,000 per unit (200 ft³)
Distillate fuel oil	5,800,000 per bbl (42 gallon)
Residual fuel oil	6,300,000 per bbl
Natural gas	1,050 per cu ft
Manufactured gas	550 per cu ft

TABLE 24-3

Nationwide Emissions of Particulates by Year (3) (10^6 tons)

Source	1966	1967	1968	Change from 1966 to 1968
Transportation	1.2	1.1	1.2	N[a]
Motor vehicles	0.7	0.7	0.8	+0.1
Other	0.5	0.4	0.4	−0.1
Fuel combustion	9.2	8.9	8.9	−0.3
Coal	8.5	8.2	8.2	−0.3
Fuel oil	0.3	0.3	0.3	N
Natural gas	0.1	0.2	0.2	+0.1
Wood	0.3	0.2	0.2	−0.1
Industrial processes	7.6	7.3	7.5	−0.1[b]
Solid waste disposal	1.0	1.1	1.1	+0.1
Miscellaneous	9.6	9.6	9.6	N
Man-made	2.9	2.9	2.9	N
Forest fires	6.7	6.7	6.7	N
Total	28.6	28.0	28.3	−0.3

[a] N = Negligible.
[b] Apparent change.

TABLE 24-4

Nationwide Emissions of Sulfur Oxides by Year (3) (10^6 tons)

Source	1966	1967	1968	Change (1966–1968)
Transportation	0.6	0.7	0.8	+0.2
Motor vehicles	0.2	0.3	0.3	+0.1
Other	0.4	0.4	0.5	+0.1
Fuel combustion	22.5	23.1	24.4	+1.9
Coal	18.7	19.1	20.1	+1.4
Fuel oil	3.8	4.0	4.3	+0.5
Natural gas	N[a]	N	N	N
Wood	N	N	N	N
Industrial processes	7.1	7.2	7.3	+0.2
Solid waste disposal	0.1	0.1	0.1	N
Miscellaneous	0.6	0.6	0.6	N
Man-made	0.6	0.6	0.6	N
Forest fires	N	N	N	N
Total	30.9	31.7	33.2	+2.3

[a] N = Negligible.

installed. This can be shown in the emission inventory to illustrate to the community the effect.

By keeping the emission inventory current, and updating at least yearly as fuel uses change, industrial and population changes occur, and control equipment is added, a realistic record for the area is obtained. It is embarassing for a county air pollution control officer to be asked at a hearing, "What percentage of the SO_2 in our county is emitted by oil refineries?" if his emission inventory figures are a few years out of date.

V. Inventory Presentation

The desirable way to present the emission inventory data is in a series of small, related tables. An attempt to combine all data for all years into one master table results in a massive, boring collection of statistics. The control agency usually ends up abstracting specific bits of information from such a table so it should be kept concise to begin with. Tables 24-3 and 24-4, while slightly dated, are examples of concise tables that present the inventory data in readily usable form. Imagine the resulting table had one tried to combine these two tables along with similar ones for hydrocarbon emissions, carbon monoxide emissions, and oxides of nitrogen emissions for just the three years.

References

1. Ozolins, G., and Smith, R. "A Rapid Survey Technique for Estimating Community Air Pollution Emissions." P.H.S. Publication 999-AP-29, U.S. Department of Health, Education, and Welfare, Cincinnati, Ohio, 1966.
2. Stern, A. C., ed. "Air Pollution," 2nd ed. Academic Press, New York, 1968.
3. "Nationwide Inventory of Air Pollution Emissions 1968," U.S. Department of Health, Education, and Welfare, Raleigh, North Carolina, 1970.
4. Duprey, R. C., "Compilation of Air Pollution Factors". P.H.S. Publication 999-AP-42, U.S. Department of Health, Education, and Welfare, Washington, D.C., 1968.

Suggested Reading

Atkisson, A., and Gaines, R. S., eds. "Development of Air Quality Standards." Merrill, Columbus, Ohio, 1970.

Questions

1. Choose a representative area (a city, county, region, etc.) and prepare a table showing the change in air pollutant emission if natural gas were used as a fuel instead of oil and coal.
2. Why are "oxides of nitrogen" and "oxides of sulfur" usually reported in emission inventory tables rather than the actual oxidation states?
3. For a given area estimate the yearly pollutants emitted by automobiles using the figures for gallons of gasoline sold supplied by (a) the gasoline dealers association, and (b) by the local taxation authorities.

Chapter 25

SOURCE SAMPLING

I. General

Air pollutants released to the atmosphere may be characterized by qualitative descriptions or quantitative analyses. A plume may be brown, dense smoke, or 60% opacity. These are qualitative descriptions made by observing the effluent as it entered the atmosphere. Probably of more concern are the quantitative data regarding the effluent. We would like to know How many parts per million? How many kilograms per hour? How many kilograms per year? To obtain these numbers it becomes necessary to actually sample the effluent. A source sample therefore is a necessity for air pollution evaluation and control. In any situation regarding atmospheric emission of pollutants source sampling is the only way to obtain accurate numbers. Figure 25-1 shows a simple source test.

The purpose of source sampling is to obtain as accurate a sample as possible, of the material entering the atmosphere, at a minimum cost (1). This statement needs to be examined in light of each source test conducted. The questions should continually be asked: (1) Is the sampling and collecting of the material representative? Is this the material entering the atmosphere? Sampling at the base of a tall stack may be much easier than sampling at the top but just because a pollutant exists in the breeching does not mean it will eventually be emitted to the atmosphere. Molecules can undergo both physical and chemical changes after leaving the stack.

FIG. 25-1. Source test.

(2) Maximum accuracy in sampling is desirable. Is maximum accuracy attainable? Decisions regarding the total effluent will be based on what was found from a relatively small sample. Only if the sample accurately represents the total will the extrapolation to the entire effluent be valid. (3) Collecting a sample is a costly and time consuming process to say the least. The economics of the situation must be considered and the costs minimized consistent with the other objectives. It really does not make a great deal of sense to spend $5000 on an extensive stack testing analysis to decide whether to purchase a $10,000 scrubber of 95% efficiency or to try to get by with a $7000 scrubber of 90% efficiency.

II. Purpose

The reasons for performing a source test are not always the same. The test might be necessary for any one, or more, of the following reasons. (1) To obtain data concerning the emissions for an emission inventory or to identify a predominant source in the area: An example of this would be to determine the hydrocarbon release from a new type of organic solvent used in a degreasing tank. (2) To determine compliance with regulations: If authorization is obtained to construct an incinerator and the permit states that the maximum allowable particulate emission is 230 mg per standard cubic meter corrected to 12% CO_2 a source test must be made to determine compliance with the permit. (3) To gather information which will enable selection of appropriate control equipment: If a source test determines that the emission is 3000 mg of particulate per cubic meter and that it had a weight mean size of 5 μ a control device must be chosen which will collect enough particulate to meet some required standard, such as 200 mg per cubic meter. (4) To determine the efficiency of control equipment installed to reduce emissions: If a manufacturer supplies a device guaranteed to be 95% efficient for removal of particulate with a weight mean size of 5 μ, the effluent stream must be sampled at the inlet and outlet of the device to determine if the guarantee has been met.

III. Statistics of Sampling

When one considers that he may be taking a sample at the rate of 0.3 of a liter per minute from a stack discharging 2000 m³/minute to the atmosphere, the chances for error become quite large. If the sample is truly representative it is said to be both accurate and unbiased. If it is not representative, it may be biased because of some consistent phenomenon (some of the hydrocarbons condense in the tubing ahead of the trap) or in error because of some uncontrolled variation (only 1.23 gm of sample was collected and our analytical technique is accurate to ±0.5 gm).

For all practical purposes, source testing can be considered as simple random sampling (2). The source may be considered to be composed of such a large population of samples that the population N is infinite. From this population, n units are selected in such a manner that each unit of the population has an equal chance of being chosen. For the sample, determine the sample mean, $\bar{y}$:

$$\bar{y} = \frac{y_1 + y_2 + \cdots + y_n}{n} \tag{25-1}$$

If the sample is unbiased, estimate the source mean, $\bar{Y}$, as being equal to the sample mean so that

$$\bar{Y} = \bar{y} \tag{25-2}$$

TABLE 25-1

Idling Internal Combustion Engine, CO Percentages

Test No.	CO, %
1	1.8
2	1.6
3	1.8
4	1.9
5	1.7
6	1.8

For example, take 6 samples of carbon monoxide from the exhaust of an idling automobile and obtain CO percentages as shown in Table 25-1. The sample mean is

$$\bar{y} = \frac{1.8 + 1.6 + 1.8 + 1.9 + 1.7 + 1.8}{6} = 1.767$$

The source mean is the same because the sample is unbiased as seen by

$$\bar{Y} = \bar{y} = 1.767 \tag{25-3}$$

The variance of the sample and the population (source) may also be assumed equal if the sample is unbiased. The variance is S^2 defined as

$$S^2 = \frac{\sum_1^n (y_i - \bar{y})^2}{n - 1} \tag{25-4}$$

The variance of the source is usually calculated by the formula

$$s^2 = \frac{1}{n - 1}\left[\sum y_i^2 - \frac{(\sum y)^2}{n}\right] \tag{25-5}$$

For example

$$\sum y_i^2 = 18.78, \qquad \sum y = 10.6 \quad n = 6,$$

$$s^2 = \frac{1}{6 - 1}\left[18.78 - \frac{(10.6)^2}{6}\right] = 0.01067$$

The standard deviation of the sample is defined as the square root of the variance. For the example

$$s = (s^2)^{1/2} = (0.01067)^{1/2} = 0.103$$

The sample represents a population (source) which if normally distributed has a mean of 1.767% and a standard deviation of 0.103%. This can be illustrated as shown in Fig. 25-2 which is the normal curve discussed in Chapter 14, Fig. 14-4.

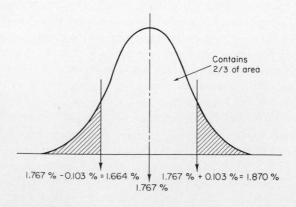

FIG. 25-2. Distribution of carbon monoxide from automotive source.

The inference from the statistical calculations is that the true mean value for the carbon monoxide from the idling automobile stands a 66.7% chance of being between 1.664% and 1.870%. The best single number for the carbon monoxide emission would be 1.767% (the mean value).

Further statistical procedures can be applied to determine confidence limits of the results. Generally, just the values for the mean and standard deviation would be reported. The reader is referred to any good statistical text to expand upon the brief analysis presented here (3).

IV. Sampling Test Preliminaries

A. Test Preliminaries

The first thing that must be done for a successful source test is a complete review of all background material. The test request may come as either a verbal or written request. If it is verbal, it should be put into writing for the permanent record. The request may contain much or little information but it is important to see that it is complete and understood. Questions to ask are (1) Why should the test be made? Is it to measure a specific pollutant such as SO_2 or is it to determine what is causing the odor problem in the new residential area? (2) What will the test results be used for? Will it be necessary to go to court or are the results for general information only? This may make a difference regarding the test method selected. (3) What equipment or process is to be tested? What are its operational requirements? (4) What methods would be preferred by the analytical group? Are the analytical methods standard or unique? A literature search regarding the process and test should be conducted unless the test crew is thoroughly familiar with the source and all possible test methods. It is important to check into the regulations regarding the process and specific test procedures as a part of the search.

When familiar with all the background material, it is time to inspect the source to be tested. This inspector should be accompanied by the plant manager or someone who knows the process in detail. It is also important that any technicians or mechanics be contacted at this time regarding necessary test holes, platforms, scaffolding, power requirements, etc. During this inspection checks should be made for environmental conditions and space requirements at the sampling site. Testing in a noisy or dusty place at elevated temperatures is certainly uncomfortable and possibly hazardous. Rough estimates of several important factors should be made at this time. These can be estimated and noted in writing during the inspection. A simple check sheet, such as shown in Fig. 25-3, should be a great aid.

The information obtained during the background search, and from the source inspection, will enable selection of the test procedure to be used. The choice will be based on the answers to several questions: (1) What are the legal requirements? For specific sources there may be only one method acceptable. (2) What range of accuracy is desirable? Should the sample be collected by a procedure that is ±5% accurate or should a statistical technique be used on data from eight tests at ±10% accuracy? Costs of different test methods will certainly be a consideration here. (3) Which sampling and analytic methods are available that will give the required

```
┌─────────────────────────────────────────────────────────────────┐
│              SOURCE TEST PRELIMINARY VISIT CHECK LIST             │
│  Plant _____ │
│  Location _____ │
│  By _____ Date _____  │
│    1.  Gas flow at test point, m/min _____ , m³/min _____ │
│    2.  Gas temperature, °C _____  │
│    3.  Gas pressure, mm of water (±)  _____  │
│    4.  Gas humidity, R. H. , % _____  │
│    5.  Pollutants of concern _____  │
│    6.  Estimate of concentration _____  │
│    7.  Any toxic materials? _____ │
│    8.  Test crew needed _____  │
│    9.  Site check:                                                │
│        Electric power _____  Test holes _____  │
│        Ambient temperature _____  Illumination _____  │
│        Platform _____  Scaffolding _____  │
│        Hoist _____  Ladders _____  │
│        Test date _____ │
│   10.  Environmental or safety gear _____  │
│   11.  Personnel involved (names)                                 │
│        Plant manager or foreman _____  │
│        Mechanic or electrician _____  │
└─────────────────────────────────────────────────────────────────┘
```

FIG. 25-3. Source test check list.

accuracy for the estimated concentration? An Orsat gas analyzer with a sensitivity limit of ±0.2% would not be chosen to sample carbon monoxide at 50–100 ppm. Conversely, an infrared gas analyzer with a full-scale deflection of 1000 ppm would not be chosen to sample CO_2 from a power boiler. (4) Is a continuous record required over many cycles of source operation or will one or more grab samples suffice? If a source only emits for a short period of time a method would not be selected which requires hours to gather the required sample.

The test must be scheduled well in advance for the benefit of all concerned. Let the plant personnel, as well as the test crew, know the intended date and time of the test. It is also a good idea to let the chemist or analytical service know when the testing will be conducted so they can be ready to do their portion of the work. It may be necessary to schedule or rent equipment in advance, such as, boom trucks, scaffolding, etc. Make sure when scheduling the test that the source will be operating in its normal manner. A boiler may be only operating at one-third load on weekends because the plant steam load is off the line and only a small heating load is being carried.

B. Gas Flow Measurement

Gas flow measurement is a very important part of source testing. The volume of gaseous effluent from a source must be determined to obtain the mass loading to the atmosphere. Flow measurement through the sampling train is necessary to determine the volume of gas containing the pollutant of interest. Many of the sampling devices used for source testing have associated gas flow indicators which must be continually checked and calibrated.

Gas flows are often measured by measuring the associated pressures. Figure 25-4 illustrates several different pressure measurements commonly made on systems carrying gases. Static pressure measurements are made to adjust the absolute pressure to standard conditions specified in the test procedure.

The quantity of gaseous effluent leaving a process is usually calculated from the continuity equation which for this use is written as

$$Q = AV \qquad\qquad (25\text{-}6)$$

where Q = flow at the specified conditions of temperature, pressure, and humidity, A = area through which the gas flows, and V = velocity of the effluent gas averaged over the area.

A is commonly measured, and V determined, to calculate Q. The velocity V is determined at several points, in the center of equal duct areas, and averaged. Table 25-2 shows one commonly accepted method of dividing stacks or ducts into equal areas for velocity determinations.

For rectangular ducts, the area is evenly divided into the necessary number of measurement points. For circular ducts Table 25-3 can be used to determine the location of the traverse points. In using Table 25-3 realize that traverses are made along two diameters at right angles to each other as shown in Fig. 25-5.

In most source tests, the measurement of velocity is made with a pitot-static tube, usually referred to simply as a pitot tube. Figure 25-6a, b illustrates the two types of pitot tubes in common use. The standard type as shown in Fig. 25-6a, does not need to be calibrated but may be easily plugged in some situations. The S type as shown in Fig. 25-6b does not plug as easily but it does need calibration to assure its accuracy. The velocity pressure of the flowing gas is read at each point of the

TABLE 25-2

Number of Velocity Measurement Points

Stack diameter or (length + width)/2 (m)	Number of velocity measurement points
0.0–0.3	8
0.3–0.6	12
0.6–1.3	16
1.3–2.0	20
2.0→	24

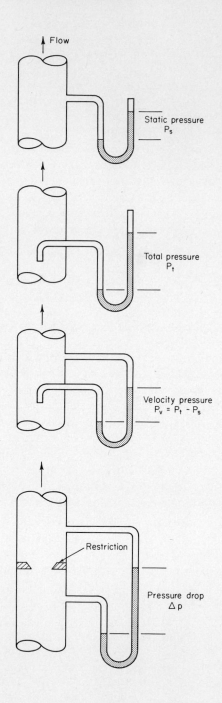

FIG. 25-4. Pressures commonly measured in flow systems.

TABLE 25-3

Velocity Sampling Locations, Diameters from Inside Wall to Traverse Point

Point number	Number of equal areas to be sampled				
	2	3	4	5	6
1	0.067	0.044	0.033	0.025	0.021
2	0.250	0.147	0.105	0.082	0.067
3	0.750	0.295	0.194	0.146	0.118
4	0.933	0.705	0.323	0.226	0.177
5		0.853	0.677	0.342	0.250
6		0.956	0.806	0.658	0.355
7			0.895	0.774	0.645
8			0.967	0.854	0.750
9				0.918	0.823
10				0.975	0.882
11					0.933
12					0.979

traverse and the associated gas velocity calculated from the formula

$$V = 420.5[(P_v/\rho)^{1/2}] \qquad (25\text{-}7)$$

where V = velocity, meters per minute, P_v = velocity pressure, millimeters of water, and ρ = gas density, kilograms per cubic meter.

The velocities are averaged for all points of the traverse to determine the gas velocity. Caution is advised not to average velocity pressures as a serious error results.

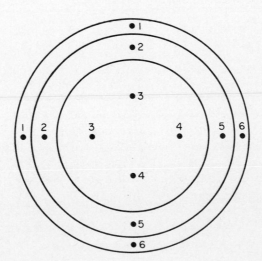

FIG. 25-5. Circular duct divided into three equal areas as described in Table 25-3. Numbers refer to sampling points.

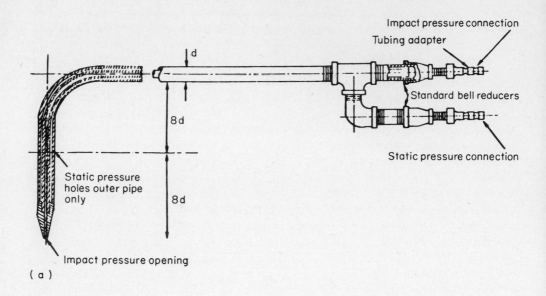

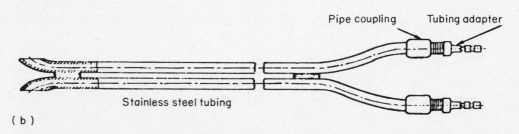

FIG. 25-6. Pitot tubes for velocity determinations. (a) Standard type; (b) the "S"-type (Stauscheibe). From "Annual Book of ASTM Standards," 1972.

Gas velocities can also be measured with anemometers (rotating vane, hot wire, etc.), from visual observations such as the velocity of smoke puffs, or from mass balance data (knowing the fuel consumption rate, air/fuel ratio, and stack diameter).

In the sampling train itself the gas flow must be measured to determine the sample volume. Particulates and gases are measured as micrograms per cubic meter. In either case, the determination of the fraction requires that the gas volume be measured for the term in the denominator. Some sample trains contain built-in flow indicating devices such as orifice meters, rotometers, or gas meters. These devices require calibration to assure that they read accurately at the time of the test and under test conditions.

To determine the volume through the sampling train a positive displacement system can be used. A known volume of water is displaced by gas containing the sample. Another system that works well and is inexpensive, is to measure the time necessary for the gas to fill a plastic bag to a certain static pressure inside the bag. The volume of the bag can be accurately measured under the same conditions and hence the flow determined by dividing the bag volume by the time required to fill it.

C. Collection of Source Sample

A typical sample train is shown in Fig. 25-7. This is the minimum number of components but in some systems the components may be combined. Extreme care must be exercised to assure that no leaks occur in the train and that the component parts of the train are identical for both calibration and sampling. The pump shown in Fig. 25-7 must be both oilless and leakproof. If, the pump and volume measurement devices are interchanged, the pump would no longer need to be oilless and leakproof, but the volume measurement would be in error unless it were adjusted for the change in static pressure. Some sampling trains become very complex as additional stages with controls and instruments are added. Many times the addition of components to a sampling train ends up making it so bulky and complicated that it becomes nearly impossible to use. A sampling train developed in an air conditioned laboratory can be useless on a shaky platform in a snow storm.

Standard sampling trains are specified for some tests. One of these standards is the system specified for incinerator testing (4). This train was designed for sampling combustion sources and should not be selected over a simpler sampling train when sampling noncombustion sources such as low temperature effluents from cyclones, baghouses, filters, etc. (5).

Before actually taking the sample train to the test site it is wise to prepare the operating curves for the particular job. With most factory assembled trains these curves are a part of the package. If a sampling train is assembled from components, the curves must be developed. The type of curves will vary from source to source and from train to train. Examples of operating curves which are useful include (1) velocity versus velocity pressure at various temperatures (6); (2) probe tip velocity versus flow meter readings at various temperatures; (3) flowmeter calibration curves of flow versus pressure drop. It is much easier to take an operating point from a previously prepared curve rather than to take out a slide rule and pad to make the calculations at the moment of the test. Remember too, that time may be a

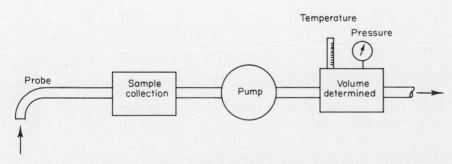

Fig. 25-7. Sampling train.

factor and settings must be made as rapidly as possible to obtain the necessary samples.

For sampling of gases the sample collection can be by any of several devices. Some commonly used methods, previously described in Chapter 14 include (1) Orsat analyzers, (2) absorption systems, (3) adsorption systems, (4) bubblers, (5) reagent tubes, (6) condensers, and (7) traps. Continuous analyzers are coming into increasing use and some of these are used for both source sampling and then source monitoring. They include (1) infrared analyzers, (2) flame ionization detectors, (3) mass spectrometers, (4) calorimetric systems, and (5) analytical systems which have been automated to continually sample the gas stream. Since the gases undergoing analysis offer no flow resistance, all that is necessary for sampling is to insert a probe and withdraw a gas sample.

For sampling particulate matter one is dealing with pollutants that have greatly different inertial and other characteristics from the carrying gas stream. It becomes important therefore to sample so the same velocity is maintained in the probe tip as exists in the adjacent gas stream. Such sampling is called *isokinetic*. Figure 25-8 shows the flow lines if anisokinetic sampling is allowed. In the first case with the probe velocity less than the stack velocity particles will be picked up by the probe which should have been carried around it by the gas streamlines. The inertia of the particles allows them to continue on their path and to be intercepted. If the probe velocity exceeds the stack velocity the inertia of the particles carries them by the probe tip even though the carrying gases are collected. Adjustment of particulate samples taken anisokinetically to corrected stack values is possible if all of the

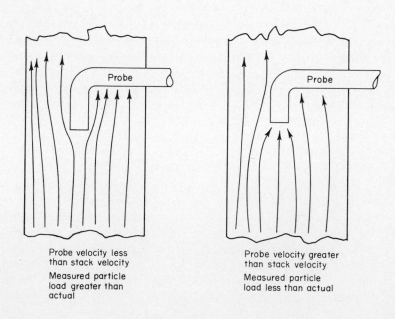

Probe velocity less
than stack velocity

Measured particle
load greater than
actual

Probe velocity greater
than stack velocity

Measured particle
load less than actual

Fig. 25-8. Anisokinetic stack sampling.

variables of the stack gas and particulate can be accounted for in the appropriate mathematical equation (7).

For the detection and intensity of odorous substances the nose is still the instrument usually relied upon. Since odors are gaseous, they may be sampled by simply collecting a known volume of effluent and performing some manipulation to dilute the odorous gas with known volumes of "pure" air. The odor is detected by an observer or a panel of observers. The odor-free air for dilution can be obtained by passing air through activated carbon or any other substance that removes all odors while not affecting the other gases that constitute "normal air." The odor free air is then treated by adding more and more of the odorous gas until the observer just detects the odor. The concentration is then recorded as the odor threshold as noted by that observer. The test is not truly quantitative as much variation between observers and between different samples is common.

If the compound causing the odor is known, and can be chemically analyzed, it may be possible to get valid quantitative data by direct gas sampling. An example would be a plant producing formaldehyde. If the effluent were sampled for formaldehyde vapor this could be related, through proper diffusion formulas, to indicate whether the odor would cause any problems in a residential neighborhood adjacent to the plant.

Several separating systems are used for particulate sampling. All rely on some principle of separating the aerosol from the gas stream. Many of the actual systems use more than one type of particulate collection device in series. If a size analysis is to be made on the collected material it must be remembered that multiple collection devices in series will collect different size fractions and size analyses must be made at each device and mathematically combined to obtain the size of the actual particulate in the effluent stream. In any system the probe itself removes some particulate before the carrying gas reaches the first separating device so the probe must be cleaned and the weight of material added to that collected in the remainder of the train.

Care should be exercised when sampling for aerosols that are condensable. Some separating systems, such as wet impingers, may remove the condensables from the gas stream while some, such as electrostatic precipitators, will not. Of equal concern should be possible reactions in the sampling system to form precipitates or aerosols which are not normally found when the stack gases are exhausted directly to the atmosphere. SO_3 plus other gaseous products may react in a water-filled impinger to form a particulate matter not truly representative of normal SO_3 release.

When sampling particulate from combustion processes it is necessary to take corresponding CO_2 readings of the effluent. Emission standards sometimes require that combustion stack gases are reported relative to either 12% CO_2 or 50% excess air. Adjusting to a standard CO_2 or excess air value normalizes the emission base. Also, emission standards require that the loadings be based on weight per standard cubic volume of air (usually at 20°C and 760 mm Hg). In most regulations the agency will require that the standard volume be dry but this is not always specified.

Once the source sample is collected it should be processed without delay. The laboratory personnel who will be analyzing the sample should be familiar with the

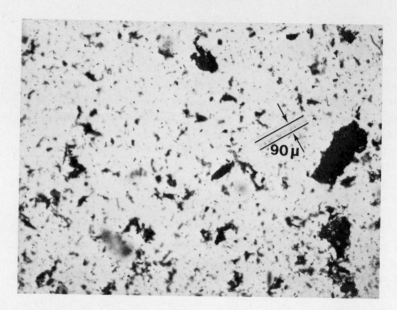

Fig. 25-9. Photomicrograph of particulate from source test of wood-fired boiler.

field phases of the test. The final analytic work should be done by the same person who obtained the original tare readings and on the same equipment. Blanks should be considered wherever changes occur due to handling or ambient temperature or humidity changes.

Extreme care should be taken in transporting and storing the samples between the time of collection and the time of analysis. Some condensible hydrocarbon samples have been lost because the collection device was subjected to elevated temperatures during shipment. Equally disastrous is placing the sample in an oven at 105°C to drive off the moisture only to discover that the particles of interest had a very low vapor pressure and also departed the sample. At times like this source sampling can be very frustrating.

A very important analytic tool that is overlooked by many source testing personnel is the microscope. A microscopic analysis of a particulate sample can tell a great deal about the type of material collected as well as its size distribution. It is a necessity if the sample was collected to aid in the selection of a piece of control equipment. All of the efficiency curves for particulate control devices are based upon fractional sizes. One would not try to remove submicron aerosol with a cyclone collector but unless a size analysis is made on the sampled material he is just guessing at the actual size range. Figure 25-9 is a photomicrograph of material collected during a source test.

V. Calculations and Report

Calculations that are repeatedly made can be made more accurately, and at lower cost, by using an electronic computer. If, for example, automotive emissions are continually tested over a standardized driving cycle, a computer program to analyze the data is a necessity. Otherwise, days would be spent calculating the data obtained in hours.

For sampling a relatively small number of individual sources a simplified calculation form may be used. Simplified calculation forms enable the office personnel to perform the arithmetic necessary to arrive at the answers, freeing the technical staff for proposals, tests, and reports. Many of the manufacturers of source testing equipment have example calculation forms included as part of their operating manuals. Some standard sampling methods have calculation forms included as a part of the method (8). Many control agencies have developed standard forms for their own use and will supply copies on request.

The source test report is the end result of a large amount of work. It should be thorough, accurate, and written in a manner understandable to the person intending to use it. It should state the purpose of the test, what was tested, how it was tested, the results obtained, and the conclusions reached as a result of conducting the test. The actual data and calculations should be included in the appendix of the source test report so that they are available to substantiate the report if questioned.

References

1. Holmes, R. G., ed. "Air Pollution Source Testing Manual." 179 pp. Los Angeles County Air Pollution Control District, Los Angeles, California, 1965.
2. Cochran, W. G. "Sampling Techniques." p. 11. Wiley, New York, 1963.
3. Parzen, E. "Modern Probability Theory and Its Applications," 464 pp. Wiley, New York, 1964.
4. "Specifications for Incinerator Testing at Federal Facilities." U.S. Dept. of Health, Education, and Welfare, Durham, North Carolina, 1967.
5. Boubel, R. W. "A High Volume Sampling Probe." *J. Air Pollut. C. A.* **21,** 783–787 (1971).
6. Marks, L. S., ed. "Standard Handbook for Mechanical Engineers," 7th ed. McGraw-Hill, New York, 1967.
7. Badzioch, S. "Correction for Anisokinetic Sampling of Gas-borne Dust Samples." *J. Inst. Fuel,* **33,** 106 (1960).
8. Marks, L. S., ed. "ASHRE Guide." American Society of Heating, Refrigerating and Air-Conditioning Engineers, Baltimore, Maryland, 1969.

Suggested Reading

Brenchley, D. L., Turley, C. D., and Yarmac, R. F. "Industrial Source Sampling," 481 pp. Ann Arbor Science Publishers, Ann Arbor, Michigan, 1973.
Cooper, H. B. H., and Rossano, A. T. "Source Testing for Air Pollution Control," 227 pp. Environmental Sciences Services Division, Wilton, Connecticut 1970.
Morrow, N. L., Brief, R. S., and Bertrand, R. R. "Sampling and Analyzing Air Pollution Sources," *Chem. Eng.* **79,** 85–98 (1972).

Questions

1. During a pitot traverse of a duct the following velocity pressures, in millimeters of water, were measured at the center of equal areas: 13.2, 29.1, 29.7, 20.6, 17.8, 30.4, 28.4, 15.2. If the flowing

fluid was air at 760 millimeters of mercury absolute and 85°C, what was the average gas velocity?

2. Would you expect errors of the same magnitude when sampling anisokinetically at 80% of the stack velocity as when sampling anisokinetically at 120% of stack velocity? Explain.

3. Suppose a particulate sample from a stack is separated into two fractions by the sampling device. Both are sized microscopically and found to be log-normally distributed. One has a count mean size of 5.0 μ and a geometric deviation of 2.0. The other has a count mean size of 10.0 μ and a geometric deviation of 2.2. Two grams of the smaller sized material were collected for each 10 gm of the larger. What would be reported for the weight mean size and geometric deviation of the stack effluent?

4. A particulate sample was found to weigh 0.0216 gm. The sample volume from which it was collected was 0.60 m³ at 60°C, 760 mm Hg absolute, and 90% relative humidity. What was the stack loading in milligrams per standard cubic meter?

5. A particulate sample was found to contain 350 mg/standard m³. The CO_2 during the sampling period averaged 7.2%. If the exhaust gas flow was 2000 m³/minute, what would be the particulate loading in both milligrams per cubic meter and kilograms per hour, corrected to 12% CO_2?

Chapter 26

ENGINEERING CONTROL

The application of control technology to air pollution problems assumes that a source can be reduced to a predetermined level to meet a regulation or some unknown minimum value. Control technology cannot be applied to an uncontrollable source, such as a volcano, nor can it be expected to completely control a source resulting in zero atmospheric emissions. The cost of controlling any given air pollution source is usually an exponential function of the percent of control and therefore becomes an important consideration in the level of the control required (1). Figure 26-1 shows a typical cost curve for control equipment.

If the material recovered has some economic value then the picture is different. Figure 26-2 shows the previous cost of control with the value recovered curve superimposed upon it.

The plant manager looking at such a curve would want to be operating in the area to the left of the intersection of the two curves while the local air pollution forces would insist upon operation as far to the right of the graph as the best available control technology would allow.

Control of any air pollution source requires a complete knowledge of the contaminant and source. The engineer controlling the source must be thoroughly familiar with all available physical and chemical data on the effluent from the source. He must know the rules and regulations of the control agencies involved and this includes not only the air pollution control agency but also any agencies which may

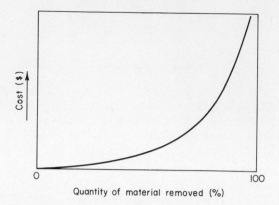

Fig. 26-1. Air pollution control equipment cost.

have jurisdiction over the construction, operation, and final disposal of the waste from the source (2).

In many cases, heating or cooling of the gaseous effluent will be required before it enters the control device. The engineer must be thoroughly aware of the gas laws, thermodynamic properties, and reactions involved to secure a satisfactory design. For example, if a gas is cooled there may be condensation if the temperature drops below the dewpoint. If water is sprayed into the hot gas for cooling, it adds greatly to the specific volume of the mixture. As the gases pass through hoods, ducts, fans, and control equipment temperatures and pressures change and hence, also, specific volumes and velocities (2).

The control of atmospheric emissions from a process will generally take one of three forms depending upon the process, fuel, types, availability of control equipment, etc. The three general methods are (1) process change to a less polluting process or to a lowered emission from the existing process through a modification or change in operation, (2) fuel change to a fuel which will give the desired level of emissions, and (3) installation of control equipment between the point of pollutant

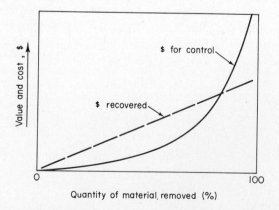

Fig. 26-2. Control equipment cost with value recovered.

generation and their release to the atmosphere. Control may consist of either removal of the pollutant or conversion to a less polluting form (3).

I. Process Change

A process change can be either a change in operating procedures for an existing process or the substitution of a completely different process. Consider a plant manager who for years has been using "Solvent A" for a degreasing operation. By past experimentation he has found that if he maintains the conveyor speed at 100 units per hour, with a solvent temperature of 80°C, he gets maximum cleaning with a solvent loss that results in the lowest overall operating cost for the process.

A new regulation is passed requiring greatly reduced atmospheric emissions of organic solvents, which includes "Solvent A." He has several alternatives open to him. He can

1. Change to another more expensive solvent, which by virtue of its lower vapor pressure would emit less organics
2. Reduce the temperature of the solvent and slow down the conveyor to get the same amount of cleaning. This may require the addition of another line or another 8-hour shift
3. Put in the necessary hooding, ducting, and equipment for a solvent recovery system which will decrease the atmospheric pollution and also result in some economic solvent recovery
4. Put in the necessary hooding, ducting, and equipment for an afterburner system which will burn the organic solvent vapors to a less polluting emission but with no solvent recovery

In some cases the least expensive control is achieved by completely abandoning the old process and completely replacing it with a new, less polluting process. Any increased production and/or recovery of material may help offset a portion of the cost. It has proved to be cheaper to abandon old steel mills and to replace them with completely new furnaces of a different type than to modify the old systems to meet pollution regulations. Kraft pulp mills found that the least costly method of meeting stringent regulations was to replace the old, high-emission recovery furnaces with a new furnace of completely different design. The Kraft mills have generally asked for, and received, additional plant capacity to partially offset the cost of the new type furnaces.

II. Fuel Change

For many air pollution control situations a change to a less polluting fuel offers the ideal solution to the problem. If a power plant is emitting large quantities of SO_2 and fly ash, conversion to natural gas may be cheaper than installing the necessary control equipment to reduce the pollutant emissions to the permitted values. If the drier at an asphalt plant is emitting 350 mg of particulate per standard cubic meter of effluent when fired with heavy oil of 4% ash, it is probable that a switch to natural gas will allow the operation to meet an emission standard of 250 milligrams per standard cubic meter.

Fuel switching based upon meteorological or air pollution forecasts is a common practice to reduce the air pollution burden at critical times. Many control agencies allow power plants to operate on residual oil during certain periods of the year when pollution potential is low. Some large utilities for years have followed a policy of switching from their regular coal to a more expensive, but lower sulfur coal, when stagnation conditions were forecast.

Caution should be exercised when considering any change in fuels to reduce emissions. Specific considerations might be

1. What are current and potential fuel supplies? In many areas natural gas is already in short supply. It may not be possible to convert a large plant with current allocations or pipe line capacity.

2. Most large boilers use a separate fuel for auxilliary or standby. An example can be made of a boiler fired with wood residue as the primary fuel with residual oil as the standby. A change was made to natural gas as the primary fuel with residual oil kept for standby. This change was made to lower particulate emissions along with a predicted slightly lower cost. Because of gas shortages, the plant now operates on

FIG. 26-3. Trojan nuclear power plant (Portland General Electric Company).

residual oil during most of the cold season and the resulting particulate emission greatly exceeds that which was previously emitted from the wood fuel. In addition, an SO_2 emission problem exists with the oil fuel that never occurred with the wood residue. Overall costs have not been lowered because natural gas rates have increased since the conversion.

3. Charts or tables listing supplies or reserves of low sulfur fuel may not tell the entire story. For example, a large percentage of low sulfur coal is owned by steel companies and is therefore not generally available for use in power generating stations even though it is listed in tables published by various agencies.

4. A very strong competition exists for low pollution fuels. While one area may be drawing up regulations to require use of natural gas or low sulfur fuels it is probable that other neighboring areas are doing the same. While there may have been sufficient premium fuel for one or two areas, if all the region changes, not enough exists. Such a situation has resulted in extreme fuel shortages during cold spells in some large cities. The supply of low sulfur fuels was simply exhausted during periods of extensive use.

The increasing number of atomic reactors used for power generation has been questioned from an air pollution standpoint as well as from other environmental points of view. A modern atomic plant, as shown in Fig. 26-3, appears to be relatively pollution free compared to the more familiar fossil fuel fired plant.

III. Pollution Removal

A great number of situations exist where sufficient control over emissions cannot be obtained by fuel or process change. In cases such as these, the level of the pollutants of concern in the exhaust gases or process stream must be reduced to allowable values before they are released to the atmosphere.

The equipment for the pollutant removal system includes all hoods, ducting, controls, fans, and disposal or recovery systems that might be necessary. The entire system should be engineered as a unit for both maximum efficiency and economy. Many systems operate at less than maximum efficiency because a portion of the system was designed, or adapted, without consideration of the other portions (4).

Efficiency of the control equipment is normally specified before the equipment is purchased. If a plant is emitting 500 kg/hr of a pollutant and the regulations only allow an emission of 25 kg/hr, it is obvious that at least 95% efficiency is required of the pollution control system. This situation requires the regulation to state, "At least 95% removal on a weight basis." The regulation should further specify how the test will be made to determine the efficiency. Figure 26-4 shows the situation as it exists.

The efficiency for the device shown in Fig. 26-4 may be calculated several ways:

$$\text{Efficiency, } \% = 100\,(C/A), \quad \text{but, since} \quad A = B + C \tag{26-1}$$

$$\text{Efficiency, } \% = 100\left(\frac{C}{B+C}\right) \quad \text{or} \quad 100\left(\frac{A-B}{A}\right) \quad \text{or} \quad 100\left(\frac{A-B}{B+C}\right) \tag{26-2}$$

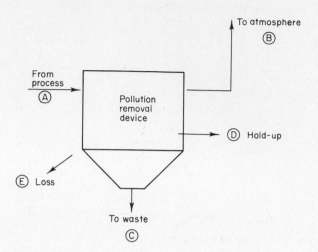

Fig. 26-4. Typical pollution control device as shown for efficiency calculations.

The final acceptance test would probably be made by measuring two of the three quantities and using the appropriate equation. For a completely valid efficiency test the effect of hold-up (D) and loss (E) must also be taken into account.

To remove a pollutant from the carrying stream requires that some property of the pollutant that is different from the carrier must be exploited. The pollutant may have a different size, inertia, electrical, or absorption properties. Removal requires that the equipment be designed to apply the scientific principles necessary to perform the separation.

IV. Removal of Particulate Matter

Aerosols, or particulate matter, differ so much from the carrying gas stream that their removal should present no major difficulties (5). The aerosol is different physically, chemically, and electrically. It has vastly different inertial properties than the carrying gas stream and can be subjected to an electric charge. It may be soluble in a specific liquid. With such a variety of removal mechanisms that can be applied it is not surprising that particulate matter, such as mineral dust, might be removed by a filter, wet scrubber, or electrostatic precipitator with equally satisfactory results.

Practical considerations in selecting particulate removal equipment include many factors other than just cost and efficiency—(1) What is the mean size and range of size of the dust? (2) What is the dust density, shape, and surface characteristic? (3) What are the abrasive or corrosive properties of the dust? (4) What are the volumes, temperature, and pressure of the gases to be handled? (5) Are condensable vapors present, including water? (6) What about the physical location of the collection unit, its size, and shape? and, (7) What provisions are needed for servicing and maintenance?

A. Filters

A filter removes particulate matter from the carrying gas stream because the particulate impinges upon and then adheres to the filter material (6). As time passes, the deposit of particulate becomes greater and the deposit then acts as a filtering media. When the deposit becomes so heavy that the pressure necessary to force the gas through the filter becomes excessive, or the flow reduction severely impairs the process, the filter must either be replaced or cleaned.

The filter media can be fibrous, such as cloth; granular, such as sand; a rigid solid, such as a screen; or a mat, such as a felt pad. It can be in the shape of a tube, sheet, bed, fluidized bed, or any other desired form. The material can be natural or man made fibers, granules, paper, metal, ceramic, glass, or plastic. It is not surprising that filters are manufactured in nearly an infinite number of different types, sizes, shapes, and materials.

The theory of filtration of aerosols from a gas stream is much more involved than the sieving action which is the case for particles in a liquid medium. Figure 26-5 shows three of the mechanisms of aerosol removal by a filter. In practice the particles and filter elements are seldom spheres or cylinders.

Direct interception occurs when the fluid streamline carrying the particle passes within one-half a particle diameter of the filter element. Regardless of the particle size, mass, or inertia, it will be collected if the streamline passes sufficiently close. Inertial impaction occurs when the particle would miss the filter element if it followed the streamline but its inertia resists the change in direction taken by the gas molecules and it continues in a direct enough course to be collected by the filter element. Electrostatic attraction occurs because either, or both, the particle or filter possess sufficient charge to overcome the inertial forces and the particle is collected instead of passing the filter element.

Other lesser mechanisms that result in aerosol removal by filters are (1) gravitational settling due to the different mass of the aerosol and carrying gas, (2) thermal precipitation due to the temperature gradient between a hot gas stream and the

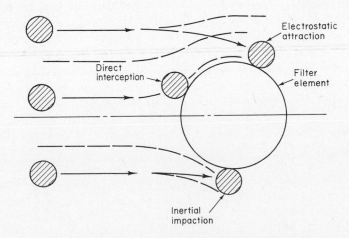

FIG. 26-5. Particulate removal by filter element.

cooler filter media which causes the particles to be bombarded more vigorously by the gas molecules on the side away from the filter element, (3) Brownian deposition as the particles are bombarded with gas molecules that may cause enough movement to permit the aerosol to contact the filter element. Brownian motion may also result in some of the particles missing the filter element because they are moved away from it as they pass by. For practical purposes, only the three mechanisms shown in Fig. 26-5 are effective as far as removal of aerosols from a gas stream.

Regardless of the mechanism which causes the aerosol to contact the filter element, it will only be removed from the gas stream if it adheres to the surface. Aerosols arriving later at the filter element will in turn adhere to the collected aerosol instead of the filter element. The result is that actual aerosol removal seldom agrees with theoretical calculations. One should also consider that there are also particles that do not adhere to the filter element even though they contact them. As time passes, the heavier deposits on the filter surface will be dislodged easier than the light deposits, resulting in increased reentrainment. Because of plugging of the filter with time, the size of the filter element increases which causes more interception and impaction. The general effect of all of these variables on the particle build-up and reentrainment is shown in Fig. 26-6.

The pressure drop through the filter is a function of two separate effects. The clean filter has some initial pressure drop. This is a function of filter material, depth of the filter, the superficial gas velocity which is the gas velocity perpendicular to the filter face, and the viscosity of the gas. Added to the clean filter resistance is the resistance caused by the adhering particles forming a cake on the filter surface.

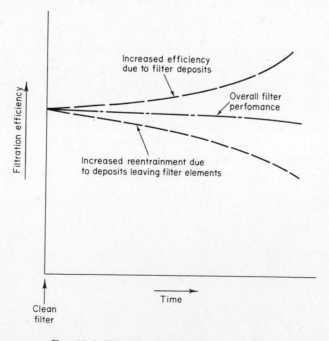

FIG. 26-6. Filtration efficiency change with time.

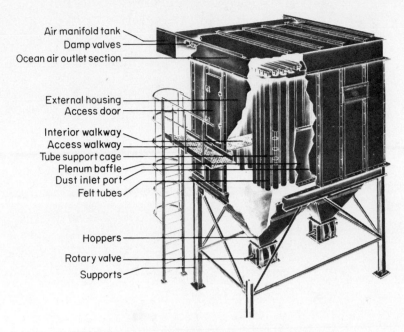

Air manifold tank
Damp valves
Ocean air outlet section

External housing
Access door
Interior walkway
Access walkway
Tube support cage
Plenum baffle
Dust inlet port
Felt tubes

Hoppers
Rotary valve
Supports

FIG. 26-7. Industrial bag house (Carborundum).

This cake increases in thickness as approximately a linear function of time and the pressure difference necessary to cause the same gas flow becomes a linear function also. The usual situation is that the pressure available at the filter is limited so that as the cake builds up the flow decreases. Filter cleaning can be based, therefore, on (1) increased pressure drop across the filter, (2) decreased volume of gas flow, or (3) time since the last cleaning.

Industrial filtration systems may be of many types. The most common type is the bag house as shown in Fig. 26-7. The filter bags are fabricated from cloth with the material and weave selected to fit the specific application. Cotton and synthetic fabrics are used for relatively low temperatures with glass cloth fabrics being used for elevated temperatures, up to 290°C.

The filter ratio for bag houses, also called the gas to cloth ratio, varies from 0.6 to 1.5 m³ of gas per minute per square meter of fabric. The pressure drop across the fabric is a function of the filter ratio; about 80 mm of water for the lower filter ratios up to about 200 mm of water for the higher. Before selecting any bag filter system a thorough engineering study should be made followed by a consultation with different bag and bag house manufacturers.

The bags must be periodically cleaned to remove the accumulated particulate matter. Bag cleaning methods vary widely with manufacturer and with bag house style and use. Methods used for cleaning include (1) mechanical shaking by agitation of the top hanger, (2) reverse flow of gas through a few of the bags at a time, (3) continuous cleaning with a reverse jet of air passing through a series of orifices on a ring as it moves up and down the clean side of the bag, and (4) collapse and pulsation cleaning methods.

The cleaning cycles are usually controlled by a timing device which deactivates the section being cleaned. The dusts removed during cleaning are collected in a hopper at the bottom of the bag house and then are removed, through an air lock or star valve, to a bin for ultimate disposal.

Other types of industrial filtration systems include (1) fixed beds or layers of granular material such as coke or sand. Some of the original proposals for cleaning large quantities of gases from smelters and acid plants suggested passing the gases through such beds, (2) plain, treated, or charged mats or pads. Common throw-away air filters used for hot air furnaces and for air conditioners are of this type, (3) paper filters of multiple plies and folds to increase filter efficiency and area. The throw-away type of dry air filters used on automotive engines are of this type, (4) rigid porous beds which can be made of metal, plastic, or porous ceramic. These materials are most efficient for removal of large sized particulate such as the 30-μ particles from a wood sanding operation, (5) fluidized beds in which the granular material of the bed is caused to act as a fluid by the gas passing through it. Most fluidized beds are used for purposes of heat or mass transfer and using them for filtration has not been thoroughly investigated.

B. Electrostatic Precipitators

High-voltage electrostatic precipitators (ESP) have been widely used throughout the world for particulate removal since they were perfected by Fredrick Cottrell early in the 20th century (7). Most of the original units were used exclusively for recovery of process materials but today, gas cleaning for air pollution control is often the main reason for their installation (8). The ESP has some distinct advantages over other aerosol collection devices—(1) it can easily handle high temperature gases which makes it a likely choice for boilers, steel furnaces, etc., (2) it has an extremely small pressure drop so fan costs are minimized, (3) it has an extremely high collection efficiency if operated properly on selected aerosols. Many cases are on record, however, where relatively low efficiencies were actually obtained because of unique or unknown dust properties, (4) it can handle a wide range of particulate sizes and dust concentrations. Most precipitators work best on particles less than 10 μ so an inertial precleaner is often used to remove the large particles, (5) if properly designed and constructed its operating and maintenance costs are lower than for any other type of particulate collection system.

A few of the disadvantages of electrostatic precipitators are (1) the initial cost is the highest of any particulate collection system, (2) a large amount of space is required for the installation, (3) it should not be considered for combustible particles such as grain and wood dust.

The principle of the electrostatic precipitator is the charging of the dust with ions and then collecting the ionized particles on a surface. For cleaning and disposal, the particles are then removed from the collection surface. Figure 26-8 illustrates a tubular type of electrostatic precipitator designed for sampling aerosols.

A high voltage (30 or more kV) DC field is established between the central wire electrode and the grounded collecting surface. The voltage is high enough that a visible corona can be seen at the surface of the wire. The result is a cascade of negative ions in the gap between the central wire and the grounded outer tube. Any aerosol entering this gap is both bombarded and charged by these ions. The aerosols then migrate to the collecting surface because of the combined effect of the bombardment and charge attraction. When the particle reaches the collecting surface, it loses its charge and adheres because of the attractive forces existing. It should remain there until the power is shut off and it is physically dislodged by rapping, washing, or sonic means.

In actual practice, the tubes are from 8 to 25 cm in diameter and 1 to 4 m long. They are arranged vertically in banks with the central wires, about 2 mm in diameter, suspended in the center with tension weights at the bottom. Many innovations, including square wires, triangular wires, and barbed wires are used by different manufacturers.

A plate-type electrostatic precipitator is similar in principle to the tubular type except that the air flows across the wires horizontally, at right angles to them. The particles are collected on vertical plates which usually have fins or baffles to strengthen them and prevent dust reentrainment. Figure 26-9 illustrates a large plate-type precipitator. These precipitators are usually used to control and collect dry dusts.

Problems with electrostatic precipitators develop because the final unit operates at something other than ideal conditions. Gas channeling through the unit can result in high dust loadings in one area and light loads in another. The end result is less than optimum efficiency because of much reentrainment. The resistivity of the dust greatly affects the reentrainment of the dust in the unit. If a high resistance dust collects on the plate surface, the effective voltage across the gap is decreased. Some power plants burning high ash, low sulfur coal have reported very low effi-

FIG. 26-8. Tubular electrostatic precipitator.

Fig. 26-9. Commercial plate type electrostatic precipitator (Joy Manufacturing Company).

ciency from the precipitator because the ash needed more SO_2 to decrease its resistivity. The suggestion that precipitator efficiency could be greatly improved by *adding* SO_2 to the stack gases has not yet been accepted. Addition of steam is also a possibility to change the resistivity, but the problems of handling the wet fly ash would have to be overcome.

Properly designed and operated precipitators with ideal dusts should operate around 96% to 98% efficiency and visible emissions should be virtually nonexistent. To secure the ultimate in cleaning the effluent of Kraft recovery boilers, two precipitators are operated in series resulting in overall efficiencies exceeding 99.5%. If necessary, the boiler can be operated on only one precipitator while the other is down for maintenance.

C. Inertial Collectors

Inertial collectors, whether they be cyclones, baffles, louvers, or rotating impellers operate on the principle that the aerosol material in the carrying gas stream has a greater inertia than the gas (9). Since the drag forces on the particle are a function of the diameter squared while the inertial forces are a function of the diameter cubed, it follows that as the particle diameter increases, the inertial (removal) force becomes relatively greater. Inertial collectors all have the property, therefore, that they are most efficient for larger particles. The inertia is also a function of the mass of the particle so the heavier particles are more efficiently removed by inertial collectors. These facts explain why an inertial collector will be highly efficient for removal of 10-μ rock dust and very inefficient for 5-μ wood particles. It would be very efficient though for 75-μ wood particles.

The most common inertial collector is the cyclone which is used in two basic forms, the tangential inlet and the axial inlet (10). Figure 26-10 shows the two types. In actual industrial practice the tangential inlet type is usually a large

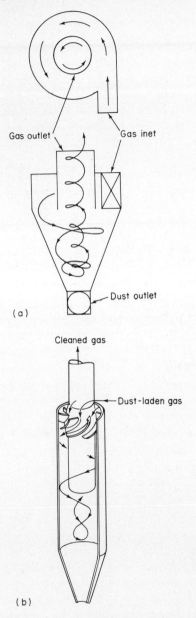

FIG. 26-10. (a) Tangential inlet cyclone; (b) axial inlet cyclone.

TABLE 26-1

Effect of Independent Variables on Inertial Collector Efficiency

Independant variable of concern	Increase or decrease to improve efficiency
Radius of curvature	Decrease
Mass of particle	Increase
Particle diameter	Increase
Particle surface/Volume ratio	Decrease
Gas velocity	Increase
Gas viscosity	Decrease

(1–5 m in diameter) single cyclone, while the axial inlet cyclone is relatively small (about 20 cm in diameter) and arranged in parallel units for the desired capacity.

For any cyclone, regardless of the type, the radius of motion (curvature), the particle mass, and the particle velocity are the three factors which determine the centrifugal force exerted on the particle. This centrifugal force may be expressed as:

$$F = MA \qquad (26\text{-}3)$$

where F = force (centrifugal), M = mass of particle, and A = acceleration (centrifugal) and

$$A = \frac{V^2}{R} \qquad (26\text{-}4)$$

where V = velocity of particle, R = radius of curvature; therefore

$$F = \frac{MV^2}{R} \qquad (26\text{-}5)$$

Other types of inertial collectors which might be used for particulate separation from a carrying gas stream depend upon the same theoretical principles developed for cyclones. Table 26-1 summarizes the effect of the common variables on inertial collector performance.

It should further be pointed out that while decreasing the radius of curvature and increasing the gas velocity both result in increased efficiency, these same changes cause increased pressure drop through the collector. Design of inertial collectors for maximum efficiency at minimum cost and minimum pressure drop is a problem which lends itself to computer optimization. Unfortunately, many inertial collectors, including the majority of the large, single cyclones, have been designed to fit a standard-sized sheet of metal rather than a specific application and gas velocity. As tighter emission standards are adopted, the role of inertial collectors will probably become that of precleaners for the more sophisticated gas cleaning devices.

D. Scrubbers

Scrubbers, or wet collectors, have been used as gas cleaning devices for many years (11). The process has two distinct mechanisms which result in the removal

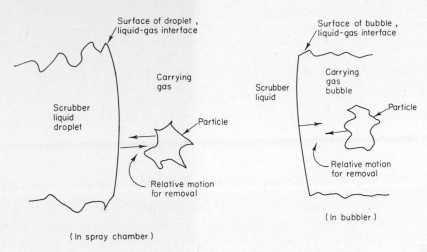

Fig. 26-11. Wetting of aerosols in spray chamber or bubbler.

of the aerosol from the gas stream. The first mechanism involves the wetting of the particle by the scrubbing liquid. As shown in Fig. 26-11 this process is essentially the same whether the system uses a spray to atomize the scrubbing liquid or a diffuser to break the gas into small bubbles. In either case, it is assumed that the particle is trapped when it travels from the supporting gaseous medium across the interface to the liquid scrubbing medium. Some relative motion is necessary for the particle and liquid–gas interface to contact each other. In the spray chamber, this is provided by spraying the droplets through the gas so they impinge upon, and contact the particles. In the bubbler, inertial forces and severe turbulence achieve this contact. In either case, the smaller the droplet or bubble, the greater the collection efficiency. In the scrubber, the smaller the droplet, the greater the surface area for a given weight of liquid and the greater the chance for wetting the particles. In a bubbler, smaller bubbles mean not only more interface area but also that the particles have a shorter distance to travel before reaching an interface where they can be wetted.

The second mechanism important in wet collectors is the removal of the wetted particles on some type of collecting surface followed by their eventual removal from the device. The collecting surface can be in the form of a bed or simply a wetted surface. One common combination is to follow the wetting section with an inertial collector which then separates the wetted particles from the carrying gas stream. Figure 26-12 illustrates an industrial scrubber suitable for removing particulate from a gas stream at a moderate pressure drop with a nominal efficiency.

Increasing either the gas velocity or the liquid droplet velocity in a scrubber will increase the efficiency because of the greater number of collisions per unit time. The ultimate scrubber in this respect is the venturi scrubber which operates at extremely high gas and liquid velocities with a very high pressure drop across the venturi throat.

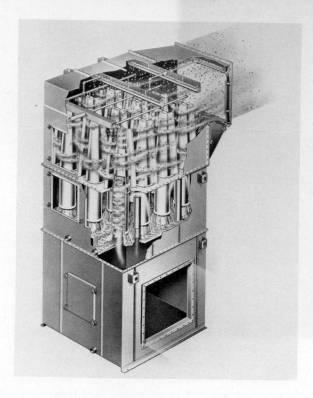

FIG. 26-12. Cyclonic scrubber (American Air Filter Company, Inc.).

E. Comparison of Particulate Removal Systems

When selecting a system to remove particulate from a gas stream it is obvious that many choices concerning equipment can be made. The selection could be made on the basis of cost, gas pressure drop, efficiency, temperature resistance, etc. Table 26-2 presents a look at these overall factors for comparative purposes. Caution must be exercised in not using the tabular values as absolute because great variations occur between types and manufacturers. No such table is a substitute for a qualified consulting engineer or a reputable manufacturer's catalog.

V. Removal of Gaseous Pollutants

Gaseous pollutants may be easier or harder to remove from the carrying gas stream than aerosols depending upon the individual situation. The gases may be reactive to other chemicals and this property can be used to collect them. Of course, any separation system relying on differences in inertial properties must be ruled out. Four general methods of separating gaseous pollutants are currently in use. These include (1) absorption in a liquid, (2) adsorption on a solid surface, (3) condensation to a liquid, and (4) conversion into a less polluting or nonpolluting gas.

TABLE 26-2

Comparison of Particulate Removal Systems

Type of collector	Particle size range (μ)	Removal efficiency	Space required	Max. temp. (°C)	Pressure drop (cm, H₂O)	Annual cost ($ per year/ m³)ᵃ
Bag house (cotton bags)	0.1– 1.0	Fair	Large	80	10	7.00
	1.0–10.0	Good	Large	80	10	7.00
	10.0–50.0	Excellent	Large	80	10	7.00
Bag house (Dacron, nylon, orlon)	0.1– 1.0	Fair	Large	120	12	8.50
	1.0–10.0	Good	Large	120	12	8.50
	10.0–50.0	Excellent	Large	120	12	8.50
Bag house (glass fiber)	0.1– 1.0	Fair	Large	290	10	10.50
	1.0–10.0	Good	Large	290	10	10.50
	10.0–50.0	Good	Large	290	10	10.50
Bag house (Teflon)	0.1– 1.0	Fair	Large	260	20	11.50
	1.0–10.0	Good	Large	260	20	11.50
	10.0–50.0	Excellent	Large	260	20	11.50
Elecrostatic precipitator	0.1– 1.0	Excellent	Large	400	1	10.50
	1.0–10.0	Excellent	Large	400	1	10.50
	10.0–50.0	Good	Large	400	1	10.50
Standard cyclone	0.1– 1.0	Poor	Large	400	5	3.50
	1.0–10.0	Poor	Large	400	5	3.50
	10.0–50.0	Good	Large	400	5	3.50
High-efficiency cyclone	0.1– 1.0	Poor	Moderate	400	12	5.50
	1.0–10.0	Fair	Moderate	400	12	5.50
	10.0–50.0	Good	Moderate	400	12	5.50
Spray tower	0.1– 1.0	Fair	Large	540	5	12.50
	1.0–10.0	Good	Large	540	5	12.50
	10.0–50.0	Good	Large	540	5	12.50
Impingement scrubber	0.1– 1.0	Fair	Moderate	540	10	11.50
	1.0–10.0	Good	Moderate	540	10	11.50
	10.0–50.0	Good	Moderate	540	10	11.50
Venturi scrubber	0.1– 1.0	Good	Small	540	88	28.00
	1.0–10.0	Excellent	Small	540	88	28.00
	10.0–50.0	Excellent	Small	540	88	28.00

ᵃ Includes: Water and power cost, maintenance cost, operating cost, capital and insurance costs.

A. Absorption Devices

Absorption of pollutant gases is accomplished by using a selective liquid in a wet scrubber, packed tower, or bubble tower. Pollutant gases commonly controlled by absorption include sulfur dioxide, hydrogen sulfide, hydrogen chloride, chlorine, ammonia, oxides of nitrogen, and low-boiling hydrocarbons.

The scrubbing liquid must be chosen with specific reference to the gas being removed. The gas solubility in the liquid solvent should be high so that reasonable quantities of solvent are required. The solvent should have a low vapor pressure to reduce losses, be noncorrosive, inexpensive, nontoxic, nonflammable, chemically

418   26. ENGINEERING CONTROL

stable, and have a low freezing point. It is no wonder that water is the most popular solvent used in absorption devices. The water may be treated with an acid or a base to enhance removal of a specific gas. If carbon dioxide is present in the gaseous effluent, and water is used as the scrubbing liquid, a solution of carbonic acid will gradually replace the water in the system.

In many cases, water is a poor scrubbing solvent for one reason or another. Sulfur dioxide, for example, is only slightly soluble in water so a scrubber of very large liquid capacity would be required. SO_2 is readily soluble in an alkaline solution so scrubbing solutions containing ammonia or amines are used in commercial applications.

Chlorine, hydrogen chloride, and hydrogen fluoride are examples of gases that are readily soluble in water so water scrubbing is very effective for their control. For years hydrogen sulfide has been removed from refinery gases by scrubbing with diethanolamine. More recently the light hydrocarbon vapors at petroleum refineries and loading facilities have been absorbed, under pressure, in liquid gasoline and returned to storage (12). All of these gases mentioned have an economic value when recovered and can be valuable raw materials or products when removed from the scrubbing solvent.

B. Adsorption Devices

Adsorption of pollutant gases occurs when certain gases are selectively retained on the surface, or in the pores or interstices of prepared solids. The process may be strictly a surface phenomena with only molecular forces involved or it may be combined with a chemical reaction occurring at the surface once the gas and adsorber are in intimate contact. The latter type of adsorption is known as chemisorption.

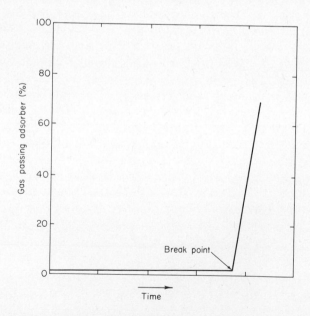

Fig. 26-13. Adsorbent breakpoint at saturation with adsorbate.

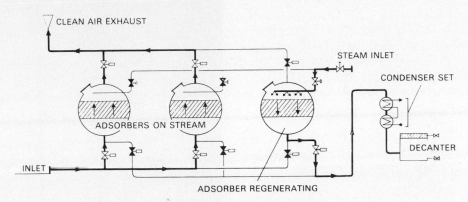

FIG. 26-14. Flow diagram for adsorber (The British Ceca Company, Ltd.).

The solid materials used as adsorbents are usually very porous with extremely large surface to volume ratios. Activated carbon, alumina, and silica gel are widely used as adsorbents depending on the gases to be removed. Activated carbon, for example, is excellent for removing light hydrocarbon molecules which may be odorous. Alumina is useful for chemisorption of SO_2 from stack gases as a conversion to aluminum sulfate occurs after adsorption. Silica gel, being a polar material, does an excellent job of adsorbing polar gases. Its characteristics for removal of water vapor are well known.

A further requirement of the solid adsorbents are that they be structurally capable of being packed into a tower, do not fracture easily, and can be regenerated and reused after saturation with gas molecules. While some small units use throw-away canisters or charges, the majority of industrial adsorbers regenerate the adsorbent not only to recover the adsorbent but also to recover the adsorbate which usually has some economic value.

The efficiency of most adsorbers is very near 100% at the beginning of operation and remains extremely high until a breakpoint occurs when the adsorbent becomes saturated with adsorbate. It is at the breakpoint that the adsorber should be renewed or regenerated. This is shown graphically in Fig. 26-13.

Industrial adsorption systems are engineered so that they operate in the region before the breakpoint and are continually regenerated by units. Figure 26-14 shows a schematic diagram of such a system with steam being used to regenerate the saturated adsorbent. Figure 26-15 illustrates the actual system shown schematically in Fig. 26-14.

C. Condensers

In many situations, the most desirable control of vapor type discharges can be accomplished by condensation. Condensers may also be used ahead of other air pollution control equipment to remove condensable components. The reasons for using condensers include (1) to recover economically valuable products, (2) to

Fig. 26-15. Adsorption system shown schematically in Fig. 26-14 (The British Ceca Company, Ltd.).

remove components that might be corrosive or damaging to other portions of the system, (3) to reduce the volume of the effluent gases.

Although condensation can be accomplished either by reducing the temperature or increasing the pressure, in practice it is done by temperature reduction only.

Condensers may be one of two general types depending upon the specific application. Contact condensers operate with the coolant, vapors, and condensate inti-

TABLE 26-3

Representative Applications of Condensers in Air Pollution Control

Petroleum refining	Petrochemical manufacturing	Basic chemical manufacture	Miscellaneous industries
Gasoline accumulator	Polyethylene gas vents	Ammonia	Dry cleaning
Solvents	Styrene	Chlorine solutions	Degreasers
Storage vessels	Copper naphthenates		Tar dipping
Lube oil refining	Insecticides		Kraft paper
	Phthalic anhydride		
	Resin reactors		
	Solvent recovery		

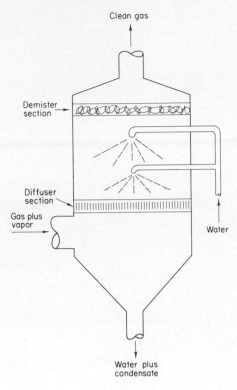

Clean gas

Demister section

Diffuser section

Gas plus vapor

Water

Water plus condensate

FIG. 26-16. Contact condenser.

mately mixed. In surface condensers, the coolant does not contact either the vapors or the condensate. The usual shell and tube type of condenser is of the surface type. Figure 26-16 illustrates a contact condenser which might be used to clean, or preclean, a hot corrosive gas (13, 14).

Table 26-3 lists several applications of condensers currently in use. For most operations listed, air and noncondensable gases should be kept to a minimum as they tend to reduce the condenser capacity.

D. Conversion to Nonpollutant Material

A widely used system, for the control of organic gaseous emissions, is the oxidation of the combustible components to water and carbon dioxide (15). Other systems such as the oxidation of H_2S to SO_2 and H_2O are also used even though the SO_2 produced is still considered a pollutant. The trade off occurs because the SO_2 is much less toxic and undesirable than the H_2S. The odor threshold for H_2S is about three orders of magnitude less than that for SO_2. For oxidation of H_2S to SO_2 the usual device is simply an open flare with a fuel gas pilot or auxiliary burner if the H_2S is below stoichiometric concentration.

Afterburners are widely used as control devices for oxidation of undesirable combustible gases. The two general types are (1) direct flame afterburners in which

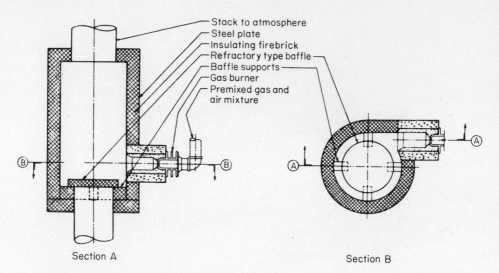

Stack to atmosphere
Steel plate
Insulating firebrick
Refractory type baffle
Baffle supports
Gas burner
Premixed gas and
air mixture

Section A   Section B

FIG. 26-17. Direct-fired afterburner (Los Angeles County Air Pollution Control District).

the gases are oxidized in a combustion chamber at, or above, the temperature of autogenous ignition, and (2) catalytic combustion systems in which the gases are oxidized at temperatures considerably below the autogenous ignition point.

Direct flame afterburners are the most commonly used air pollution control device where combustible aerosols, vapors, gases, and odors are to be controlled. The components of the afterburner are shown in Fig. 26-17. They include the combustion chamber, gas burners, burner controls, and exit temperature indicator. Usual exit temperatures for the destruction of most organic materials are in the range of 650°C to 825°C with retention times at the elevated temperature of 0.3–0.5 seconds.

Direct flame afterburners are nearly 100% efficient when properly operated. They can be installed for approximately $175–$350 per cubic meter of gas flow. Operating and maintenance costs are essentially those of the auxiliary gas fuel. On larger installations, the overall cost of the afterburner operation may be con-

TABLE 26-4

Comparison of Gaseous Pollutant Removal Systems

Type of equipment	Pressure drop (cm, H_2O)	Installed cost ($ per m³)	Annual operating cost ($ per m³)
Scrubber	10	2.45	3.50
Absorber	10	2.60	7.00
Condenser	2.5	7.00	1.75
Direct flame afterburner	1.2	2.10	2.10 + gas
Catalytic afterburner	2.5	2.90	7.00 + gas

FIG. 26-18. Afterburner with heat recovery. Labeled parts: A, Fume inlet to insulated forced draft fan (11,000 SCFM at 450°F). B, Regenerative shell and tube heat exchanger (55% effective recovery). C, Automatic bypass around heat exchanger for temperature control (required for excess hydrocarbons in fume stream under certain process conditions). D, Fume inlet and burner chamber internally insulated (fume stream raised to 880°F by heat exchanger). E, Combustion chamber, refractory lined for 1500°F duty (operating at 1400°F for required fume oxidation to meet local pollution regulations). F, Discharge stream leaving regenerative heat exchanger at 970°F enters ventilating air heat exchanger for further waste heat recovery. G, Ventilating air fan and filter (11,000 SCFM of outside air at −10°–100°F). H, Automatic bypass with dampers for control of ventilating air temperature. I, Heated air for winter comfort heating requirements leaves at controlled temperature. J, Discharge stack (890°F). K, Combustion safeguard system with dual burner manifold and controls for high turndown. L, Remote control panel with electronic temperature controls (Hirt Combustion Engineers).

siderably reduced by using heat recovery equipment as shown diagrammatically in Figure 26-18. In many industrial situations, boilers or kilns are used as entirely satisfactory afterburners for gases generated in other areas or processes.

Catalytic afterburners are currently used primarily for the control of solvents and organic vapor emissions from industrial ovens. Their use as emission control devices for gasoline powered automobiles has been investigated and this may become by far their greatest use.

The main advantage of the catalytic afterburner is that the destruction of the pollutant gases can be accomplished around 315°–485°C, which results in considerable savings as far as fuel costs. However, the installed costs of the catalytic systems are higher than for the direct flame afterburners because of the expense of the catalyst and associated systems so the overall annual costs tend to balance out.

In most catalytic systems there is a gradual loss of activity due to contamination or attrition of the catalyst so the catalyst must be replaced at regular intervals. Other variables that effect the proper design and operation of catalytic systems involve gas velocities through the system, amount of active catalyst surface,

TABLE 26-5

Possible SO₂ Control Systems

Method	Remarks
Limestone—dolomite injection (dry)	Calcined limestone reacts with sulfur oxides. Then, removal with dry particulate control system.
Limestone—dolomite injection (wet)	Calcined limestone reacts with sulfur oxides. Removal by wet scrubbers.
Alkalized alumina sorption	Sulfur oxides removed by sorption on solid metal oxide. Metal oxides then removed with particulate recovery system and regenerated.
Catalytic oxidation	SO_2 is catalytically oxidized to SO_3 and then scrubbed and recovered as sulfuric acid.
Caustic scrubbing	Caustic neutralizes sulfur oxide compounds. Only in use on small processes.

residence time, and preheat temperature necessary for the complete oxidation of the emitted gases.

E. Comparison of Gaseous Removal Systems

Just as with particulate removal systems, it is apparent that many choices are available for removal of gases from effluent streams. Table 26-4 presents some of the factors that should be considered in making a selection of equipment.

For control of SO_2, for example, several systems are currently in various stages of development (13). Table 26-5 briefly explains these systems.

References

1. Stern, A. C., ed. "Air Pollution," 2nd ed. Academic Press, New York, 1968.
2. "Technical Progress Report, Control of Stationary Sources," Vol. I. Los Angeles County Air Pollution Control District, Los Angeles, California, 1960.
3. Strauss, W. "Industrial Gas Cleaning," 471 pp. Pergamon, Oxford, 1966.
4. Danielson, J. A., ed. "Air Pollution Engineering Manual." U.S. Department of Health, Education, and Welfare, 999-AP-40, Cincinnati, Ohio, 1967.
5. "Control Techniques for Particulate Air Pollutants." U.S. Department of Health, Education, and Welfare, AP-51, Washington, D.C., 1969.
6. "Filtration," American Petroleum Institute, New York, 1959.
7. "Electrostatic Precipitators." American Petroleum Institute, New York, 1958.
8. White, H. J. "Industrial Electrostatic Precipitation," 376 pp. Addison-Wesley, Reading, Massachusetts, 1963.
9. "Gravity, Inertial, Sonic, and Thermal Collectors." American Petroleum Institute, New York, 1959.
10. "Cyclone Dust Collectors." American Petroleum Institute, New York, 1955.
11. "Wet Collectors." American Petroleum Institute, New York, 1961.
12. "Control Techniques for Hydrocarbons and Organic Solvent Emissions from Stationary Sources." U.S. Department of Health, Education, and Welfare, AP-68, Washington, D.C., 1970.

13. "Control Techniques for Sulfur Oxide Air Pollutants." U.S. Department of Health, Education, and Welfare, AP-52, Washington, D.C., 1969.
14. "Control Techniques for Nitrogen Oxide Emissions from Stationary Sources." U.S. Department of Health, Education, and Welfare, AP-67, Washington, D.C. 1970.
15. "Control Techniques for Carbon Monoxide Emissions from Stationary Sources." U.S. Department of Health, Education, and Welfare, AP-65, Washington, D.C., 1970.

Suggested Reading

Lund, H. F., ed. "Industrial Pollution Control Handbook." McGraw-Hill, New York, 1971.
Ridker, R. G. "Economic Costs of Air Pollution," 214 pp. Praeger, New York, 1967.
Weisburd, M. I., ed. "Air Pollution Control Field Operations Manual." U.S.P.H.S. Publication 937, Washington, D.C., 1962.
Wolozin, H., ed. "The Economics of Air Pollution." Norton, New York, 1966.

Questions

1. For a given process at a plant, the cost of control can be related to the equation: dollars for control $= 10,000 + 10e^x$, where x = percent of control $\div 10$. The material collected can be recovered and sold and the income determined from the equation: dollars recovered = (1000) (percent of control).
   (a) At what level of control will the control equipment just pay for itself?
   (b) At what level of control will the dollars recovered per dollars of control equipment be the maximum?
   (c) What would be the net cost charged to the process for increased control from 97.0% to 99.5%.
2. You wish to design a bag house to clean 3000 cubic meters at a filter ratio of 3 m^3/m^2 of cloth. The filter bags are 15 cm in diameter by 3 m long. If you design a "square" bag house, with the bags on 30 cm centers, what would be the exterior dimensions neglecting ductwork? An alternate system uses 15-mm-diameter porous plastic tubes 1 m long on 25-mm centers. For the same filter ratio and flow, what would be the exterior dimensions for a "square" enclosure?
3. For a given cyclone collector, plot centrifugal force as a function of particle specific gravity (0.50–3.00), gas velocity (from 175 m/minute to 1750 m/minute), and radius of curvature (from 30 cm to 250 cm).
4. List the advantages and disadvantages of using a bag house, wet scrubber, or electrostatic precipitator for particulate collection from an asphalt plant drying kiln. The gases are at 250°C and contain 4500 mg/m^3 of rock dust in the 0.1–10 μ size range. Gas flow is 2000 m^3/minute. Consider initial and operation cost, space requirement, ultimate disposal, etc.
5. Suppose a gaseous process effluent of 30 m^3/minute is at 200°C and 50% relative humidity. It is cooled to 65°C by spraying with water that was initially at 20°C. What volume of saturated gas would you have to design for at 65°C? How much water per cubic meter would the system require? How much water per cubic meter would you have to remove from the system?
6. If an electrostatic precipitator is 90% efficient for particulate removal, what overall efficiency would you expect for two of the precipitators in series? Would the cost of the two in series be double the cost of the single precipitator? List two specific cases where you might use two precipitators in series.
7. The gaseous effluent from a process is 30 m^3/minute at 65°C. How much natural gas at 8900 kg cal/m^3 would have to be burned, per hour, to raise the effluent temperature to 820°C? Natural gas requires 10 m^3 of air for every cubic meter of gas at a theoretical air/fuel ratio. Assume the air temperature is 20°C and the radiation and convection losses are 10%.
8. For question 7, if heat recovery equipment were installed to raise the incoming effluent to 425°C, how much natural gas would have to be burned per hour?

Chapter 27

REGULATORY CONTROL

Regulatory control is governmental imposition of limits on emission from sources. In addition to quantitative limits on emissions from chimneys, vents, and stacks, regulations may limit the quantity or quality of fuel or raw material permitted to be used; the design or size of the equipment or process in which it may be used; the height of chimneys, vents, or stacks; the location of sites from which emissions are or are not permitted; or the times when emissions are or are not permitted. Regulations usually also specify acceptable methods of test or measurement.

One instance of an international air pollution control regulation is the cessation of atmospheric testing of nuclear weapons by the United States, the U.S.S.R. and other powers signatory to the cessation agreement. Another is the Trail Smelter Arbitration (1) in which Canada agreed to a regulatory protocol affecting the smelter to prevent flow of its stack emissions into the United States.

National air pollution control regulations in some instances are pre-emptive in that they do not allow subsidiary jurisdictions (states, provinces, counties, towns, boroughs, cities or villages) to adopt different regulations. In other instances, they are not pre-emptive in that they allow subsidiary jurisdictions to adopt regulations that are not less stringent than the national regulations. In the United States, the regulations for mobile sources are an example of the former (there is a provision of a waiver to allow more stringent automobile regulations in California), those for stationary sources are an example of the latter.

426

In many countries, provinces or states have enacted air pollution control regulations. Unless or until superseded by national enactment, these regulations are the ones currently in force. In some cases, municipal air pollution control regulatory enactments are the ones currently in force, and will remain so until superceded by state, provincial, or national laws or regulations.

A regulation may apply to all installations; to new installations only; or to existing installations only. Frequently new installations are required to meet more restrictive limits than existing installations. Regulations which exempt from their application installations made before a specified date are called "grandfather clauses," in that they apply to newer installations but not to older ones. Regulatory enactments sometimes contain time schedules for achieving progressively more restrictive levels of control.

I. Emission Limits

A. Subjective Limits

Limits on stack emissions arc of both the subjective and objective type. Subjective limits are based upon the visual appearance or smell of an emission. Objective limits are based upon physical or chemical measurement of the emission. The most common form of subjective limit is that which regulates the optical density of a stack plume, measured by comparison with a Ringelmann Chart (Fig. 27-1). This form of chart has been in use for over 70 years and is widely accepted for grading the blackness of black or gray smoke emissions. Within the past two decades, it has been used as the basis for "equivalent opacity" regulations for grading the optical density of emissions of colors other than black or gray.

The original Ringelmann Chart was a reflectance chart, in that the observer viewed light reflected from the chart. More recently light transmittance charts have been developed for both black (2) and white (3) gradations of optical density which correlate with the Ringelmann chart scale. It is now common practice in the United States to send air pollution inspectors to a "Smoke School" where they are trained and certified as being able to read the density of black and white plumes to an accuracy that is acceptable for court testimony (4).

Before the widespread acceptance of the Ringelmann Scale, smoke was regulated by prohibiting the emission of black smoke. Now regulatory practice is to prohibit the emission of smoke whose density is greater than a specified Ringelmann number. Over recent decades, there has been a progressive decrease in the Ringelmann number thus specified, culminating in the specification of Number 1 for large new steam power plants in the United States (5). In addition to subjective observation of smoke density, systems have been developed to objectively measure, by means of a photocell, the decrease in intensity of a beam of light projected through the plume prior to its emission (Fig. 27-2).

Subjective evaluation of odor emission is made difficult by the phenomenon of odor fatigue, which means that after a person has been initially subjected to an odor, he loses the ability to perceive the continued presence of low concentrations of

Spacing of Lines on Ringelmann Chart

Ringelmann chart no.	Width of black lines (mm)	Width of white spaces (mm)	Percent black
0	All white		0
1	1	9	20
2	2.3	7.7	40
3	3.7	6.3	60
4	5.5	4.5	80
5	All black		100

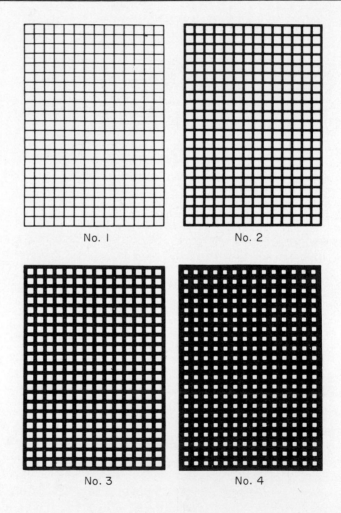

FIG. 27-1. Ringelmann smoke chart.

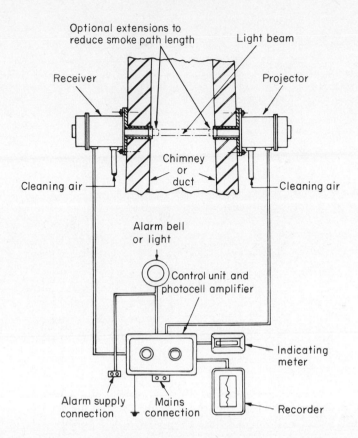

FIG. 27-2. Smoke-density meter. Reproduced from BS 2811 Smoke Density Indicators and Recorders, 1969, by permission of the British Standards Institution, 2 Park Street, London W1A.

that odor. Therefore all systems of subjective odor evaluation rely upon preventing olfactory fatigue of the observer by letting him breathe odor-free air for a sufficient time prior to his breathing the odorous air and evaluating its odor content. Usually an activated charcoal bed is used to clean up the air to provide the odor-free air required by the observer (Fig. 27-3) (see also Chapter 12).

B. Objective Limits

There are two major categories of objective emission limits; those which limit the emission of a specific pollutant regardless of the process or equipment from which it is emitted; and those which limit the emissions of a specific pollutant from a specific process or type of equipment. Regulations may require the same emission limit for all sources, regardless of size or capacity; or they may vary the allowable emission with size or capacity of the source. Limits may be stated in absolute terms, i.e., not more than a specified mass of pollutant per unit of time; or in relative terms, i.e., not more than a specified mass of pollutant per unit mass of fuel burned; material being processed; or product produced; or per unit of heat released in a furnace. In

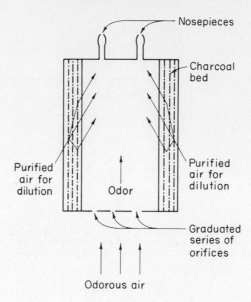

Fig. 27-3. Scentometer. Odorous air passes through graduated orifices and is mixed with air from the same source, which is purified by passing through charcoal beds. Dilution rates are fixed by orifice selection.

the case of gaseous pollutants, limits may be stated in volumetric rather than gravimetric terms. Emission limits are sometimes stated as mass of pollutant per unit volume of effluent. Effluent volume varies with gas temperature and pressure, the presence or absence of diluting air, and the amount thereof. Therefore volumes must be reduced to a specified temperature, pressure and percent of diluting air. In the case of flue gases from fuel combustion dilution is usually expressed as percent excess air or percent of carbon dioxide in the flue gas. As an example of how such limits are stated, Table 27-1 gives the limits adopted by the United States for new installations of large fuel-fired steam power plant boiler furnaces, cement kilns, municipal incinerators, sulfuric acid plants and nitric acid plants (5).

The most common rationale for developing emission limits for stationary sources is the application of best practicable means for control. Under this rationale, the degree of emission limitation achievable at the best designed and operated installation in a category sets the emission limits for all other installations of that category. As new technology is developed, what was best practicable means in 1970 may be short of the best attainable in 1980. Since there is thus a moving target, means must be provided administratively to allow the plant which compiled with a best practicable means limit to be considered in compliance for a reasonable number of years thereafter. It is important to note that best practicable means limits are set without regard to present or background air quality or the air-quality standards for the pollutants involved, the number and location of sources affected by the limit and the meteorology or topography of the area in which they are located. However, an additional limit on minimum stack height or buffer zone, based upon the above

TABLE 27-1

United States National Standards for Maximum 2-Hour Emission from New Sources (5)

Particulate matter

Steam power plants	0.1 lb/million BTU fired
Portland cement plants	
Kiln	0.3 lb/ton feed
Clinker cooler	0.1 lb/ton feed
Municipal incinerators	0.08 grain/scf at 12% CO_2
Sulfuric acid plants	0.15 lb/ton acid (as H_2SO_4)

Sulfur dioxide

Steam power plants	
Liquid fuel	0.8 lb/million BTU fired
Solid fuel	1.2 lb/million BTU fired
Sulfuric acid plants	4.0 lb/ton acid

Nitrogen oxides (as NO_2)

Steam power plants	
Gaseous fuel	0.2 lb/million BTU fired
Liquid fuel	0.3 lb/million BTU fired
Solid fuel	0.7 lb/million BTU fired
Nitric acid plants	3.0 lb/ton acid

noted factors, may be coupled with a best practicable means limit on mass of emission.

The other major rationales for developing emission limits have been based upon some or all of the above noted factors—air quality, air-quality standards, number and location of sources, meteorology, and topography. These include the roll-back approach which involves all these factors except source location, meteorology, and topography; and the single-source mathematical modeling approach which con-

TABLE 27-2

Emission Limits Adopted by the United States for New Light-Duty Automobiles (gm/vehicle mile)

	Vehicle (model year applicable)				
Pollutant emission	1972	1973	1974	1975	1976
Carbon monoxide	39	39	39	3.4	3.4
Hydrocarbons[a]	3.4	3.4	3.4	0.41	0.41
Oxides of Nitrogen	—	3	3	3	0.4

[a] Crankcase emissions: No crankcase emissions shall be discharges into the ambient atmosphere. Fuel Evaporative Emissions: 2 gm per test procedure for vehicles beginning with 1972.

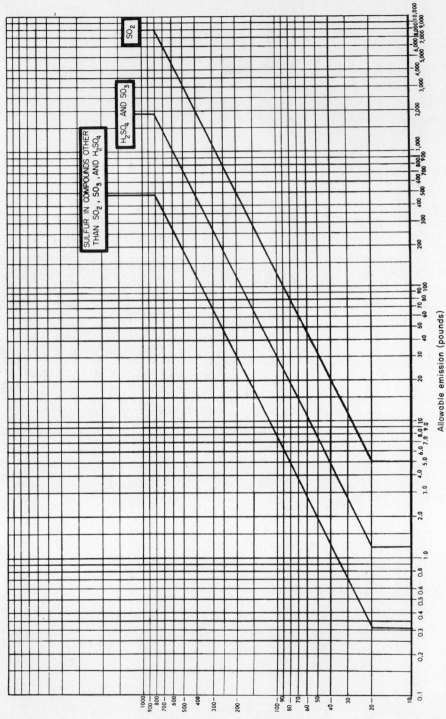

FIG. 27-4. Allowable emissions for sulfur compounds—New Jersey.

siders only the air quality standard and meteorology, ignoring the other factors listed. In the decade of the 1970's emission limits based upon a multiple source mathematical modeling approach involving all these factors will come to the fore.

An example of a set of emission limits based upon the roll-back approach are the limits adopted by the United States for carbon monoxide, hydrocarbons and oxides of nitrogen emissions from new automobiles (6) (Table 27-2). An example of an emission regulation based upon the single-source mathematical modeling approach is that for sulfur compounds in New Jersey (Fig. 27-4).

II. Control of New Installations

In theory, if starting at any date, all new sources of air pollution are adequately limited, all sources installed prior to that date eventually will disappear, leaving only those adequately controlled. The weakness of relying solely on new installation control to achieve community air pollution control is that installations deteriorate in control performance with age and use; and that the number of new installations may increase to the extent that what was considered adequate limitation at the time of earlier installations may not prove adequate in the light of these increased numbers. Although, in theory, old sources will disappear, in practice they take such a long time disappearing, that it may be difficult to achieve satisfactory air quality solely by new installation control.

One way of achieving new installation control is to build and test prototype installations and only to allow the use of replicates of an approved prototype. This is the method used in the United States for the control of emissions from new automobiles (see Table 27-2 above).

Another path is followed where installations are not replicates of a prototype, but rather are each unique, e.g., cement plants. Here two approaches are possible. In the one, the owner assumes full responsibility for compliance with regulatory emission limits. If at testing the installation does not comply, it is the owner's responsibility to rebuild or modify it until it does comply. In the alternate approach, the owner makes such changes to the design of his proposed installation as the regulatory agency requires after inspection of his plans and specifications. The installation is then deemed in compliance, if when completed, it conforms to the approved plans and specifications. Some regulatory agencies require both testing and approval. In this case, the testing is definative and the plan filing is intended to prevent less sophisticated owners from investing in installations they later would have to rebuild or modify.

In following the approved replicate route, the regulatory agency has the responsibility to randomly sample and test replicates to assure conformance with the prototype. In following the owner responsibility route, the agency has the responsibility of locating owners, particularly those who are not aware of the regulatory requirements applicable to them, and then of testing their installations. Locating owners is a formidable task requiring good information communication between governmental agencies concerned with air pollution control and those which perform inspections of buildings, factories and commercial establishments for other purposes.

Testing installations for emissions is also a formidable task and is discussed in detail in Chapter 25. Plan examination requires a staff of well qualified plan examiners.

Because of the above requirements, it takes a sophisticated organization to do an effective job of new installation control. Prototype testing is best handled at the national level; installation testing and plan examination at the state, provincial or regional level. Location of owners is a local operation with respect to residential and commercial structures, but may be a state or provincial operation for industrial sources.

III. Control of Existing Sources

Existing installations may be controlled on a retrospective, a present or a prospective basis. The retrospective basis relies upon the receipt of complaints from the public and their subsequent investigation and control. It follows the theory that the squeaking wheel gets the grease. Complaint investigation and control can become the sole function of a small air pollution control organization to the detriment of its overall air pollution control effectiveness. Fire-fighting activities of this type may improve an agency's short-term image by appeasing the more vocal elements of the public, but this may be accomplished at the expense of the agency's ability to achieve long range goals.

On a present basis, an agency's staff can be used according to plan to investigate and bring under control selected source categories, selected geographic areas, or both. The selection is done, not on the basis of the number or intensity of complaints, but rather on an analysis of air quality data, emission inventory data, or both. An agency's field activity may be directed to the enforcement of existing regulatory limits, or the development of data from which new regulatory limits may be set and later enforced.

On a prospective basis, an agency can project its source composition and location and their emissions into the future and by the use of mathematical models and statistical techniques attempt to determine what control steps have to be taken now to establish future air-quality levels. Since the future involves a mix of existing and new sources, decisions must be made as to the control levels required for both categories, and whether these levels should be the same or different.

Regulatory control on a complaint basis requires the least sophisticated staffing and is well within the capability of a local agency. Operation on a present basis requires more planning expertise and a larger organization and therefore is better adapted to a regional agency. To operate on a prospective basis an agency needs a still higher level of planning expertise, such as may be available only at the state or provincial level. To be most effective, an air pollution agency must operate on all three bases, retrospective, present, and prospective, because no agency can afford to ignore complaints, to neglect special source or area situations, nor to fail to do long-range planning.

IV. Control of Mobile Sources

Mobile sources include railroad locomotives, marine vessels, aircraft and automotive vehicles. There is over a hundred years of experience in regulating smoke and cinder emission from steam locomotives and steam operated marine craft. Methods of combustion equipment improvement, fireman training and smoke inspection for these purposes are well documented (7). Cooperation of the railroads and steamship lines must be obtained, voluntarily if possible; by punitive measures, if necessary. Best results are obtained when the railroads and steamship lines hire their own smoke inspectors and train their own fireman. However, the control agency must have its own inspectors to make the operation effective. This type of control is best at the local level.

Regulation of aircraft engine emissions has been made a national level responsibility by law in the United States. The Administrator of the Environmental Protection Agency is responsible for emission limits of aircraft engines, while the Secretary of Transportation is required to prescribe regulations to insure compliance with these limits.

In the United States, regulation of emissions from new automotive vehicles has followed the prototype-replicate route already discussed in Section II of this chapter. The direction which will be taken for the regulation of used car emissions during the 1970's is still unclear. The options are essentially inspection or no inspection. The argument for no inspection, which is, in effect, no regulation of used cars, is that to be acceptable as prototypes, cars must meet emission limits after 50,000 miles of driving. The argument for routine annual or semiannual inspection is that cars should be routinely inspected for safety (brakes, lights, steering and tires) and that the additional time and cost to check the car's emission during the same inspection will be minimal. There is no doubt but that such an inspection will pinpoint cars whose emission control system has been removed, altered, damaged or deteriorated, and force such defects to be remedied. The question is whether the improvement in air quality that would result from correcting these defects would be worth the cost to the public of maintaining the inspection system. Another way of putting the question is whether the same money would be better invested in making the prototype test requirements more rigid with respect to the durability of the emission control system (with the extra cost added to the cost of the new car) than in setting up and operating an inspection system for automotive emissions from used cars. A final question in this regard is whether or not the factor of safety included in the new car emission standards is sufficient to allow a percentage of all cars on the road to exceed the emission standards without jeopardizing the attainment of the air-quality standard. These questions have yet to be answered.

V. Air-Quality Control Regions

Workers in the field of water resources are accustomed to thinking in terms of watersheds and watershed management. It was these people who introduced the term airshed to describe the geographic area requiring unified management for achieving air pollution control. The term airshed was not well received because its physical connotation is wrong. It was next followed by the term air basin, which was closer to the mark but still had the wrong physical connotation, since unified air pollution control management is needed in flat land devoid of valleys and basins. The terminology that finally evolved was the air-quality control region, meaning that geographical area including the sources significant to production of air pollution in an urbanized area and the receptors significantly affected thereby. If long averaging time isopleths (i.e., lines of equal pollution concentration) of a pollutant, such as suspended particulate matter are drawn on the map of an area, there will be an isopleth that is at essentially the same concentration as the background concentration [Fig. 4-2 ($40\mu/g/m^3$ isopleth)]. The area within this isopleth meets the description of an air-quality control region.

For administrative purposes it is desirable that the boundaries of an air-quality control region be the same as those of major political jurisdictions. Therefore when the first air-quality control regions were officially designated in the United States by publishing their boundaries in the Federal Register, the boundaries given were those of the counties all or part of which were within the background concentration isopleth.

When about a hundred such regions were designated in the United States, it was apparent that only a small portion of the land area of the country was in officially designated regions. For uniformity of administration of national air pollution legislation, it became desirable to include all the land area of the nation in designated air-quality control regions. The Clean Air Amendments of 1970 therefore gave the states the option of having each state considered an air-quality control region or of breaking the state into smaller air-quality control regions mutually agreeable to the State and the U.S. Environmental Protection Agency. The regions thus created need bear no relation to concentration isopleths, but rather represent contiguous counties which form convenient administrative units. Therefore for purposes of air pollution control, the United States is now a mosaic of multicounty units, all called air-quality control regions, some of which were formed by drawing background concentration isopleths and others of which were formed for administrative convenience. Some of the former group are interstate, in that they include counties in more than one state. All of the latter are intrastate.

References

1. Dean, R. S., and Swain, R. E. Report Submitted to the Trail Smelter Arbitral Tribunal, U.S. Department of the Interior, Bureau of Mines Bulletin 453, U.S. Government Printing Office, Washington, (1944), 304 pp.
2. Rose, A. H., Jr., and Nader, J. S. Field evaluation of an improved smoke inspection guide, *J. Air Pollut. Contr. Assoc.* **8,** 117–119 (1958).
3. Connor, W. D., Smith, C. F., and Nader, J. S. Development of a smoke guide for evaluation of white plumes. *J. Air Pollut. Contr. Assoc.* **18,** 748–750 (1968).
4. "Air Pollution Control Field Operations Manual—A Guide for Inspection and Enforcement" (compiled and edited by M. I. Weisburd) U.S. Public Health Service Publication 937. Washington, D.C., 1962.

5. "Standards of Performance for New Stationary Sources," Environmental Protection Agency, Federal Register. Thursday, December 23, 1971, Washington, D.C. Volume 36, Number 247. Part II.
6. "Exhaust Emission Standards and Test Procedures," Environmental Protection Agency, Federal Register, Friday, July 2, 1971, Washington, D.C., Volume 36, Number 128, Part II.
7. "Railroad Locomotives" *in* "Air Pollution—A Bibliography" (S. J. Davenport and G. C. Morgis, eds.), pp. 57–71. Bureau of Mines Bulletin 537, U.S. Government Printing Office, Washington, D.C., 1954.

Suggested Reading

Lund, H. F., ed. "Industrial Pollution Control Handbook." McGraw-Hill, New York, 1971.
Strauss, W., ed. "Air Pollution Control," Part I, 451 pp. Wiley, New York, 1971.

Questions

1. What are the geographic boundaries of the Air Quality region (or its equivalent, if you are not in the United States) in which you reside?
2. Are there regulatory limits in the jurisdiction where you reside which are different from your national air quality or emission limits? If so, what are they?
3. Discuss the application of the prototype testing—replicate approval approach to stationary air pollution sources.
4. Figure 27-2 shows a relatively simple light-beam device. Investigate and report on the availability of more sophisticated devices to accomplish this purpose, which will maintain their calibration over a longer period of time.
5. There has been substantial objection, on the part of certain segments of industry, to the use of "equivalent opacity" regulations. Discuss the nature and validity of these objections.
6. One form of air pollution control regulation limits the pollution concentration at the owner's "fence line". Find an example of this type of regulation and discuss its pros and cons.
7. How are pollutant emissions to the air from used cars (automobiles) regarded where you reside? Discuss the merits of this extent of regulation.
8. Limitation of visible emission was the original form of air pollution control a century ago. Has it outlived its usefulness? Discuss this question.
9. Discuss the use of data telemetered to the office of the air pollution control agency from automatic instruments measuring ambient air quality and automatic instruments measuring pollutant emissions to the atmosphere as air pollution control regulatory means.

Chapter 28

ORGANIZATION FOR CONTROL

The best organizational pattern for an air pollution control agency is that which most effectively and efficiently performs all its functions. There are many functions a control agency or industrial organization could conceivably perform. The desired budget and staff for the agency or organization is determined by listing the costs to perform all desired functions. The actual functions performed by the agency or organization is determined by limitations on staff, facilities and services imposed by its actual budget.

I. Functions

The most elementary function of an air pollution control agency is its *control* function, which breaks down into two subsidiary functions: *enforcement* of the jurisdictions air pollution control laws, ordinances, and regulations, and *evaluation* of the effectiveness of existing regulations and regulatory practices and the need for new ones.

The enforcement function may be subdivided several ways. One is into control of *new sources* and *existing sources*. New source control can involve all or some of the following functions.

1. *Registration* of new sources
2. *Filing* of plans and specifications
3. *Review* of plans and specifications
4. *Issuance* of certificates of approval for construction
5. *Inspection* of construction
6. *Testing* of installation
7. *Issuance* of certificates of approval for operation
8. *Receipt* of required fees for the above services
9. *Appeal and variance* hearings and actions
10. *Prosecution* of violations

Control of existing sources can involve all or some of the following functions.

1. *Visible emission* inspection
2. *Complaint* investigation
3. *Periodic* inspection
4. *Special* industrial category or geographic area inspection
5. *Fuel* and fuel dealer inspection and testing
6. *Testing* of installations
7. *Renewal* of certificates of operation
8. *Receipt* of required fees for above services
9. *Appeal and variance* hearings and actions
10. *Prosecution* of violations

It will be noted that items 6 through 10 of the above two lists are the same.

The evaluation function may be subdivided into *retrospective* evaluation of existing regulations and practices and *prospective* planning for new ones. Retrospective evaluation involves

1. *Air-quality monitoring* and surveillance
2. *Emission inventory*
3. *Statistical analysis* of air-quality and emission data and of agency activities
4. *Analytic evaluation*
5. *Recommendation* of required regulations and regulatory practices

Prospective planning involves

1. *Prediction* of future trend
2. *Mathematical modeling*
3. *Analytical evaluation*
4. *Recommendation* of required regulations and regulatory practices

Meteorological services are closely related to both retrospective evaluation and prospective planning.

The last two items on both the above lists are the same.

Functions such as those already noted require extensive *technical* and *administrative* support. The *technical* support functions required include

1. *Technical* information services—library, technical publication, etc.
2. *Training* services—technical
3. *Laboratory* services—analytical instrumentation, etc.
4. *Computer* services
5. *Shop* services

An agency requires, either within its own organization or readily available from other organizations, provision of the following *administrative* support functions.

1. *Personnel*
2. *Procurement*
3. *Budget*—finance and accounting
4. *General* services—secretarial, clerical, reproduction, mail, phone, building maintenance, etc.

Since an air pollution control agency must maintain its *extramural* relationships, these functions must be provided and include

1. *Public relations* and information
2. *Public education*
3. *Liason* with other agencies
4. *Publication* distribution

An agency needs *legal* services for prosecution of violations, appeal and variance hearings, and the drafting of regulations. In most public and private organizations these services are provided by lawyers based in organizational entities outside the agency. However, organizationally, it is preferable that the legal function be provided within the agency organization.

One category that has been excluded from all the above lists is Research and Development. R and D is not a necessary function for an air pollution control agency at the state, provincial, regional or municipal level. It is sufficient if the national agency, e.g., the Environmental Protection Agency in the United States, maintains an R and D program sufficient for the nation's needs and encourages each industry to undertake the R and D required to solve its particular problems. National agencies tend to give highest priority to problems common to a number of areas in the country. However, where major problems of a state, province, region or municipality are unique to its locality, and not likely to have a high national priority, the area may have to undertake the required R and D.

It is apparent that some of the functions listed overlap. An example is the overlap of the laboratory function in technical support and the testing and air-quality monitoring functions in control. It is because of such overlaps that different organizational structures arise.

Many of the functions listed for a control agency are not required in an industrial air pollution control organization. A control agency must emphasize its enforcement function, sometimes at the expense of its evaluation function. The converse is the case for an industrial organization. Also, in an industrial organization, the technical and administrative support, legal and research and development functions are likely to be supplied by the parent organization.

II. Organization

In the foregoing section functions were grouped into categories. The most logical way to organize an air pollution control or industrial organization is along these categorical lines (Figs. 28-1 and 28-2), deleting from the organizational structure those functions and categories with which the agency or organization is presently not concerned. When budget and staff are small one person is required to cover all the agency's or organization's activities in more than one function; and in very small agencies or organizations, in more than one category.

Agencies and organizations which cover a large geographic area are faced with the problem of the extent to either centralize or decentralize functions and categories. Centralization consolidates the agency's or organization's technical expertise facilitating the resolution of technical matters particularly as they relate to large or complex sources. Decentralization facilitates the agency's ability to deal with a larger number of smaller sources. The ultimate decentralization is delegation of certain of an agency's functions to lesser jurisdictions. This can be lead to a three-tiered structure, with certain functions fully centralized; certain ones delegated; and certain ones decentralized to regional offices or laboratories.

In an industrial organization the choice is between centralization in corporate or company headquarters or decentralization to the operating organizations or individual plants. The usual organization is a combination of headquarters centralization and company decentralization.

A major organizational consideration is where to place an air pollution control agency in the hierarchy of government. As state, provincial, and municipal governmental structure evolved during the 19th century, smoke abatement became a function of the departments concerned with buildings and with boilers. In this century, until the 1960's, air pollution control shifted strongly to national, state, provincial, and municipal health agencies. However, since the mid-1960's there has been an increasing tendency to locate air pollution control in agencies concerned with natural resources and the environment.

III. Finance

A. Fines and Fees

Governmental air pollution control agencies are primarily tax supported. Most agencies charge fees for certain services, such as plan filing, certificate issuance, inspection and tests. In some agencies, these fees provide a substantial fraction of the agency's budget. In general, this is not a desirable situation because when the

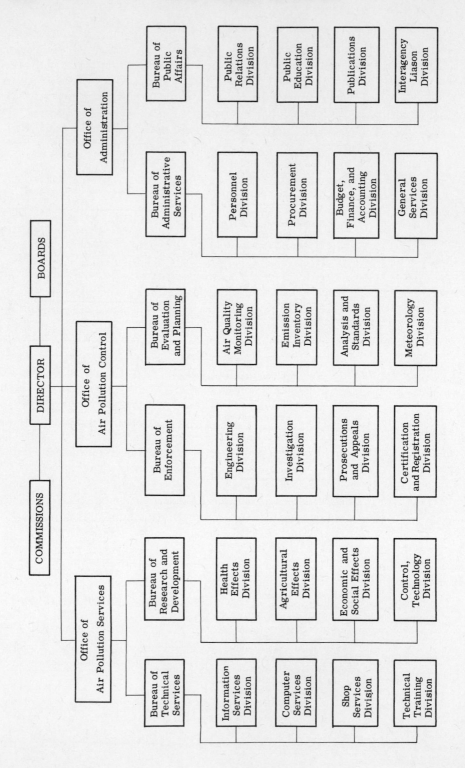

FIG. 28-1. Organization plan which encompasses all functions likely to be required of a governmental air pollution control agency.

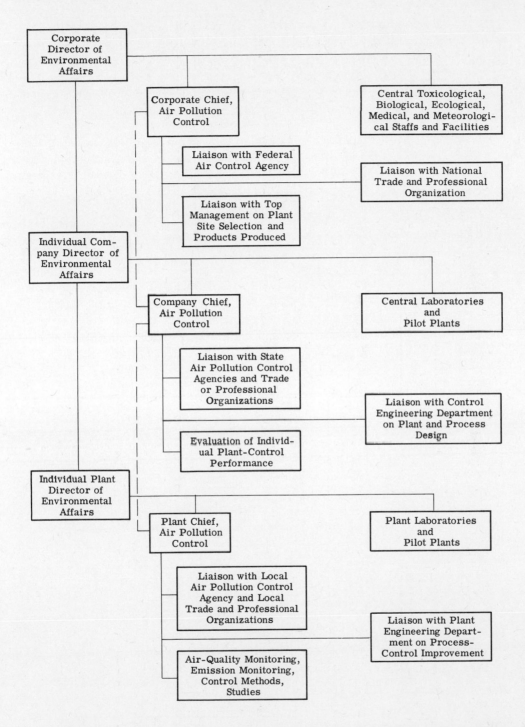

Fɪɢ. 28-2. Organization plan which encompasses all functions to be required of an industrial air pollution organization.

continuity of employment of an agency's staff depends upon the amount of fees collected, there is an understandable tendency to concentrate effort on fee producing activities and resist their deemphasis. This makes it difficult for an agency to change direction with changes in program priorities. Even when these fees go into the general treasury, rather than being retained by the agency, such collection gives some leverage to the agency in its budget request and has the same effect as if the fees were retained by the agency.

In most jurisdictions violators of air pollution control laws and regulations are subject to fines. In the past, it has been considered good administrative practice for these fines to go into the general treasury, to discourage imposition of fines as a means to support the agency's program, which might result if the agency retained the fines. Collection of fines could become an end in itself, if the continuity of employment of the agency staff were dependent upon it. Moreover, as in the case of fees, an agency can use its fine collection to give it leverage in budget allocation.

There has been extensive recent rethinking of the role of fees and fines as means to influence industrial decision-making with regard to investment in pollution control equipment and pollution-free processes. In their new roles, fees and fines take the form of tax write-offs and credits for pollution control investment; taxes on the sulfur and lead content of fuels; continuing fines based on pollution emission rate; and effluent fees on the same basis. Tax write-offs and credits tend to be resisted by treasury officials since they diminish tax income. Air pollution control agencies tend to look with favor on such write-offs and credits since they result in air pollution control with minimal effort on the part of their staffs and with minimal effect on their budget.

One problem with fuel taxes, continuing fines, and effluent fees is how to use the funds collected most effectively. Among the ideas which have developed are to use the funds as the basis for loans or subsidies for air pollution control installations by industrial or domestic sources; and for financing air pollution control agency needs of jurisdictions subsidiary to those which collect and administer the fines or fees. By ensuring that the continuity of employment of agency personnel is divorced from the fine and fee setting, collection and administrative process, it is argued that the use of fines and fees can be a constructive rather than a stultifying process.

B. Budget

An agency's budget consists of the following categories.

> Personnel
>   Salaries
>   Fringe benefits
>   Consultants fees
> Space, furniture and office equipment
>   Acquisition or rental
>   Operation
>   Maintenance
> Technical and laboratory equipment
>   Acquisition or rental
>   Operation
>   Maintenance

Transportation equipment
   Acquisition
   Operation
   Maintenance
Travel and sustenance
Supplies
Reproduction services
Communication services
Contractual services
Other—miscellaneous

The principal precaution is not to have too high a percentage of an agency's budget in the personnel category. When this happens, the ability of the agency staff to perform its functions is unduly restricted.

In an industrial organization, the decision has to be made as to whether to charge-off air pollution control costs as a corporate charge, so that the plant manager does not include them in his accounts; or the reverse, so that the plant manager must show a profit for his plant as a cost center, after including costs of air pollution control.

IV. Personnel

The personnel needs of an agency or industrial organization can be assessed qualitatively and quantitatively (1). Qualitatively, the professional categories required by an agency or organization staff are determined by the functions included in their scope of work. Table 28-1 lists the types of professional training usually specified for persons to exercise the principal functions. For smaller agencies or organizations where one person handles several functions, or categories, the type of person trained by institutions with curricula in air pollution, or by internship in a large air pollution control agency or organization, is specially valuable.

Quantitative personnel needs are determined by the answer to either How many people should there be ideally to perform all functions? or Within specific budget restrictions, how should personnel be allocated among the several functions? There have been several studies of manpower needs in air pollution control from which Tables 28-2 and 28-3 have been developed to help answer these two questions.

Training of personnel is of two types: to perform their assigned job and to allow them to be promoted to a higher job classification. The means to accomplish such training also follows two paths—on-the-job (in-service or intramural) training; and extramural training, which ranges from intensive certificate-granting short courses to extensive training in degree-granting graduate programs. In the United States, the principal organization offering short courses is the Office of Manpower Development, Office of Air Programs, Environmental Protection Agency (2).

V. Advisory Groups

It is common practice, at least in the United States, that air pollution control agencies have associated with them nonsalaried groups who meet from time to

TABLE 28-1

Professional Categories Required in Air Pollution Control Agencies

Control agency function	Professional category of persons usually responsible for functions
Administrative services	Business managers
Air-quality monitoring	Chemists
Analytic evaluation	Public administrators
Budget, finance, accounting	Accountants
Certificate issuance and renewal	Business managers
Computer services	Computer scientists
Construction inspection	Engineers
Education	Educators
Emission inventory	Engineers
Fee receipt	Accountants
Hearings and prosecution	Lawyers
Inspection (supervision)	Engineers
Inspection (field force)	Sanitarians
Installation testing	Chemists; engineers
Investigation	Chemists; engineers
Laboratory services	Chemists
Legal	Attorneys
Liason	Public administrators
Mathematical modeling	Meteorologists
Meteorological services	Meteorologists
New source registration and filing	Business managers
Personnel	Business managers
Plan review	Engineers
Procurement	Business managers
Public relations	Public relation specialists
Regulation recommendation	Public administrators
Research and development	
Agricultural effects	Biological scientists
Atmospheric effects	Physical scientists
Control technology	Engineers
Economic and social effects	Social scientists
Health effects	Medical scientists
Shop services	Technicians
Statistical analysis	Statisticians
Technical information	Librarians
Training services	Educators
Trend prediction	Statisticians

time in capacities ranging from official to advisory. The members of such groups may be paid fees or expenses for their days of service, or they may contribute their time and expenses without cost to the agency. The members may be appointed variously by elected officials, legislative bodies, department heads, program directors, or directors of program categories. They may have to be sworn into office and the records of their meetings may be official documents; or their appointments and meetings may be quite informal.

TABLE 28-2

Manpower Levels in U. S. Air Pollution Control Agencies (1970)

Manpower category	Total number of persons employed[a]				Percent of agencies employing persons in category				Median number of persons per agency utilizing category[b]			
	State and territorial agencies	County and regional agencies	City agencies	All agencies	State and territorial agencies	County and regional agencies	City agencies	All agencies	State and territorial agencies	County and regional agencies	City agencies	All agencies
Administrators	157	246	165	568	96	99	100	98	2	2	2	2
Attorneys	20	32	20	72	57	41	45	46	1c	1c	1c	1c
Chemists	132	125	90	347	93	57	58	65	2	1	1	1
Engineers	425	237	177	839	93	50	58	61	4	2	1	2
Meteorologists	17	14	7	38	32	9	12	15	1	1	1	1
Sanitarians	110	195	97	402	30	54	41	45	4	2	1	2
(Inspectors)	60	373	368	801	30	44	64	47	4	3	3	3
Specialists–data processing	14	22	26	62	28	16	12	17	1	1	1	1
Specialists–public information	18	29	13	60	42	27	19	27	1	1	1c	1
Technicians	272	172	114	558	68	51	39	50	5	2	1	3
Total	1225	1445	1077	3747								
Number of agencies	53	121	83	257								
Average number of persons/agency	23	11	13	15								
Median number of persons/agency	16	7	6	8								

[a] Equivalent full-time employees – a part-time employee counted as $\frac{1}{2}$ person.

[b] This is not the median number of persons in all agencies, since some agencies do not utilize category.

[c] Part-time.

TABLE 28-3

Manpower Levels in Ideal Governmental Air Pollution Control Agencies[a]

Manpower category	Professional category	Positions per million population
Administrators	Accountants	1
	Business managers	4
	Clerks	4
	Public administrators	2
Attorneys	Attorneys	1
Chemists	Chemists	2
Engineers	Engineers	8
Meteorologists	Meteorologists	1
Sanitarians[b]	Sanitarians	8
Specialists (Data processing)	Computer scientists	1
	Statisticians	1
Specialists (Public information)	Educators	1
	Librarians	1
	Public relations specialists	1
Technicians	Technicians	4
Total		40

[a] Excluding Research and Development.
[b] Including inspectors.

In general, when such bodies are official, there is a requirement of law or ordinance which sets a statutory base for their creation. This base may be as specific as the statutory requirement that a specified group be appointed in a specified manner with specified duties and authorities, or as broad as a general statutory statement that advisory groups may be created, without specifying any details of their creation or duties. Such groups may provide their own secretariat and have their own budget authority; or be completely dependent upon the agency for services and costs.

The highest level of such groups are the Commissions or Boards, which among possible authorities, can promulgate standards and regulations, establish policy, issue variances and award funds. The next highest level are the Hearings or Appeals Boards which can issue variances, but which neither promulgate standards or regulations nor set policy. Next in rank are those Boards, Panels or Committees which formally recommend standards or regulations but which do not have authority to promulgate them. Next in line are formally organized groups which review requests for funds, and make recommendations; but lack final award authority. These latter groups are particularly prevalent in federal programs, such as that of the United States, where large amounts of money are awarded as grants, fellowships, or contracts. These review groups tend to be specialized, e.g., a group of cardiologists to review requests for funds to do cardiovascular research. Although such groups do not have final award authority, the award process tends to follow their recommendations, thereby making their real authority greater than their statutory authority.

At the lowest level are groups that review program content and make recommendations. The impact of groups in this category depends upon whether their reports and recommendations are or are not published, and the ground rules for determining whether or not they shall be published. If the groups are creations of the agency whose programs they review, and the decision as to whether or not to publish rests with the agency, minimal impact can be expected. Where review groups are relatively independent of the agency whose programs they review and have independent authority and means for publication, maximum impact can be expected. An example of the latter is the statutory requirement in Section 202 of the U.S. Clean Air Act Amendments of 1970, that automobile exhaust standards be reviewed by the National Academy of Sciences.

References

1. "Manpower and Training Needs for Air Pollution Control." Report of the Secretary of Health, Education and Welfare to the Congress of the United States, June 1970. Senate Document No. 91-98, 91st Congress, 2nd Session, U.S. Government Printing Office. Washington, D.C. (44 pp).
2. Institute for Air Pollution Training, Research Triangle Park, North Carolina 27711.

Suggested Reading

Davies, J. C., III. "The Politics of Pollution," 231 pp. Pegasus, New York, 1970.
Sax, J. L. "Defending the Environment: A Strategy for Citizen Action," 252 pp. Knopf, New York, 1971.

Questions

1. How is the Air Pollution Control Agency in your state or province organized? Where is its principal office? Who is its head?
2. Is there a local air pollution control agency in your city, county or region? How is it organized? Where is its principal office? Who is its head?
3. What is the function and role of the pilot plants shown in Fig. 28-2?
4. Draw an organization chart for a governmental air pollution control organization which is limited by budget to ten persons, including clerical and secretarial staff.
5. Write a job specification for the chief of the Bureau of Enforcement (see Fig. 28-1).
6. Discuss the relative roles of the staff of an air pollution control agency and its advisory board in the development and promulgation of air quality and emission standards.
7. Discuss the alternatives mentioned in last paragraph of Section III regarding allocation of costs of air pollution control in an industrial organization.
8. Prepare a brief report on the feasibility of employing effluent fees in air pollution control.
9. How does an air pollution control agency organize to insure that its registration of new sources does not miss significant new sources?

Appendix A

RELATIVE SUSCEPTIBILITY OF PLANTS TO VARIOUS AIR POLLUTANTS

Relative Susceptibility of Plants to Various Air Pollutants[a,b]

Plant	Sulfur dioxide	Ozone	Fluoride	PAN	Ethylene	Oxidant	Nitrogen dioxide	2, 4 D	Chlorine	Ammonia	Hydrogen chloride	Mercury vapor	Hydrogen sulfide
Alder		S											
Alfalfa	S	I	R	I		S			S				
Aloe												R	
Apple	I	S	I					S	S	R			R
Apricot	I		S										
Arborvitae			I		I								
Ash			R					R					
Asparagus			R				R						
Aspen		S	I										
Aster	I		I					I					S
Azalea			I	R	I		S		I			I	
Bachelor's button	S												
Barberry			I										
Barley	S	I	I	I									
Basswood												I	
Bean	S	S	R	S	S	S	S	R	I			S	S
Beech											R		
Beet	S	S	I	I	R	S					S		
Begonia	I		I	R					R			I	I
Bidens			R										
Bindweed	S												
Birch	S		R					S			R		
Blackberry			R						S				
Black mustard	I												
Blueberry			S										
Blue grass	I			S			I		I	I			I
Boxelder	R	S	S					S	S				
Bridal wreath		S	R										
Brittlewood							S						
Broccoli	S		R	R									
Brussels sprout	S		R										
Buckwheat	S		I						S	I			S
Burdock	I		R										
Butterfly weed												S	
Cabbage	I	R	R		R			R					
Calliopis													S
Camellia			R									I	
Canterbury bells			R										

Plant	Sulfur dioxide	Ozone	Fluoride	PAN	Ethylene	Oxidant	Nitrogen dioxide	2, 4 D	Chlorine	Ammonia	Hydrogen chloride	Mercury vapor	Hydrogen sulfide
Cantaloupe		R											
Careless weed	S												
Carissa			R				R						
Carnation		S			S								R
Carrot	S		I	I	I								
Caster bean													I
Catalpa	S	S											
Catnip	I												
Cattail			S										
Cauliflower	I		R										
Cedar								I					
Celery		S	R										
Cheeseweed				I			I		I	I			R
Chenopodium murale		S	I										
Cherry	I		I					I	I		S	R	R
Chestnut								S	S				
Chickweed		S	I	S			I		S	R			I
Chicory	S												
Chrysanthemum	R	S	R	R									
Cinquefoil												S	
Citrus			I			S	I		I				
Clover	S		I		R	S							S
Cockleburr	I	I											
Coleus									S	I			R
Columbine			R									I	
Corn (sweet)	R	S	S	R				I	S				
Cornflower													I
Cosmos	S								S			I	S
Cotoneaster												I	
Cotton	S		R	R	S								
Cowpea					S				I				
Crabgrass			I										
Croton							R					R	
Cucumber		R	R	R	S				I				S
Curly dock	S												
Currant		R											
Dahlia			I	S					I				
Dandelion	I		R				I		I	R			I
Dock		I	R	I									

Plant	Sulfur dioxide	Ozone	Fluoride	PAN	Ethylene	Oxidant	Nitrogen dioxide	2, 4 D	Chlorine	Ammonia	Hydrogen chloride	Mercury vapor	Hydrogen sulfide
Dogwood			R					S					
Eggplant		I	R					S	R				
Elderberry			R					S					
Elm	S		R										
Endive	S	S			R	S							
Fern									I			S	R
Fir, Douglas			S								R		
Filaree		R											
Fleabane	S												
Forsythia			R					S				I	
Four o'clock	S												
Fuchia												I	
Galinsoga sp.		S	R										
Gardenia					I								
Gaura parviflora	S												
Geranium			I						I			I	
Gladiolus	I	R	S					I					I
Goldenrod			I										
Gomphrena									S				
Grape	I	S	I					S	I				
Grass		R			R					I			
Gum									I				
Halesia									I				
Heath							R						
Hedge mustard	I												
Hemlock								I	R				
Hibiscus							S						
Hickory								S					
Holly					I				R			I	
Honeylocust		S											
Honeysuckle			I										
Hydrangea												S	
Iris	I		S										
Ivy												R	
Ixora							R						
Johnny-jump-up			R						S				
Johnson grass			S						S				
Jimson weed		S											
June grass	S												

Plant	Sulfur dioxide	Ozone	Fluoride	PAN	Ethylene	Oxidant	Nitrogen dioxide	2, 4 D	Chlorine	Ammonia	Hydrogen chloride	Mercury vapor	Hydrogen sulfide
Juniper			R										
Lamb's-quarter	I	S	I	I			R	S	R	I			S
Larch	S		S								S		
Lettuce	S	S	I	S			S						
Lilac	R	I	I										
Lily												I	
Linden			I					S					
Lobelia			R										
London plane tree			R					S					
Mallow	S	S											
Manzanita	R		R										
Maple		S	I					S	S		I	I	
Marigold	I		R		S								
Mexican tea		I											
Milkweed	R		R										
Mimosa												S	
Mock orange			I										
Morning glory	S		R										
Mulberry	S		I					I					
Muskmelon		S				S							
Mustard		R		S			S		S	S			R
Narcissus			I										
Nasturtium			R						I				I
Nettleleaf goosefoot			I	S			R		I	R			S
Nightshade	I		R										
Oak			R					S	I		R	I	
Oat	S	S		S	R	S							
Okra	S												
Olive									R				
Onion	R	I	I	R	R				S				
Orchard grass	I	S	I										
Orchid		I			S								
Oxalis			I						R			S	
Parsley	I	I											
Parsnip	I		I										
Patient dock	I												
Pea			I		S								
Peach	I		I		S		I	I	R			I	R
Peanut		S											

Plant	Sulfur dioxide	Ozone	Fluoride	PAN	Ethylene	Oxidant	Nitrogen dioxide	2, 4 D	Chlorine	Ammonia	Hydrogen chloride	Mercury vapor	Hydrogen sulfide
Pear	I		R					R			R		
Peony			I					R					
Pepper	S		R						R				I
Periwinkle			R										
Persimmon												I	
Petunia		S	R	S		S			I				
Philodendron					S								
Phlox			R										
Pigweed			I				R		R	R			R
Pine	S	S	S			S			I			I	
Pinus remorata		S											
Plantain	S		R										
Plum	I		S										
Poa annua		S											
Pokeweed			I										
Poplar	S		I										
Poppy	S												S
Potato	R	S	I					I					
Prickly lettuce	S		R										
Primrose									S				
Privet		S	R		S			I	S			I	
Prune (Italian)			S										
Pumpkin	S	R											
Purslane	I		R										R
Pyracantha			R										
Radish	S	I	R	R	R				S				S
Ragweed	S	R	R					I					
Raspberry			R										
Red goosefoot		I											
Redroot	I												
Rhododendron			I					I					
Rhodotypos									I				
Rhubarb	S	I	R					R					
Rose	R	R	I		S			I	S			I	I
Russian thistle	R		R										
Rye	I	S	I				I						
Rye grass	S		I										
Safflower	S												
Salsify	S												

Plant	Sulfur dioxide	Ozone	Fluoride	PAN	Ethylene	Oxidant	Nitrogen dioxide	2, 4 D	Chlorine	Ammonia	Hydrogen chloride	Mercury vapor	Hydrogen sulfide
Saltgrass	R												
Salvia	S											I	S
Sarcococca												R	
Sassafras									S				
Saxifrage												I	
Serviceberry			I										
Shepherd's purse	R												
Smartweed	I		I						R				
Snapdragon			R										
Snowberry	R	S											
Sorghum			I	R	R			R					
Sorrel								S					
Soybean	S		R	I	I				R				S
Spinach	S	S	I	I		S							
Spruce			I					I			I		
Squash	S	R	R		I				I				
Squirrel tail	S												
Stock			R										
Strawberry			R									I	R
Sumac			I					S					
Sunflower	S	R	I				S		S	S		S	I
Sweet basil		S											
Sweet clover	I		I										
Sweetgum			R					I	S				
Sweet pea	S		R										
Sweet potato	S		I		S								
Sweet William			I										
Swiss chard	S	S		S									
Sycamore		S	I										
Tobacco	S	S	R	I		S	S	S	S	I		I	S
Tomato	I	I	I	S	S				S	I	I	I	S
Touch-me-not				R									
Tree of heaven			R					S	S				
Tulip			S						S				
Tumbling mustard	I												
Turnip	S	I											
Velvet weed	S												
Venus-looking-glass									S				
Verbena	S		I										

Plant	Sulfur dioxide	Ozone	Fluoride	PAN	Ethylene	Oxidant	Nitrogen dioxide	2, 4 D	Chlorine	Ammonia	Hydrogen chloride	Mercury vapor	Hydrogen sulfide
Viburnum			R								S	I	
Vinca			R									I	
Violet	S		I										
Virginia creeper	R		R						S				
Walnut			I										
Wandering Jew									I				
Watermelon	I												
Weeping willow		S	R									S	
Wheat	S	S	I	I									
Wisteria								S					
Witch Hazel									S				
Yellowwood								S					
Yew			I					I	R				
Zinnia	I		R					S	S				

[a] S = Sensitive; I = intermediate; and R = Resistant.

[b] The susceptibility rating is a compilation of published data (see Chapter 10). The authors decided upon a rating when published information differed.

Appendix B

PSEUDOADIABATIC CHART

Appendix C

PSYCHROMETRIC CHART

To obtain wet bulb temperature, T_w, associated with values of T, R, and e, follow upward to the left, parallel to the diagonal straight line until it intersects the saturation curve. The set of diagonals is drawn for sea level pressure. For higher elevations, construct alternate diagonals using proportions indicated by the line marked 5000 ft.

To obtain dewpoint temperature, T_d, associated with values of T, R, and e, follow horizontally to the left parallel to lines of constant vapor pressure until intersection with the saturation curve.

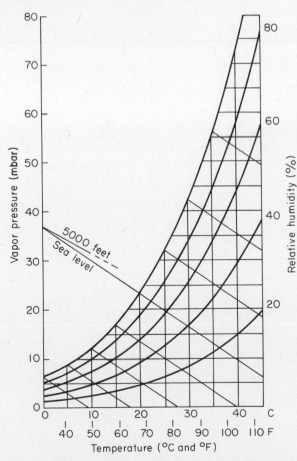

Appendix D

THE GAUSSIAN PLUME MODEL

Basic Form of the Model

In Chapter 20, Section III, it was noted that the form for a model of downwind concentration of a pollutant should be

$$\chi(x, y, z) = (Q/U) F(y) G(z)$$

where the x axis lies along the mean wind direction, y is crosswind, and z vertical. The "normalized" form is $(\chi U/Q) = F(y)G(z)$ so that the product of the crosswind dilution term $F(y)$ and the vertical dilution term $G(z)$ gives a relative number that is independent of the particular source strength or wind speed, and dependent only on the intensity of the crosswind and vertical mechanisms of dilution.

In the Gaussian plume model, the dilution terms are both given the form of the normal distribution:

$$F(y) = (\sqrt{2\pi}\sigma_y)^{-1} \exp{-\tfrac{1}{2}(y/\sigma_y)^2} \quad \text{and} \quad G(z) = (\sqrt{2\pi}\sigma_z)^{-1} \exp{-\tfrac{1}{2}(z/\sigma_z)^2}$$

The model also allows for perfect reflection of a pollutant from the ground surface by adding an imaginary mirror image source located beneath the actual source, which is in turn H units above the origin of coordinates. Since reflections are assumed to have only vertical components, the sum of the equations for the aboveground and the mirror sources is the Gaussian plume model:

$$(\chi U/Q)(x, y, z; H)$$

$$= (2\pi\sigma_y\sigma_z)^{-1} \exp{-\frac{1}{2}\left(\frac{y}{\sigma_y}\right)^2}\left[\exp{-\frac{1}{2}\left(\frac{z-H}{\sigma_z}\right)^2} + \exp{-\frac{1}{2}\left(\frac{z+H}{\sigma_z}\right)^2}\right] \quad \text{(D-1)}$$

The model and the following discussions of it are included, along with a great many other aspects of the use of the model, in Turner's workbook,* as well as in other places.†

Special forms of the model follow directly, such as the concentration on the ground level plane:

$$(\chi U/Q)(x, y, O; H) = (\pi\sigma_y\sigma_z)^{-1} \exp{-\tfrac{1}{2}(y/\sigma_y)^2} \exp{-\tfrac{1}{2}(H/\sigma_z)^2} \quad \text{(D-2)}$$

and the ground level concentration beneath the plume centerline:

$$(\chi U/Q)(x, O, O; H) = (\pi\sigma_y\sigma_z)^{-1} \exp{-\tfrac{1}{2}(H/\sigma_z)^2} \quad \text{(D-3)}$$

And again, since $\exp(0) = 1$, the ground level concentration of the centerline of a plume from a ground level source,

$$(\chi U/Q)(x, O, O; O) = (\pi\sigma_y\sigma_z)^{-1} \quad \text{(D-4)}$$

TABLE D-I

Pasquill's Classification of Atmospheric Stability as a Function of Ambient Variables, with Notes Based on Comments of Turner

Surface wind speed (10 m) (m sec^{-1})	Day[a] Incoming solar radiation is ...			Night[a] Low level cloud cover is ...	
	Strong[b]	Moderate[c]	Slight	$\geq 4/8$[d]	$\leq 3/8$[d]
Less than 2	A	A,B	B	(X)	(X)
2 to 3	A,B	B	C	E	F
3 to 5	B	B,C	C	D	E
5 to 6	C	C,D	D	D	D
Above 6	C	D	D	D	D

[a] For central urban environments, classes A,B, and C become B,C, and D; and class D is always used at night.

[b] "Strong" refers to clear skies with zenith angles, $Z \leq 30°$.

[c] "Moderate" refers to clear skies with $30° < Z < 55°$.

[d] In the United States, cloud cover is in tenths rather than oktas. (X) Not defined; "plume" will not be identifiable.

* Turner, D. B., "Workbook of Atmospheric Dispersion Estimates." Public Health Service Publication No. 999-AP-26. U.S. Department of Health, Education and Welfare: Washington. 1969. 84 pp.

† See References No. 5a and 5b of Chapter 20.

Aids to Computation

In any of the forms (D-1)–(D-4), a value of $(\chi U/Q)$ may be obtained by the multiplication of several expressions in which the only variables are the two standard deviations σ_y and σ_z, and various exponentials all of the same form. Table D-1, together with Figs. D-1, D-2, and D-3 provide the means to obtain all these variables nomographically. Thus solutions of the Gaussian plume may be obtained by desk-top calculations, even with a slide rule.

Figures D-1 and D-2 must be entered through Table D-I, which contains Pasquill's classification of atmospheric stability according to several environmental elements, with notes taken from discussions of Turner.* The rationale for this table may be obtained in Chapter 21, Sections I and III.

Once the stability class is obtained from Table D-I, values of σ_y and σ_z are obtained as functions only of x from Figs. D-1 and D-2, respectively. With these two parameters evaluated, it is then possible to express the departures from centerline

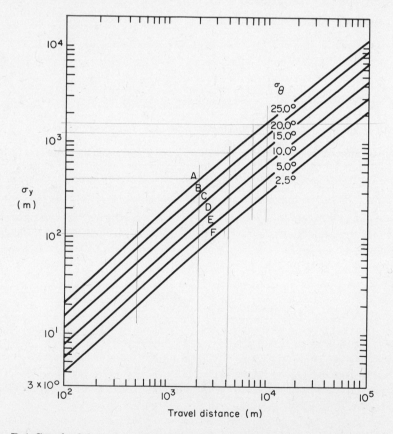

FIG. D-1. Standard deviation of crosswind dilution parameter σ_y as a function of travel distance x and the Pasquill stability class. Values shown associating stability classes with standard deviations of crosswind angular departure are only suggestive. (From Slade, Reference 3 of Chapter 20.)

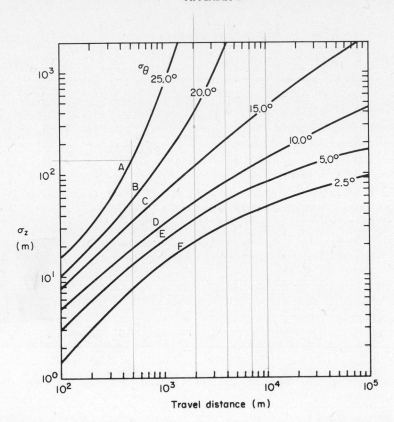

FIG. D-2. Standard deviation of vertical dilution parameter, σ_z, as a function of travel distance x and the Pasquill stability class. Values shown associating stability classes with standard deviations of vertical angular departure are only suggestive. (From Slade, Reference 3 of Chapter 20.)

as fractions of these values: (y/σ_y) and $(z \pm H)/\sigma_z$. If any of these fractions is designated A, the value of the necessary exponential $\exp -\frac{1}{2}A^2$ may be obtained from Fig. D-3. With values of σ_y, σ_z, and all exponentials, calculation of the normalized concentration $(\chi U/Q)$ is made by simple steps.

Sample Calculations

Of the myriad sample calculations which might be made to illustrate the use of the Gaussian plume model, only two have been selected as being basic to most of the others. First, calculate the relative concentration at point (0.8 km, 0.1 km, 0) with a source height of 50 m and stability class C. The point is at ground level, off the centerline, and far downwind relative to the source height.

From Fig. D-1, $\sigma_y = 86$ m; from Fig. D-2, $\sigma_z = 50$ m. Thus $(y/\sigma_y) = 1.16$ and $(H/\sigma_z) = 1.0$. From Fig. D-3, $\exp -\frac{1}{2}(y/\sigma_y)^2 = 0.51$, and $\exp -\frac{1}{2}(H/\sigma_z)^2 = 0.61$. With these values

$$(\chi U/Q) = [(\pi)(86)(50)]^{-1} \cdot (0.51)(0.61) = 2.3 \times 10^{-5}\ \mathrm{m}^{-2}$$

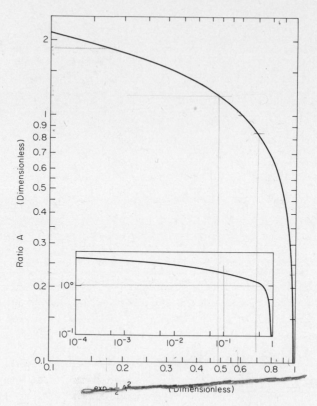

FIG. D-3. Nomograph relating the dimensionless ratio A to its exponential $(\exp - \frac{1}{2}A^2)$. The inset extends the nomograph to larger values of A and smaller values of the exponential.

For a wind speed of 5 m sec^{-1} and an emission rate of 100 g sec^{-1}, for example, the estimate of the true concentration at the point is $(2.3)(10^{-5})(100)/(5) = 4.6 \times 10^{-4}$ g m^{-3}.

With the given conditions, the profile of relative concentration at $x = 0.8$ km would appear as in Fig. D-4. The profile shows the result of ground reflection clearly,

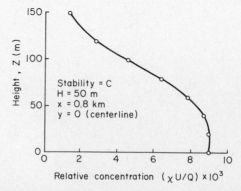

FIG. D-4. Vertical cross section of relative concentration in the plume defined in the first sample calculation.

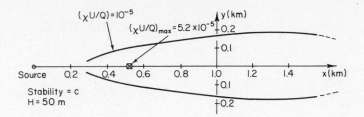

FIG. D-5. Ground plane isopleth of relative concentration $= 10^{-5}$ m^{-2}, for the plume defined in the second sample calculation. The location and magnitude of the maximum relative concentration on the ground plane is taken from Fig. D-6 and shown on the centerline at $x = 0.52$ km.

and is produced by values for various heights, z, from the product $(\pi\sigma_y\sigma_z)^{-1}$ with the sum of two exponentials in $(z \pm H)$.

The second calculation concerns the plotting, on the ground plane, of an isopleth of equal concentration. Suppose, for the conditions of the first example, the plot is to be of the isopleth of relative concentration 10^{-5} m^{-2}. The key to the calculations is the ratio $(\chi U/Q)(x, y, 0; H)/(\chi U/Q)(x, 0, 0; H) = \exp -\tfrac{1}{2}(y/\sigma_y)^2$, which requires the value of the isopleth chosen, $(\chi U/Q)(x, y, 0; H)$ and the centerline concentration at a chosen distance x, $(\chi U/Q)(x, 0, 0; H)$. For each value of x, a value of $\exp -\tfrac{1}{2}(y/\sigma_y)^2$ is thus determined, and from Fig. D-3 the corresponding value of (y/σ_y) is B. The off-centerline distance of the isopleth at x is $y = B\sigma_y$. Figure D-5 shows the result for the specified conditions.

Special Forms and Conditions

Figure D-6 shows an empirical result which is often useful. It is the downwind distance to the maximum ground level concentration, which is of course on the centerline. Also given is the value of the maximum relative concentration at that point as a function of source height H and stability class. The maximum is plotted in Fig. D-5 for the conditions given.

If the plume is contained, or "trapped," beneath an inversion layer whose effective base is L units above the surface, it may be assumed (a) the plume is unaffected out to the distance at which $\sigma_z \doteq L/2$, and (b) the plume is distributed uniformly in the *vertical* beyond the distance at which $\sigma_z \doteq L$. Beyond the latter distance, concentrations in the trapped plume may be estimated with

$$(\chi U/Q)_T = (\sqrt{2\pi}\sigma_y L)^{-1} \exp -\tfrac{1}{2}(y/\sigma_y)^2 \qquad \text{(D-5)}$$

Between the two distances, concentrations may be obtained by interpolation.

The plume from an infinite line source (for example a freeway or auto route) is uniformly distributed *crosswind* with the relative concentration downwind being estimated by

$$(\chi U/Q)_L = (\sqrt{2}/\sin \lambda \cdot \sigma_z) \exp -\tfrac{1}{2}(H/\sigma_z)^2 \qquad \text{(D-6)}$$

at ground level, where Q_L is the source strength in rate per unit line length (e.g.,

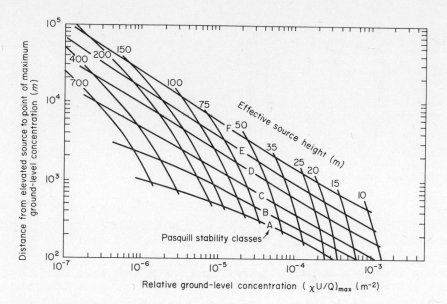

FIG. D-6. Downwind distance to the maximum ground plane concentration x_{max} and value of that relative maximum, $(\chi U/Q)_{max}$, as a function of source height and Pasquill stability class. (From Slade, Reference 3 of Chapter 20.)

$g\,sec^{-1}\,m^{-1}$) and λ is the acute angle between the line source and the wind direction x. This equation becomes very unreliable unless λ exceeds $45°$.

The specific form of the Gaussian puff—an instantaneous point source—was given in Eq. (20-7), where it is seen that the puff has the general form $(\chi U/Q)_I = F(y)G(z)J(x - Ut)$. Now y and z are measured from the puff center and x along the trajectory from the source. The reader is cautioned that the standard deviations, σ_I, in this equation are less than their analogs in Eqs. (20-6) and (D-1). The sources previously cited on page 461 should be consulted for values of the σ_I's.

Concluding Remarks

Every writer who discusses the Gaussian plume model in any detail is careful to leave the reader with the reminder that, though it gives a comforting feeling of rigor, accuracy, and precision, the model is in fact very much an approximation to reality.

Short distances downwind (up to 0.4 km, for example), with steady winds and uniform terrain without local obstacles, the estimates may be within a factor of 2. Departures from ideal conditions, i.e., those specified in the derivation, quickly make the results at best order-of-magnitude estimates. Turner has generally good advice on this and other points of accuracy and precision.

Appendix E

WIND ROSES AND TRAJECTORIES

Wind Roses

Table E-I gives an imaginary set of wind frequencies, based on eight compass points and upon the four speed classes of Fig. 20-13a. In what follows, of course, the number of compass points and of speed classes is at the investigator's choice.

The table says, for example, that of all the wind observations in the record which the table represents, 5% are for winds in the speed class 1–4 knots from the southeast. Further, 40% of all observations were of winds in class 1–4 kt, and 6% were from the southeast.

In a graphic form, Table E-1 assumes the identity of the wind rose in Fig. E-1. Here the bars extending from the center toward the southeast, for example, are for winds *from* the southeast; the lengths of bars are proportional to the frequency according to the scale which is part of the rose. The speed classes are separated by the kind of bar symbol, also part of the rose. The frequency of calms is given in the central circle.

Although other variations on this basic construction have been proposed,* the procedure suggested here is the most frequently used in the air pollution literature. As suggested in Chapter 20 and 22, however, its utility for many types of transport

---

* Munn, R., Reference 2 (page 103) of Chapter 22; and Court, A. *Weather* **18**, 106–110 (1963).

TABLE E-1

An Imaginary Set of Wind Frequencies at a Station [a]

Wind speed class	Wind direction								
	N	NE	E	SE	S	SW	W	NW	Total
Calm	.	.	.	.	.	.	.	.	5
1–4 kt	10	5	3	5	2	8	2	5	40
5–9 kt	3	—	—	—	4	15	3	10	35
≥10 kt	—	—	1	1	3	10	3	2	20
Total	13	5	4	6	9	33	8	17	100

[a] The entries are in percent of all wind observations of speed and direction.

problems involving actual wind sequences is not as great as that of the constructed trajectory.

Construction of Trajectories

In Table E-2 are given wind data for ten consecutive hours at a station. These are the data used in constructing Fig. E-2. During the first hour, for example, the data say the average wind was at 5 speed units (the units are not important for this discussion) *from* a direction of 270°; i.e., due west. During the second hour the speed was 4 and the average direction was 280°, i.e., 10° north of west, etc.

The principal trajectory in Fig. E-2 was constructed in the following manner. From an origin (source) a vector was laid out so as to "fly with the wind" for 5 units. From the end point of that vector (the first "node") a second vector was laid out to fly with the wind for 4 units; and so on. The nodes are labeled with

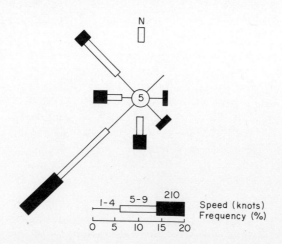

Fig. E-1. Wind rose of the wind frequency data in Table E-1.

numbers denoting the expected position, after the indicated number of hours travel, of a parcel which left the source at time zero. This procedure produced the principal trajectory shown by the heavy, solid line beginning at the origin.

The principal trajectory was drawn on the assumption that during the first full hour the average wind was at speed 5 from 270°. If, on the other hand, the hour designation "1" means, for example, that the average wind was determined by a short sample (say 2 minutes as in standard station routine) *as observed at clock hour 1*, then a slightly different trajectory could reasonably be drawn as follows. The average wind is now assumed to be speed 5, from 270° for the hour from 30 minutes before hour 1 to 30 minutes after, then the speed was 4 from 30 minutes before hour 2 to 30 minutes after, etc. The sequence of vectors based on this assumption is shown in Fig. E-2 by the sequence of triangles, labeled with numbers in parentheses. The meaning of these numbers, although analogous to those on the principal trajectory, is now slightly changed.

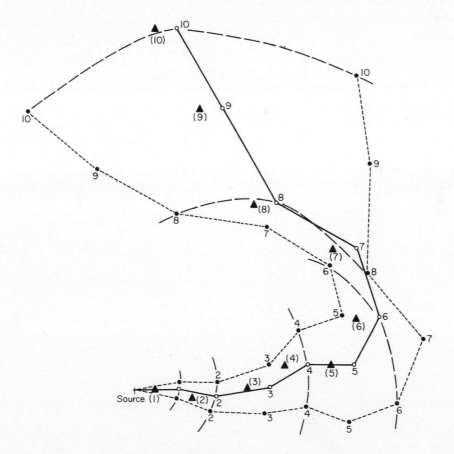

Fig. E-2. Principal trajectory (heavy solid line), extreme trajectories (dotted lines), and alternate trajectory (unconnected triangles) derived from the sequence of wind data in Table E-2. Dashed lines connect points of equal travel time on the principal and two extreme trajectories.

Meaning of a Constructed Trajectory

In Table E-2 are given indications of the directional variability of the wind during each hour. Such data are not customarily available in climatic records, but the reader should keep in mind that they hold the key to the true meaning of any trajectory constructed in the manner described above.

The standard deviation of the variability of wind direction, σ_θ, specifies, when added to and subtracted from the mean direction, the directional spread within which the least variable two thirds of all the instantaneous directional values occur for the time period specified. This statement is based on the theory of the normal distribution, for which the reader is referred to the discussion of Figure 25-2.

The average direction $\pm 2\sigma_\theta$ specifies the directional spread within which 95% of all instantaneous wind directions occur. In Fig. E-2 extreme trajectories are constructed and shown in dotted lines, as follows.

Beginning at the origin, lines of length 5 are drawn for wind directions $270° \pm 2\sigma_\theta$; namely, at 280° and 260°. These two vectors specify two additional end points. Subsequently, vectors have been added to that of the 280° vector at directions only of mean direction *plus* $2\sigma_\theta$ for each hour, and similarly for additions to the 260° vector leaving the origin. The resulting extreme trajectories suggest the envelope in plan within which 95% of all parcels must remain if they departed from the origin at time zero.

The three trajectories—principal and two extremes—are connected in the following way. There are three segments for any sequential pair of time markers, for example 4-5. Since $2\sigma_\theta$ for hour 5 is 20° in Table E-2, the vector of the left extreme trajectory lies 20° left of the vector segment in the principal trajectory, while the vector in the right extreme trajectory is 20° to the right of the segment in the

TABLE E-2

An Imaginary Set of Wind Data for Ten Consecutive Hours at a Station [a]

Hour	Mean speed	Mean direction (tens of degrees)	Crosswind variability ($2\sigma_\theta$ in degrees)
1	5	27	10
2	4	28	10
3	6	26	10
4	5	24	20
5	5	27	20
6	6	23	40
7	8	16	40
8	10	14	20
9	12	15	30
10	10	15	20

[a] The units of the speed are unspecified to preserve generality.

principal trajectory. All three of these vectors show speed 5. Finally, points of equal travel time are suggested by the arcs through times 2, 4, 6, 8, and 10.

Thus, although the principal trajectory contains much more information than a wind rose or frequency table, it depicts only the mean, or expected, trajectory from the source. Inclusion of directional variability makes for a more complex, but greatly more realistic picture of the likely transport of any pollutants leaving the source at time zero. For all its complexity, the trajectory envelope and travel time arcs do not tell the story for pollutants leaving the source at any other time, and furthermore, they only pretend to account for 95% of the emissions from the source at time zero. Finally, they are based on the questionable assumption that the data of Table E-2 represent a uniform wind field over the entire map of Fig. E-2. Even these more complex graphic analytic tools are at best only rough approximations to reality.

Appendix F

SYMBOLS AND DEFINITIONS

A	Cross-sectional area of raindrop, length^2
	Basal area of "box" [Eq. (20-10)], length^2
A_c	Acceleration, centrifugal
B, B'	Luminance of an object, no units
B_L, B_L'	Apparent luminance at distance L, no units
B_0, B_z	Turbidity coefficient at standard level (0), and at height z, length^{-1}
C	Concentration of chemical species
C_0, C_L	Visual contrast $= \Delta B/B$, no units
D	Drop diameter, length
E	Total particle energy
	Evaporation rate, mass/time-area
	Collision efficiency, no units
E_ν, E_λ	Monochromatic radiant emission of black body, frequency base (ν), wavelength base (λ), energy/area-time-bandwidth
E'	Monochromatic radiant emission, any radiator, energy/area-time-bandwidth
E^*	Energy per photon, energy
F	Force
F_a, F_ϵ	All-wave radiant flux from the atmosphere (a), from earth-atmosphere interface (ϵ), energy/time-area
F_p	Plume rise parameter [Eq. (20-9)], length^4/time^3
G	Flux of heat beneath earth-air interface, energy/area-time
H	Convective flux of sensible heat above earth-air interface, energy/area-time
	Stack height (height of pollu-

I_i, I_o tant source above ground level), length

I_i, I_o Infrared radiant flux incoming (i) and outgoing (o) relative to earth–air interface, energy/area-time

I_1, I_2 Indices of air pollution potential [Eq. (22-1) and (22-2)], no units

J Unimolecular chemical reaction ($J \rightarrow$ Products)

K Empirical constant, especially in Table 17-I, no units

K_d Dissociation constant
Eddy diffusivity, area/time

K_e Equilibrium constant for chemical species

L Length of fluid flow path, length
Path length through absorbing medium, length
Mixing height, length
Distance seen through the atmosphere, length

L_o Visual range with a constant illumination of 0.05 lumens, length

L_m Meteorological range, length

L_s Slant visual range, length

L_v Latent heat of vaporization, energy/mass

M Mass of substance
Particle mass

M_v, M_d Sample mass of vapor (v), dry air (d), mass

M_w, M_s Sample mass of water (w), solute (s), mass

N Number of units in a population (statistical)

$N(D)$ Space density of raindrops, volume^{-1}

$N(d)$ Space density of particles, volume^{-1}

N_i Space density of particles in the i-th size interval, volume^{-1}

N_L Mole fraction of solvent

O_m Molality

Q Point source strength, mass/time

Q_L Source strength per unit length, line source, mass/time-length

Q_v Volume rate of flow (length^3/time)

Q' Area source strength, mass/area-time

R, R_0 Radiant intensity of any wavelength, initial (0) and after transmission, length

R_c Radius of curvature in circular motion, length

R_d Gas constant for dry air, length^2/time^2—°K
Distance from sun to earth, length

R_e Radius of earth, length

$R_{n,t}$ Residual fraction of pollutants remaining in a "box" [Eq. (20-10)], no units

R_p Rainfall rate, length/time

R_v Gas content for water vapor, length^2/time^2—°K

R_λ Monochromatic radiant intensity, energy/area-time-solid angle

R_ω Radius of sun, length

S The solar constant, or the flux of solar energy through a unit area normal to the solar beam above the atmosphere. Energy/area-time = 2.0 cal/cm^2 min

S_c Downwind dimension of a city, length

S_h, S_s Flux of solar radiation on a unit area horizontal (h), on slope (s), above the atmosphere, energy/area-time

S_i, S_o Flux of solar radiation per unit area of earth–atmosphere interface, incoming (i), outgoing (o), energy/area-time

S_N^2 Variance of a population (statistical)

S_n^2 Variance of distribution of means of samples of size n (statistical)

S_n Flux of solar radiation on a unit area normal to the solar beam, beneath the atmosphere, energy/area-time

S_t Stimulus intensity

T Temperature of emitter
Air temperature
Air temperature in sounding

T_a Temperature of air or of atmosphere

T_d	Dewpoint temperature	a_s	Absorptivity with respect to shortwave radiation (solar), no units
T_e	Temperature of parcel environment	b, b_a, b_s	Extinction coefficient for absorption (a), scattering (s), length^{-1}
T_0	Air temperature at standard pressure altitude		
T_p	Temperature of air parcel	c	The speed of light, length/time
T_s	Temperature of stack effluent	c_p, c_{pd}	Specific heat of air (dry air = pd) at constant pressure, energy/mass—°K
T_w	Wet bulb temperature		
T_ϵ	Temperature of earth–air interface	c_v	Specific heat of air at constant volume, energy/mass —°K
T_π	Effective radiant temperature of earth–atmosphere system	d	Particle diameter, length
T_ω	Temperature of sun's surface	d'	droplet diameter, length
U	Mean horizontal wind speed in mixed layer, length/time	dH	Change in sample heat energy per unit mass
U_c	Velocity of particle on circular path, length/time	dU	Change in sample internal energy per unit mass
U_g	Geostrophic wind speed, length/time	dW	Work done on or by unit mass of gas sample
U_0, U_z	Mean horizontal wind speed at standard height (0), arbitrary height (z), length/time	e	Ambient vapor pressure, force/length^2
V, V_t	Mean vertical velocity, and terminal velocity (t), length/time	e_s, e_w	Vapor pressure at saturation (s), at the wet bulb temperature (w), force/length^2
V_g	Velocity of gas in duct or stack, length/time	e_s	Vapor pressure above a pure solvent surface
V_L	Volume of liquid, length^3	e_{sa}	Vapor pressure above a solution surface, force/length^2
V_{st}	Velocity of stack effluent, length/time	g	Acceleration of gravity, length/time^2
W	Valley width, length	h	Planck's constant = 6.55×10^{-27} erg-sec
W_p	Washout coefficient, time^{-1}		
Y	Diffusivity of gas, area/time	h_c	Height of a capillary column, length
$\bar{Y}$	Population mean (statistical)		
Z	Zenith angle (measured from directly overhead to the solar position), degrees or radians	i	Angle of inclination of a slope, degrees or radians
		j	Initial concentration of chemical reactant
Z_g	Number of colliding molecules per unit volume	k	Boltzmann's constant = 1.37×10^{-16} erg/°K
Z_i	Height of inversion base above earth–air interface, [Eq. (22-1)], length		Absorption coefficient, length^{-1}
a	Absorptivity with respect to radiation, no units	k_s, k_λ	Absorption coefficient, monochromatic at wavelength λ, for solar spectrum (s), length^{-1}
	Empirical exponent in Eq. (20-1), no units		
a_g	Adsorbed gas per unit mass	k'	Absorption coefficient for unit concentration
a^+_M, a^-_M, a_{MA}	Ionic activity of chemical species	m	Ambient mixing ratio, no units
		m_s	Saturation mixing ratio, no units

n	Moles of chemical species	y	Horizontal coordinate normal to x
	Number of units in a sample (statistical)		Distance in the crosswind direction, length
n_c	Amount of cloud cover in tenths, no units	$\bar{y}$	Sample arithmetic mean (statistical)
n_i	Refractive index, no units	$\bar{y}_g$	Sample geometric mean (statistical)
n_v	Number of molecules per unit volume	z	Vertical distance from earth–air interface, length
p	Atmospheric pressure		Vertical coordinate normal to x
p_s	Static pressure, force/length^2	z_a	Valence, or charge on ion
p_t	Total pressure, force/length^2	ΔH	Correction applied to stack height to account for momentum and buoyancy of effluent stream, length
p_v	Velocity pressure, force/length^2		
q	Advected pollutant flux into a "box" [Eq. (20-10)] in unit time step, mass	ΔT_b	Boiling point elevation, °C
		ΔT_f	Freezing point lowering, °C
r	Radius of a capillary tube, length	ΔS_t	Differential threshold
	Reflectivity with respect to radiation, no units	ΔZ	Thickness of inversion layer [Eq. (22-1)], length
	Relative humidity, no units	Δj	Decrease in concentration of chemical reactant
	Particle radius, length		
r_c	Equilibrium relative humidity over a curved surface, no units	$\Delta\Theta$	Difference in potential temperature between top and bottom of inversion layer [Eq. (22-1)], °K
$r_{n,t}$	Residual fraction of pollutant remaining in a "box" [Eq. (20.10)], no units		
		Θ	Potential temperature, °K
r_i	Radius of particle in i-th size class, length	Π	Osmotic pressure
		X	Spatial concentration of pollutant, mass/volume
r_{sa}	Equilibrium relative humidity over a salt solution, no units	Ω	Rate of rotation of the earth about its axis, time^{-1}
r_{st}	Inside radius of a stack, length	α	$2\pi r/\lambda$, where r is the particle radius
r'	Radius of curvature of a surface, length	β	Sample volume per unit mass (specific volume) of air, volume/mass
s	Surface tension, force/length		
s_g	Sample standard geometric deviation (statistical)	γ_e, γ_p	Lapse rate of temperature $(-dT/dz)$ in a sounding (e), of an air parcel (p), °C/length
s_n	Standard deviation of a sample (statistical)		
s_n^2	Variance of a sample (statistical)	γ_d	The dry adiabatic lapse rate $= 1.0$ °C/100 m
t	Time	δ	Solar declination, degrees or radians
	Travel time $= x/U$	ϵ	Emissivity, no units
u	Root-mean-square velocity of molecules, length/time	θ	Angle of incidence (measured from the normal to a surface). Equivalent to the zenith angle, Z, when the surface is horizontal, degrees or radians
v	Crosswind eddy component of wind speed, length/time		
w	Particle mass		
x	Horizontal coordinate parallel to mean wind vector		
	Distance in downwind direction, length	η	Hour angle, degree or radians
			Coefficient of viscosity

κ Scattering area ratio, no units

κ_i Scattering area ratio for particles in the i-th size interval, no units

λ Wavelength of radiation, length

λ_m Mean free molecular path, length

μ_s Ionic strength of chemical species

ν Frequency of radiation, length^{-1}

ρ Density of sample, mass/volume

$\rho_{n,t}$ Residual fraction of pollutant remaining in a "box" [Eq. (20-10)], no units

ρ_e, ρ_p Density of air in a parcel (p) and in its environment (e), mass/volume

ρ_v, ρ_d Density of dry air (d), water vapor (v), mass/volume

σ The Stefan–Boltzmann constant $= 0.817 \times 10^{-10}$ cal/cm^2 min ($^\circ$K)^4

σ_m Molecular diameter, length

$\sigma_x, \sigma_y, \sigma_z$ Standard deviation (as a Gaussian measure of spread) from plume centerline or from puff centroid in the x, y, and z directions, length

σ_v Standard deviation of the variation of the crosswind eddy component about its mean value, length/time

$\sigma_\theta, \sigma_\psi$ Standard deviations of crosswind (θ) and vertical (ψ) angular departures from zero, degrees or radians

τ Transmissivity with respect to radiation, no units

φ Latitude (N, positive), degrees or radians

ω Solar azimuth angle, degrees or radians

ω' Slope azimuth angle, degrees or radians

SUBJECT INDEX

A

Absorption,
  atomic, 180, 181–182
  infrared, 180–181
    limits of detection for, 187
  of pollutant gases, 417–418, 422
  of solar radiation, 201
Absorption coefficient, 215
Absorptivity, 212–214
  for black body, 210
Accuracy,
  in air sampling, 170
  in pollutant analysis, 175
Acetylene, heat of formation of, 91
Acid(s),
  equilibrium constants of, 89
  manufacture of, 361, 362–363
    emission limits for, 431
Acid rain, 39–40
Acoustical analysis, 183–184
Acrolein, from varnish cooking, 361
Activation energy, 90, 91
Addition, relative, for pollutant concentration, 282
Adiabatic process, 228

Adiabats,
  dry, 231, 238
  wet, 231, 238
Adsorption, 85–86
  in removal of gaseous pollutants, 418–419, 420
Aerosols, *see also* Particulate matter
  meteorological range and, 309–311
  natural sources of, 353
  particle size, 255–256
  photochemical, 94
  sampling for, 397
Aesthetics, 139, 307–308
  air-quality standards and, 158
Afterburners, 421–424
Ageing,
  in animals, ozone and, 120
  in man, 128
    ozone and, 136
Agglomeration, 23
Air,
  ambient, 67
  definition of, 19
  interrelations with water and soil, 11–13
    balancing, 14
  nondegradation policy for, 17
  thermodynamics of,